SPACE - A NEW COMMUNITY OF OPPORTUNITY

CONFERENCE COMMITTEE

Executive Chairman	Aaron Cohen, NASA-JSC
Genera l Chairman	Henry N. Bowes, Lockheed
Technical Chairman	William G. Straight, Lockheed
Secretary	Carolyn Fine, Lockheed
Arrangement Chairmen	C. E. Gist, Lockheed Harry C., Consultant
Publicity Chairman	Joseph P. Militano, IBM
Banquet Chairman	John R. Crenshaw General Dynamics
Corporate Relations Chairman	Benjamin N. Boykin Rockwell
External Relations Chairman	Dr. Bill J. Chen Pennington-Chen Real Estate
Pre-Print Chairman	M. A. Jowid, Jr. McDonnell Douglas
Student Participation	Oran Dial Computer Sciences Corporation
Treasurer	Chris A. Minton, Lockheed
Registration	Marilyn Cavalier, Lockheed
NASA Coordination	Daniel A. Nebrig, NASA-JSC

COOPERATING AGENCIES:

American Institute of Aeronautics and Astronautics
Amateur Radar Satellite Corporation
Clear Lake Area Chamber of Commerce
Clear Lake Area Economic Development Foundation
Houston Chamber of Commerce
Institute of Electrical and Electronics Engineers
International Space University
National Space Club
National Space Society
National Contract Management Association
Space Foundation
Space Studies Institute
Southwestern Aerospace Professional Representatives Association
U.S. Space Foundation

Thanks are due Carmen Martinez-Ivins and Doreen L. Linnan, Univelt, Inc., for transcribing the talks of speakers and for final preparation of the manuscript for publication.

Front Cover Illustration:

Apollo 17 view of Earth - A fantastic view of the sphere of the Earth as photographed from the Apollo 17 spacecraft during the final lunar landing mission in NASA's Apollo program. This outstanding photograph extends from the Mediterranean Sea area to the Antarctica south polar ice cap. Note the heavy cloud cover in the Southern Hemisphere. Almost the entire coastline of the continent of Africa is clearly delineated. The Arabian Peninsula can be seen at the northeastern edge of Africa. The large island off the northeastern coast of Africa is the Malagasy Republic. The Asian mainland is on the horizon toward the northeast (NASA Photo Nos. 72-HC-928, 72-H-1578, AS17-148-22727).

Frontispiece:

Current U.S. Launch Vehicles. See paper AAS 87-625 in this volume for details.

Titan II

Titan III

Titan IV

Delta Medium Launch Vehicle

Shuttle

Atlas/Centaur

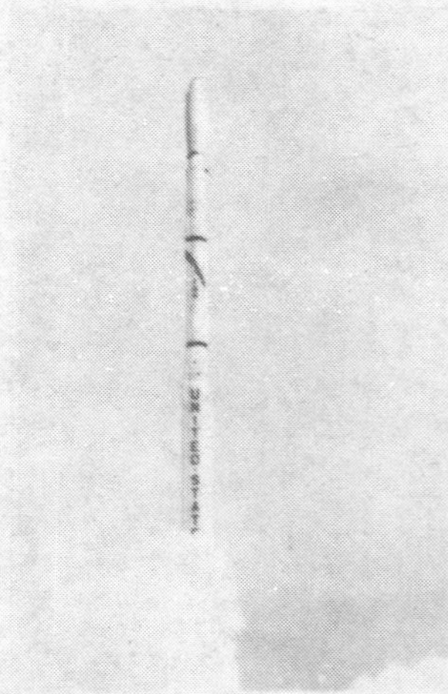

Scout

AN AMERICAN ASTRONAUTICAL SOCIETY PUBLICATION

SPACE - A NEW COMMUNITY OF OPPORTUNITY

Volume 67
ADVANCES IN THE ASTRONAUTICAL SCIENCES

Edited by
William G. Straight
Henry N. Bowes

This proceedings volume is based on the 34th Annual AAS International Conference held November 3-5, 1987, Houston, Texas

Published for the American Astronautical Society by Univelt, Incorporated, P.O. Box 28130, San Diego, California 92128

AAS Publications Office
P.O. Box 28130
San Diego, California 92128

Affiliated with the American Association for the Advancement of Science
Member of the International Astronautical Federation

First Printing 1989

Library of Congress Card No. 57-43769

ISSN 065-3438

ISBN 0-87703-297-1 (Hard Cover)
ISBN 0-87703-298-X (Soft Cover)

Published for the American Astronautical Society
by Univelt, Inc., P.O. Box 28130, San Diego, California 92128

Printed and Bound in the U.S.A.

FOREWORD

Over the years, four major AAS conferences have been held in Texas, most of them notable for their innovative themes and coverage. For all of them, proceedings volumes were published in *Advances in the Astronautical Sciences*. Following are the conferences and proceedings volumes:

- **Commercial Uses of Space**, Dallas, May 1967 (Volume 23 *Advances* and Volume 3 *AAS Microfiche Series*). We believe this to be the first conference ever held on this subject.
- **International Congress of Space Benefits**, June 1973, Dallas (Volume 30 *Advances*). To the best of our knowledge this was the first conference covering spin-offs from the space program.
- **The Future United States Space Program**, October - November 1978, Houston (Volume 38 I & II *Advances* and Volume 30 *AAS Microfiche Series*)
- **Developing the Space Frontier**, October 1982, Houston (Volume 52 *Advances*)

Two of the conferences were held in Dallas and two in Houston. The current volume covers the proceedings for the third major AAS conference held in Houston in November, 1987, **Space - A New Community of Opportunity** (Volume 67 *Advances*).

The volume covers ten panels which deal with a wide range of space technology, space science, and applications. Of particular interest are the Keynote Address by Willis Hawkins and the Banquet Address by Dr. Thomas O. Paine. Both speakers discuss many problems of the U.S. space program and what needs to be done to get things on track again. Both of these stimulating papers appear in the introduction to this volume. A closing recap summarizing many of the major issues which were brought up during the conference appears near the end of this book.

Nearly all presentations were transcribed and edited; however, each speaker was given the opportunity to review and revise his/her paper before publication. The reader will therefore find that many papers do not appear in formal style but rather reflect the informality of the presentations. Every effort has been made by the Series Editor and Associate Editor to ensure accuracy and freedom from error; however, for a volume of this size based on transcribed recordings, responsibility for technical content and expression must rest with the individual speakers or authors.

The editors wish to express their thanks to all those who have contributed to the success of this conference and to authors for their effort in finalizing material for publication.

Horace Jacobs
Series Editor

PREFACE

For me it all began in December 1985 - at a meeting of the American Astronautical Society officers, directors and technical committee chairmen - with President David Honhart presiding. I was there as a newcomer, courtesy of Al Naumann - a director and Phil Culbertson - Vice President, Technical. The primary business of the meeting was the Annual Conference to be held in Houston, Texas, in the late Fall of 1987.

In addressing the group President Honhart put forward two challenges - one, to Phil Culbertson and his Technical Committees, that they assume the responsibility for each of the technical sessions. (In the past the host committee had planned the entire program.) The second to Al Naumann and Hank Bowes, that they form a "Houston Area" Section of the American Astronautical Society. The challenges were accepted by us - and we were off to get the job done.

After a slow and painful start, all of the technical chairmen put together what I consider one of the best programs that AAS has sponsored! The results are clearly evident in the proceedings that follow.

In working with Dr. Aaron Cohen, Director of the NASA/Johnson Space Center and Executive Chairman of the AAS Annual Conference, Bill Straight, the Program Chairman, and I agreed that we would leave a finite written Summary and Recommendations that would be passed on to the appropriate civilian and defense agencies. Thus, in the program the first and last sessions were dedicated to bringing this aspiration into reality.

Now that I have the perspective of over a year to review the results, it is clear that the 34th Annual Conference did meet the two challenges of President Honhart. The technical committees presented a fine program and the Southwest Section of the AAS was formed. Furthermore, the Summary and Recommendations of the Conference are now a matter of record for the appropriate Government agencies to review and consider. I now consider my job finished! Thanks to all who made it possible.

Henry N. Bowes
Conference General Chairman

CONTENTS

Page

Page

Page

Page

Page

Page

INTRODUCTION

AAS 87-601 to -602

SPACE: A NEW COMMUNITY OF OPPORTUNITY - WELCOMING REMARKS

Henry N. Bowes*, Aaron Cohen†,
Kathryn J. Whitmire‡ and Gerry Griffin**

Henry N. Bowes: Good morning, ladies and gentlemen. I am Hank Bowes, General Chairman for the 34th National Symposium for the American Astronautical Society. I'm the one who ultimately has to answer to Aaron for how well this conference goes. So, if it's going well as far as you're concerned, please tell Aaron. If it's not, please tell me or any of my host chairmen. The hosts can be identified by a yellow badge. And they are there to serve you and make this event a very successful event. With that I'd like to introduce the Executive Chairman for this symposium. He is Director of the Johnson Space Center and I am very pleased to be working with one whom I've considered to be my friend for many years. Would you welcome Aaron Cohen.

Aaron Cohen: Thank you very much, Hank. The last time we had the American Astronautical meeting here was, I think, in 1982. And we're really pleased and proud to have it here. On behalf of the American Astronautical Society and the host committee for this symposium, it is my pleasure and honor to welcome you to this 34th Annual Symposium. With such a prestigious group of speakers, panelists, and participants that we have gathered today at this symposium, I feel certain that the goal of this conference, which is to strengthen the United States' space policy and further international cooperation, will be achieved.

It is truly a great pleasure for me to introduce the Honorable Kathy Whitmire, Mayor of the City of Houston. Mayor Whitmire has taken time from a busy schedule, and we all know it's a busy schedule, to be with us, to extend a welcome to Houston. From my personal experience, Mayor Whitmire understands the need for a strong space program in the United States. I can tell you that first hand. She is a

---

* General Chairman, 34th AAS Annual Conference. He is Vice President and Program Manager, Lockheed Engineering and Management Services Company, Inc., 2400 NASA Road One, Houston, Texas 77058.

† Dr. Cohen is Executive Chairman, 34th AAS Annual Conference. He is Director, NASA Johnson Space Center, Mail Code AA, Houston, Texas 77058.

‡ Mayor of the City of Houston.

** President and Chief Executive Officer, The Houston Chamber of Commerce.

great supporter of the space program. Please join me in welcoming a great supporter, Kathy Whitmire.

The Honorable Kathryn J. Whitmire: Thank you very much Dr. Cohen. It is a pleasure to be here to welcome this symposium of the American Astronautical Society back to our city again. We are all so very pleased that you have come to spend a few days with us, and particularly to focus on issues that are so important to the future of this nation, particularly the future of this city, and certainly the future of this entire planet. Now I have to ask, looking at this prestigious group here today, how many of you have come from outside of Houston to this meeting? Can I just get you to raise your hands so I can see how many people I am welcoming to Houston? Thank you very much. We are so pleased that you are here. Now, the real reason I wanted to find out - someone may have told you that today is election day in Houston - and I just wanted to see how many people are actually from Houston so that I might be able to encourage you to go out and vote before the day is over because my name is on that ballot and I'd just love to have your support while we're working on these endeavors of the long-term future. When you're in the business that I'm in, the future ends very quickly after election day if you don't get your constituents out to vote.

This has been an exciting time for Houston. As Dr. Cohen has indicated, it is an exciting time for the space program in Houston. We in the city of Houston are very excited and enthusiastic about the future of the Johnson Space Center, and of NASA, and of the role that space plays in our daily lives and in our future. In this year, because it's an election year, I have been trying to focus a specific message to the people of Houston. And that is the opportunities that we have for the future. We have been, for those of you who are not from Houston and may not have heard, through some difficult economic times in this city. Because through the years our primary industry has been the energy field, oil and gas, and related fields, and during these past couple of years that particular industry has fallen on some difficult times, we as a community have begun to pay more attention to diversification of our economy and to preparation for the economy of the future.

I think in a very real way each of our cities in the United States, and around the world, is looking to that future - looking to the changes that are going to occur economically and in the daily lives of all our citizens, and trying to be at the forefront of preparing for that future.

All of us at Houston are very, very fortunate because we have the Johnson Space Center here which has given us a clearer view of the future. It has given us a clearer view of the opportunities that are going to be emerging in the next decade and in the next century. So we have adopted a philosophy or a slogan called, "Reaching the Future First", that we are identifying where we can go, how we can get there, and how we can be first to arrive. I think that this not only applies to us in Houston, which we like to call "Space City", but certainly to everyone associated with the space industry.

You are the people on the cutting edge of what has not yet been dreamed about - of the new ways of doing things in the business world, the new way of doing things in the industry and in the daily lives of all citizens. You are the people who are dreaming what others have been afraid to dream, but not only are you dreaming, you are making it happen.

We have identified the economy of the future from the standpoint of Houston. And we have developed a five point program for our future. That five point program, of course, continues to include the energy and petrochemical field. But in addition to that, it includes the commercialization of space, the new opportunities that are emerging in the private sector, surrounding the activities at the Johnson Space Center. It also includes international trade as we look far beyond the boundaries of this country and see ourselves involved in a global economy. It includes the biomedical technologies that are developing in our Texas Medical Center and have become more and more closely linked with what is going on at the Johnson Space Center. And certainly it includes one other area that's very important to me and that's the convention and tourist field - another reason I'm so excited to see all of you here because we have been putting a much greater emphasis on attracting people from outside of our community to come to Houston to enjoy our people, our climate, and our enthusiasm for the future.

That's the strategy that we have in Houston, and as you can see the Johnson Space Center and all of the companies that work with the Johnson Space Center are very important to that future strategy. I believe that the ideas that will develop from the plans for the Space Station, the ideas that will develop for the use of space technology, in other processes, in biomedicine, in manufacturing processes, are going to make such a change in our entire society, that it is sometimes amazing that so few people are aware of the potential. And yet you are the people that are aware of the potential and that's why I'm so happy that you're with us here in Houston today. I hope that while you're here that you will not spend all of your time in these very exciting meetings. I hope that you will have a chance to get out and see a little bit more of the city, find out about the cultural arts in this community, enjoy the out-of-doors, perhaps even venture down to Galveston Island and get to know a community that cares about space and cares about the future. We're so happy to have you and want to give you a very warm welcome to Houston. Thank you so much.

Aaron Cohen: Thank you very much, Mayor Whitmire. And as a token here's a little something to remember us by.

The next speaker I have really needs no introduction to the space community. He's the President and Chief Executive Officer of the Houston Chamber of Commerce. Gerry Griffin knows space as few do. He came up through the ranks in the space program first as a flight controller, then in the mission control center as a flight director. He served in Washington as Associate Administrator. He has been Deputy Director of two different space centers - Dryden and Kennedy Space Centers. And he is a former Director of the Johnson Space Center. Please welcome Gerry Griffin.

Gerry Griffin: Thank you Aaron. It's always nice to step in front of an audience and see so many friendly faces. That is the case this morning. I recognize some of the faces. Some of you I haven't seen since I left NASA a couple of years ago. And so hopefully in these next couple of days I'll be able to renew some of these acquaintances. I'd like to add my welcome to Houston to that of the mayor's, and I want to say something about our mayor. We have a unique situation here in Houston with the cooperation between the public and private sector. Sometimes, I guess, necessity is the mother of invention, but I think in the last several years, particularly since Mayor Whitmire has been in office, we've seen the cementing of that relationship better than in any other city of our size in this country. She leads the charge quite often into matters, I think, most mayors would shy from. But she is not at all loathe to make that tough decision to get the city government to work with the other elected officials. And I just want you to know we really appreciate it.

The conference that you're starting now, and I notice its theme is, "A New Community of Opportunity", is very timely. I don't have to tell this group that the last year and a half have been a very tough time for the American space program. I think we have seen some light at the end of the tunnel. And I believe we are on the road now to get this most important program back on track. It's people like yourselves, those people in and around the industry, that are going to have to make it happen. But, for sure, the American policy that drives the space program has to be straightened out for the long haul. I believe conferences such as yours, where you have amongst you decision makers and people that know what the issues are, can lead to good sorts of decision analyses and so forth and I think the AAS is very good in making sure that they get their views before the people in Washington and indeed to the other elected officials around the country.

As the mayor said, Houston is poised to take advantage of this explosion which I believe we'll see in the use of space technology over the next decade. Space has been a key component of our economy here for some time, but it's not nearly the size, as the mayor mentioned, as energy or some of our other major mainstream economy drivers. I think we will see a change in that in this part of the country in the next ten years. I know that many of the companies that you represent are looking at Houston as a potential for growth. I believe Houston in its space community will grow as the American space program comes back into focus and gets going again hitting on all cylinders in a way that we have never imagined. I believe that it's because we do have the setting here in Houston to attract the kinds of businesses which many of you represent. Obviously there are other parts of the country that are very important to the space program and their growth will also be enhanced as the space program continues and we wish them well. A rising tide lifts all ships. We just hope Houston's ship sits a little higher in the water as that tide rises.

I want to welcome you to the conference. I've looked at the topics. I notice that the theme is related in part to that decision-making, and policy making that I talked about earlier and I think it's very timely to address it.

You know that I seldom take credit for anything. I want to say that Aaron asked me to make sure that there would be no rain today and tomorrow. I've arranged that, I don't know what else I can do for you.

By the way, I have to make my other plug. We are still interested in selling a few boxes in the Astrodome and I was very saddened to see that the Space Station decision was going to be delayed for a while as to which team was going to win that, because I had notified the people at a speech last week that I expect to be on the winner's doorstep immediately. Those boxes are reasonably priced; you'll enjoy them; it's a good place to watch a ball game. So, I'll be around to see you winners. Thank you very much and welcome to Houston.

Aaron Cohen: Gerry, a little token of appreciation for being with us today.

AAS 87-603 to -604

SPACE: A NEW COMMUNITY OF OPPORTUNITY OPENING REMARKS

David C. Honhart* and Aaron Cohen†

Aaron Cohen: Now to officially open the symposium it's a privilege to introduce to you Captain David Honhart of the United States Navy. Captain Honhart is the President of the American Astronautical Society, in addition to his other responsibilities as the Deputy Director for Ocean Science and Technology and Director of Space Oceanography on the Staff of the Chief of Naval Research. Ladies and Gentlemen, Captain Honhart.

David C. Honhart: Let me say quickly on behalf of the National Society, welcome to Houston. I want to tell you a few things that we're doing differently this year. Certainly the quality of the AAS as a national organization has to do with the events that it puts on: its Annual Meeting, here in Houston this year, the Military Space Symposium which occurs in Washington, and the Goddard Memorial Symposium. But since we are a cooperative society, it is the corporations and the individual members who are the ones that really make it happen. What I thought was important for myself to do this morning was to make sure that we give recognition to those companies and to those individuals who made this conference a reality. I don't want any applause because I'm a bit of a rebel rouser and in the end we're going to give them three cheers and wake everybody up in the hotel.

We made a decision last night. The next meeting is going to be in St. Louis, Missouri, about this time next year. And so we're looking for those corporate members, those corporations who have offices in the St. Louis area to step forward and help us by working with us, if you will.

The leader, without question in setting up this meeting here in Houston is the Lockheed Corporation. I'm going to read the names and then I'll read the names of the other companies and those who participated. Certainly there's Hank Bowes who's General Chairman here, and Bill Straight who's Technical Chairman, Carolyn Fine, C. E. Gist, Marilyn Cavalier, C. A. Minton, Al Naumann, Don Cravey, Sonya Yoder and Wayne Cope, all from the Lockheed Corporation who played a major role here. From IBM, Joseph Militano and J. C. Heberlig, from General Dynamics, John

* Captain Honhart is President of the American Astronautical Society. He is affiliated with the Office of the Chief of Naval Research, 800 N. Quincy Street, Code 13D, Arlington, Virginia 22217.

† Dr. Cohen is Executive Chairman, 34th AAS Annual Conference. He is Director, NASA Johnson Space Center, Mail Code AA, Houston, Texas 77058.

Crenshaw, and Carol O'Donohue, and I think it's Celeste Savagio and from Boeing Corporation, Julia Sorrells and Godfrey Bottomley; from Rockwell Corporation, Benjamin Boykin; from McDonnell-Douglas Corporation, M. A. Jowid, from Computer Sciences Corporation, Oran Dial and Consultants Harry Claggett and Bev Steadman, and the external relations here was handled by Dr. Bill Chen. And NASA played a key role in this also. Of course, headed by Aaron Cohen, Brian Welch, Steve Nesbitt, and Daniel Nebrig and a good friend of mine, Commander Don Mautner. So, let's give them three cheers. Have a good time.

Aaron Cohen: It is a pleasure for me to be Executive Chairman of the 34th AAS Conference and to have this meeting in Houston. The theme of this symposium, "Space - A New Community of Opportunity," is very appropriate. The opportunities that space provides have just come to be realized. We are really in the early phase of capitalizing on space and what space offers us. As we all know, there is so much to do. The tools we need to provide these opportunities are on the horizon. First, it's a matter of returning the shuttle to a flight status and we know that's important. A replacement orbiter. The Space Station, and the orbital maneuvering vehicle. To develop Shuttle C or the Advanced Launch System. To encourage private endeavors such as IAS in Spacehab, the more futuristic programs, such as the aerospace plane, a potential lunar base or a manned Mars mission. All those visionary type projects are becoming a reality. All of these are needed and progress is being made in all areas.

The technical and program aspects of all these ideas and all these programs will be covered in this symposium. We will also cover the scientific and commercial aspects in this symposium. And the final session will project the future. The people we have here for the next three days are some of the best in the country. A great deal of information is going to be brought forward along with some very knowledgeable ideas. It is meetings such as this that bring academia, government, and industry together that allow us to take advantage of the opportunities in space.

This, the 34th Annual Conference of the American Astronautical Society was conceived at the peak of introspection on U.S. space programs. Examination of the shuttle process is, of course, my first priority. And I want to take time to focus on the interrelationships of the United States' international space policies with the essential aerospace technologies.

The vision of space exploration has been with us for centuries. When most of us were growing up it was a vision or something we read about in the comic strips. It was not a reality. Today it is a reality. Then the technology of World War II has challenged us to make this vision a reality. The energies of governments through their policies was focused on breaking our earthly bonds.

To me policy and technology are intimately intertwined. It is appropriate for those of us that represent the technological capability of aerospace to examine the issues we face and express our opinions to the governing agencies of space policy. Our goal of this conference is to provide a forum for presenting a perspective on both policy and technology and to flush out the true issues about U.S. space policy - the interaction with technological arenas and the ever-present economics of the world.

The format of the panel is planned to allow the panelists and presenters to provide a background of information and put forth a set of issues and recommendations. Time is then provided for the audience to have floor discussions. We will be using questionnaire forms to allow each chairman and host to facilitate the process. Each panel host is charged with preparing a summary of these issues and recommendations. This summary then will be made available to the closing recap panelists on Thursday afternoon to suggest, "Where from here?" The AAS President, Dave Honhart, has committed to take the results of our final session and present them to the U.S. and international space agencies as the considered opinion of this symposium. I ask that you join in this opportunity to have our ideas heard by those who can make a difference in our space policy. As future evidence that the promise will be fulfilled, Captain Honhart and I agreed that a Houston focus is essential to make the AAS an on-going process. So, in this regard, our Houston section of the AAS is being chartered as a part of our symposium's closing ceremonies. I hope that you will attend.

Now, it gives me great privilege and pride to introduce our keynote speaker, Mr. Willis Hawkins. Mr. Hawkins, again, needs very little introduction to the space community. He's a very highly respected aerospace and military consultant. He has made major contributions to the design and development of airplanes, missile systems and space vehicles for which he has received numerous distinguished service awards for his achievement in those areas. Most notable are the U.S. Navy Distinguished Public Service Medal for his contribution to the Polaris Program; the U.S. Army Distinguished Award, which he has received twice; and the NASA Distinguished Public Service Medal. He was chairman of the NASA Aerospace Safety Advisory Panel and I had a lot of reviews with Mr. Hawkins that certified the first shuttle flight in 1982. He currently serves as a member of the National Research Council on the Shuttle Criticality Review and Hazard Analysis Audit Committee. Ladies and gentlemen, Mr. Willis Hawkins.

AAS 87-605

SPACE: A NEW COMMUNITY OF OPPORTUNITY - KEYNOTE ADDRESS

Willis Hawkins*

Thank you very much, Aaron. This is indeed a privilege. I think the purpose of your meeting is something that could be extremely important to all of our futures. I have elected today to do something that is a little different for me. I'd like to join with you in sharing your dreams. I do indeed share them and I hope I'll get a chance to demonstrate my enthusiasm in our later panel meeting. As the keynote speaker I suggest that we all take a new look at ourselves. If this turns into a sermon, I hope you'll excuse me. When you get as old as I am, you can deliver sermons I am told. I'll try to make it as creative as I can.

Those of us who are involved in space programs, particularly those who are involved in man in space, have experienced a rather major change in their technological momentum over the last few years. And technological momentum is important. We've had breathing room. We didn't want it, we didn't ask for it, but I think it's up to us to be sure that that breathing room was well spent. I think we all have had to rethink what we do and what our responsibilities really are toward those who have been supporting us in the space business over the last three decades. Really, we need new eyes. We've been living in a technological heaven that we've made for ourselves. There's nothing the matter with that, but the environment has changed, and we should recognize it and we should recognize why it has. The other evening I had a chance to listen to T. Allan McArtor, the present head of the FAA. He said, "We're in an era of inhibition of technology by the environment."

I think you all recognize that we're in that same environment. There are many of you here who can lead us into any kind of technological triumph we would like to have. And those you represent can help because I think you are the core of a very powerful technological family. What we haven't thought about is what we must do to avoid this inhibition because it is not good for us and it's not good for our nation. Perhaps we're the only ones that understand it well enough to remove it.

A couple of years ago, I had the occasion to do research for another speech on how the United States approached the problem of risk management. This research was fascinating. For instance, in the early days of the automobile business, there were a number of people who thought the automobile was the worst hazard that had ever been foisted on the public. There were those who sincerely said and believed

* Senior Advisor, Lockheed Corporation, 4500 Park Granada Blvd., Calabasas, California 91399.

that your head would blow off at 30 miles a hour. Now, the technologists knew better. They were patient with this kind of thing, but nevertheless as far as the environment was concerned, their experiments were pretty risky. But they took them. They had the judgment to take those risks, sort out the ones that were real, and the ones that weren't, and look at what they built. The technological judgment that was expressed in those days by the people who were the inventors built a sound, magnificent industry.

In the early days of the airplane things were much the same. With a few differences. The inventors were the pilots, in many cases, and therefore they recognized these risks because they were risking themselves. And they proceeded on the basis of reducing those risks as they went along as best they could. But it wasn't long before they were risking not only their own necks, but the necks of passengers and of the community. They continued to reduce these risks, but this kind of process -- inevitably I suppose -- developed a third party to be sure that these risks were rational.

The beginning agencies that became the FAA were formed. There was, in this process, a miracle of legislative judgment that took place. The legislation said that the FAA or it's predecessor, the CAA, was to regulate the safety of the product and sponsor the industry. Strangely enough in today's environment Congress has raised the question as to whether or not this is a conflict of interest. I hope they forget they asked the question because it is a wonderful piece of legislation and we could use more of it. Look at what has happened to that industry. It is one of the safest modes of transportation and the most efficient that has ever been created.

There's another example, and in this case, I didn't have any history of experience myself, but it was an equally interesting development. I refer to the nuclear power industry. In this case, the inventors of nuclear power were in an entirely different community from those who distributed the power. Therefore, they had different views of risks. One, just sort of ignored the problem. And the other had such a big problem that they didn't know how to express it to the rest of the world. Inevitably another third party was set up to regulate this industry. But this legislation didn't have the word 'sponsor' in it. Pretty soon it was recognized by everybody, and is still recognized, that the cold hand of government, if really given the task of controlling something without having to sponsor it too, could be a detriment to progress and that's what's happened.

I'm a little fearful that if we're not careful the same thing can happen in space. We need the judgment of an audience like this and those that work for you and with you to explain the risks we're taking and get the acceptance of those risks.

The nuclear power industry was very nearly dead. They've come back a little bit and they've developed some tools to help them stay alive. They've gone into the business of analytically looking at their risks, trying to see if there is a way of expressing these risks to the public. And they've been reasonably successful. They don't have very much statistical evidence to go on because they've actually been the safest power development organization that the world has ever known. Nevertheless, they're working at it and they're setting and establishing standards of safety and risk. And they're selling these standards to the regulators.

Let's look at our space history. Strangely enough we've got quite a lot of it, although it seems we've just started. When we first went into space, the public was completely with us. They recognized the dream of space and they accepted the risks that went with our first efforts. And our first efforts weren't exactly beautiful -- we scattered hardware all over the sky. But the public was still with us. Then we began to carry human beings into space. The public was still with us, accepted what we were doing, and somehow we were so successful that our success blinded us and the public to the real risks that we were taking.

I was amazed to find that in the time we have been in this business the public has permitted us to spend $116 billion dollars in non-military space. In today's environment we have to take a very sobering look at what we have done in the past. I have a few suggestions that we ought to ponder. I think we've listened poorly to what's been going on outside the space business all the while that we've been in it. When we started in the space business, airplanes weren't all that sophisticated and when you went on an airplane in those days, it was a little risky. But the things they've done between then and now are remarkable. Much of what they have done we might have been able to apply to the space business. Our success did something else to us. We sort of accepted a constant risk level, I believe. We made few changes from the time we started to where we are now, relying on retention rationale, not change, to aid us. The reception of the public was still good while we were successful. We were also blinded by the adverse public reaction to schedule changes and budget changes. That kind of fear of the public led us to do things, I believe, that we wouldn't have done had we used our own judgment. We then resorted to committees, boards, boards on boards, thus taking the responsibility away from the individual. I think we lose the judgment of real individuals when we have a democratic vote on whether to take a risk or not. And finally, we didn't really recognize how powerful the public reaction could be. There are those who will say, and absolutely rightly, that this isn't public reaction we're living with, it's the reaction of the *Washington Post*, and the *Los Angeles Times* and Congress. Believe me, that is public reaction. There's nothing you can do to avoid it. And I'm sure that the public itself is a lot more supportive than it sounds like in the headlines. But we're going to have to recognize that the *Washington Post* is the kind of public that we have to live with.

Let's look at the downside of this history and let's really go to work on it and see what we can do to reestablish the momentum and the leadership that we're all capable of taking and giving. We have the inspiration and judgment and we must not be stopped in our tracks. What about the future? Are there tasks worth doing? You can bet your bottom dollar. We're talking today, sitting on tenderhooks as to who is going to do it and asking ourselves how to establish a permanent entity -- a permanent habitation in space. It will be a center for learning, it will be a center for production, it will be a stepping stone to other orbits and other space adventures. We need ingenuity and real sensitivity to public perceptions to do this trick. I assure you from what I hear that the public really wants us to do it. But we need to do it as safely as possible. We need to do it in thoughtful steps so that we don't destroy our technical momentum, while we respond to what the pressures are, in particular, respond to the budget pressures that are bound to build. We have to select building blocks so that we can work at this a little at a time. There are many who will say this is sort of like shooting a cannon off a little at a time, it's not easy to do. I know it's not easy to do. But we're going to have to figure out how to do it so that we don't get stopped -- worse yet, stop ourselves.

We're starting another adventure -- the aerospace plane. This, too, we should do. The business of being hypersonic between airports is an exciting goal. But we need to define the milestones to do this, so that we can do the critical things first. And again, like in the Space Station, we can't expect a total national focus. We're going to have to do it while other people are doing other things around us, which also need national support. Here again we're going to have to look for the building blocks to get there. And we mustn't promise the Orient Express tomorrow, because headlines don't build budgets.

There is yet a third effort that needs our attention -- almost in desperation. How do we get the cost to orbit down? Strangely enough all of the good things that you folks are doing in trying to recover from the *Challenger* aren't really helping. I've been on a number of review boards and I have to say that I've never seen a group of people like NASA's organization and the contractors that support NASA who are more dedicated to finding out what really is wrong with the Space Transportation System and trying to fix it. But some of these are not really fixes and we're going to regret them tomorrow The business of having signatures on pieces of paper sometimes ten, sometimes 12, is difficult to handle now. But it will be much, much worse when we try to stop all that stuff and simplify the process by which we get into space. Our expanded critical items list isn't going to help. We're going to have to figure out which are the really critical ones and which really aren't. And redone rationales, logical as they may be, aren't going to help, because what we have to do is to get rid of the rationales. The things we are doing today are going to be very difficult to stop tomorrow and stopping those things tomorrow is going to be one of the keys -- it's part of the environment that inhibits our technology. We'll fix how man gets to orbit if we can fix how we get material to the same orbit for less cost.

I'm a little discouraged as to how we're going about this. We're entranced by the fact that SDI needs an awful lot of payload in orbit. You need an awful lot of payload to start up the Space Station. But we can't go about launching things the same way we always have, just tripling or quadrupling the payload to get the cost per pound down by the amortization. We're going to have to save dollars on every launch, whether it's a big launch or a little launch. I've heard a lot of the concepts on what we should do about boosters, and I've heard a few small voices saying, "Why don't we put together a lot of little pieces to make a big one, instead of making a big one to start with?" That makes a great deal of sense. If you have a lot of little pieces that all look alike, you can put three or four of them together and launch a little payload, or you can put 10 or 15 or 30 together and launch a big payload. You have the dollars saved through production rate. You have the dollars saved by having identical elements, all of which are handled the same way. We've changed our technologies quite a lot, maybe we don't need to stretch for performance the way we're doing. Do we need pump-fed engines, for instance? What about the margins we fly with? Shouldn't we put in just a little more structure so that one of these days we get to the place where we launch on the crew chief's word instead of holding a national symposium? And then, of course, if you want to dream a little while we're going off into space, why not dream of a booster that comes back all by itself, and lands itself, right where you want it--on the launching pad, perchance? I mean, if you've been in the vertical rising airplane business for a while, you ought to take a look at the thrust-to-weight ratio of one of these boosters when it's only got a 10 percent fuel load on board. I'm telling you that thrust-to-weight ratio would make your heart sing if you were in a vertical rising mode, and you certainly ought to be able to come back and land. After all, with the guidance accuracy we have in the missile business these days -- matters of feet and meters -- we certainly ought to be able to find the airport. And furthermore, we did it on the Moon 20 years ago. So, why don't we dream about things like that?

What have I said? I hope you've recognized, whatever I've said, gloomy or otherwise, that I'm talking to a roomful of miracle makers. You're all still pretty young, as near as I can tell from here. And Mayor Marilyn looks like she's 17 years old. What you want to do is good for you, and I assure you it's good for the nation. Your efforts could lead us to entirely new worlds, maybe some of them of our own making. Certainly we've stumbled. We didn't explain to the world what we were doing. Maybe we really didn't recognize some of the risks we were taking, and we weren't sufficiently alert to the opportunities to reduce the risks and to do something about them. Let's not allow the budget and schedule pressures to blind us. Let's get at our dreams with all the judgment we can bring to bear. Because those dreams aren't going to let us alone--they're sound dreams. We haven't lost the skills, we haven't lost the tools and space is out there waiting. Now it's time to use your tools, use our chastened judgment, listen to what is going on around us, explain to the world what we're doing and what we're risking, and plot a course that will thrill our most caustic critics, with truly economic steps to infinite space.

QUESTION AND ANSWERS

Henry N. Bowes: All of you when you came into the room should have received an audience participation form. If you want one and don't have one, raise your hand and we'll get your copy to you. Seeing no hands says one of two things, either you already have one, or you're not willing or able to put a question down. I like to think it's the former. Those of you who have already participated in the process, would you pass the forms to the outside? We will pick them up and we hope we will be able to sort them out if we get a flood. Do we have any yet? Too early? While you're writing, perhaps there are questions from the floor that we could take? I have one for you, Willis, maybe to get the ball started. We have been looking at the probabilistic risk assessment you mentioned in your presentation. My concern about the process is that we have very little data about 17-inch valves and other such things that are created specially for our business. Do you see any encouragement from the fact that the nuclear industry had to do the same thing in that they had little or no data on some of their machinery? Could you comment on that please, Willis?

Willis Hawkins: I do see encouragement. That's why I mentioned the nuclear industry. The real problem is that when a group of people take a look at a complicated system, such as a 17-inch valve, I think that both the assembled multitude and the individual in charge can ask, where is the critical part of that system? And there may be several critical parts but you can sort out the one that is most critical. The only trouble is you can't explain that to anyone else. There are ways however to define what you believe to be the probabilities that are involved. You can define them mathematically and if something is really catastrophic, it has a very high likelihood of causing disaster. If it isn't quite catastrophic, it suggests a different kind of definition of disaster. By putting these differences together mathematically, you can actually see what the total system is like. Furthermore, you can take a look at that mathematical combination in such a way that you can lead yourself to doing tests to remove some of the unknowns -- devise the test to remove some of the disparities that show up in this mathematical combination. Certainly if the combination really causes a disaster, then you've done it wrong in the first place. I think it's a way of mathematically putting yourself independent of total judgment in such a way that you can explain it to somebody else. That's half the problem, because whatever we do, our judgment may be pretty good, though we can't explain it very well to other people. This is what the nuclear industry has tried to do in order to explain to other people what risk level they're actually operating to. I think it's worth an exercise for all you folks in this business because it's one mechanism that has worked.

Henry N. Bowes: Willis, we have a question here:

Question: "What is your perspective on the use of used external tanks for research labs and/or a Space Station as proposed by UCARF's Space Phoenix Program?", and the writer goes on to say:

Why Important and of Interest: "It appears to be the quickest and cheapest way to proceed and it will be done by private industry".

Recommendation: And the recommendation from Pat Jackson is that it "should be supported." Pat Jackson is a freelance writer and video maker.

Willis Hawkins: In the first place,I'm going to have to state that I'm sort of ignorant as to how we can best use this kind of hardware. It always hurts my soul to throw those beautiful things away and I'm sure that's the incentive for everybody else. It sounds like a nice, simple solution, but if the solution can't be sold to the people in charge of the Space Station, then I hope it's because they're objective and they can figure out a better way of doing it. This is one of those situations, perchance, that I alluded to: "listening to the outside world while you're doing your job". I think, if this is a good idea to save dollars and actually save time in getting a Space Station put together, then it ought to be presented and analyzed by the people who are in charge of the Space Station and if they agree with it, fine, and if they don't, then I hope they've used good judgment.

Henry N. Bowes: Thank you Willis. We have a another one. This question is from Dr. Joe Kerwin of LMSC. He put a title in here which I'll let you share, Willis.

Question: "What did you mean by the statement, "We have to get rid of the rationale"?

Why Important and of Interest: "Knowing why we made a decision is useful, you must have meant something else."

Henry N. Bowes: And I'm not sure I read the question, the initial statement.

Willis Hawkins: I know what he's talking about. And I think most of the audience does. The way we handle critical items in space, particularly in the Space Transportation System is to define what it is and then, if we accept the fact that we're going to keep that critical item in the system, we have to explain to somebody else why. This is what the "rationale" is. It's written up to tell headquarters, if you will, and hopefully headquarters accepts and can stand off the *Washington Post*. But the rationale is why you have kept a critical item within the system when you're getting ready to launch. I'm trying to suggest that it would be much better not to have to write any of those rationales. In other words, I'm suggesting that we have probably not made as many changes as we should to remove critical items. It's much better not to have to explain them and be able to say we've gotten rid of the critical item.

Henry N. Bowes: All right, we have a couple more. Again, addressed to you, Willis.

Question: "The government is strongly urging increased commercialization of space, in part to remove government investment requirements. Is the space industry ready to be 'cut loose' commercially?"

Why Important and of Interest: "The pace of development is determined by the level of interest."

Recommendation: "Government should sustain their support." And this question is from S. E. Doyle, Director of Marketing, Aerojet TechSystems Company.

Willis Hawkins: The business of commercialization of space is a national quandary. In the first place, the commercialization of anything relies upon a market of some sort that you can reach for. The market to enable you to make money in space isn't at all obvious. One of the reasons for commercializing space is to take the burden off the government for having to make all expenditures for space. But you can't take everything off the government in view of the size of the market. I'm no expert in all these analytical exercises that people have been doing, but space involves quite a few things. It involves real estate because you have to launch from somewhere. It involves responsibility, and I mean responsibility in the legal sense rather than in the spiritual sense, because you have to take a lot of risks that aren't only risks of money, but risks of people that are not involved, as in overflying other countries, or launching in such a way that you can scatter hardware over some supposedly uninvolved area. That has to be handled and it is not normally handled in any commercial program except through insurance. Therefore, the government is going to be involved no matter what and it's a new role of the United States Government. They have never gotten into a program -- well, I shouldn't say never -- but they have never done a program that was half commercial and half government. I think some really good judgment and some really good ideas on how this process could take place are needed -- it's still a vacuum -- and when we hear the right idea that's salable to the government, then I think we'll have made some real progress. But the ideas still don't exist, as nearly as I can tell. Maybe we'll hear some of them later on in this program. We'll have the Department of Commerce and the Department of Transportation here, and they've been wrestling with it and I hope we'll hear of some progress. But it's going to take government and it's going to take industry, working hand-in-hand, and the *Washington Post* hates that like the plague.

Henry N. Bowes: Thank you. Another question,

Question: "As more cargo is lifted per launch, its value increases, perhaps to the point of exceeding the risk tolerance for the vehicle - where are we in regard to this? What is your perspective?"

Willis Hawkins: I don't think that that is necessarily the critical question to ask about our future in boosters. If we had decided that we're going to take 200,000 pounds to orbit, we'd probably need it and there's no other way to get it there. I think we will have taken the necessary care to produce a booster that will do this with a reasonable risk. What I really worry about is that every payload can't be 200,000 pounds and by the time you've collected a bundle of payloads to make up 200,000 pounds to get the cost per pound down, you will have had a planning exercise that may last five years. And we can't wait five years between launches and keep up any momentum. So I think the real problem is how we get a standard reliability, if you will, for all launches, from the smallest payload to the biggest. That's possible if we have the right launch concepts. So, I don't think the problem is really whether we can risk putting 200,000 pounds together, it's how we can get the risk so low for any launch that it doesn't make much difference what payload we put on it. Hopefully, the cost will go along with the payload, so we have about the same kind of cost per pound for the little payload as we have for a big one. After all, if we really approach it on a blind cost basis it would be like using 747s between Washington and Boston, and New York. And we certainly don't do that because there's a more economical way of doing it. You've got to adjust to the amount of payloads that are going to be available at a reasonable rate.

Henry N. Bowes: Willis, this is another question close to what you were just talking about.

Question: "Since the Russians have done so well with 'big dumb motors', why do you think that many small ones would be as good or better?" I'm not sure but I think you've almost answered that.

Willis Hawkins: The Russians are doing better than we are, I think. They're using a lot of small pieces to get their launch off rather than having great big pieces. They all add up to a big launcher, but I think their elements are substantially more alike which probably reduces their costs of launch.

Henry N. Bowes: Here's one on risk management.

Question: "Address application of risk management as an integral part of the design process."

Why Important and of Interest: "Building safety and reliability in 'up front' is cost effective compared to fixing things later."

Willis Hawkins: That wasn't so much a question as it was a small lecture. I can guess what organization put that in. I think one of the problems that we ought to face and one the organization of the whole space transportation business ought to face is the fact that using quality assurance, reliability and maintenance is all good. We have in effect produced these organizations and they do their work with the best tools they have. But we've produced them as staff to the doing organization. They advise the man who's running the place, and he has to decide how many of their ideas he can afford to incorporate. It's a little like what I told you about the FAA. They have two responsibilities -- one to regulate and one to sponsor. A quality assurance organization winds up saying, "You've got to do this, you've got to do that", but not having to solve the problem of how do I do all of this and still get off the pad? So what I was trying to suggest, and I think we ought to look at very hard, is to how does the program management utilize the inputs, keep track of his decisions on how much of this input he uses. It's that interconnection, I think, that needs a lot of building up because of the fact that this advice is all very good and you can stand behind it. But you can't afford it all. So how do you make those decisions and how do you make those decisions apparent to anybody else? That's the problem in management.

Henry N. Bowes: OK. We have one from one of the people that Willis referred to as the youngsters here. This is Randall Briggs, a graduate student from Worcester Polytech. A good old Yankee like myself from the North -- at least he's living in Yankee country.

Question: "What is industry doing to support educational programs at grade school, high school, and college levels?"

Why Important and of Interest: "Education of the population in the space field is critical to the development of future personnel."

Recommendation: "(1) Increased funding to these institutions; (2) development of goals for educational programs; (3) funding for fundamental research at the college levels."

Henry N. Bowes: And, Aaron, my thought on this question is, I'd like to have you express your opinion, too, since I know you do work with these kinds of programs, but Willis first.

Willis Hawkins: That's a question that could be the subject of a whole symposium, but there are two problems that industry has. Industry recognizes the necessity of having a constant influx of really well educated individuals to help solve problems so they do need to support the universities extensively. But, remember that most of the industries that you're talking about -- if you refer to space -- are industries that work for the government and the government decides how much you can give away, what you can spend your money on, or else it gets disallowed and it comes out of the stockholders' pocket. So this turns out to be an interesting tightrope for the industry to walk. Actually we do quite a few other things. I just saw a list, I believe, of 12 invited lecturers, who were selected out of one of the Lockheed divisions. And I know other companies do the same thing. But these lecturers are made available to any school on demand. They deliver lectures on how the engineering and technological inputs go into industry and how we use them. This brings the industrial experience into the schools. They're made available each year to go to various schools as lecturers for those schools at no cost. But there are lots of programs like that and I think we're doing reasonably well. Of course, everybody in industry wishes they could afford to do more.

Aaron Cohen: I am very interested in the program of working with academia. As I said in my opening remarks, I think industry, academia, and government are really the three--if you can get those three areas working properly together, you've got a winning combination. In terms of the government supporting the universities and academia we've got a number of programs. One fundamental program is for grants and contracts to universities in high technology areas that will give the students the benefit of working in those areas and working with the government on these various technologies and that does two things: Not only does it give the students something to work on officially, it also provides better students to the aerospace community. So, that's very important. We also have started very recently -- or it is coming out soon -- a university program where there will be contracts given to universities. They will be on a competitive basis, but it will be long-term contracts given to universities from our OAST organization headquarters which will stimulate research by the students. So, there are a number of things that are being done. Can it ever be enough? I don't know if it can ever be enough, but there are things that are being done, because we do want to get good students out of the universities. And there's another thing: Working with the universities, you do get the bright ideas that come out that industry can capitalize on. So I think all that's very important and we are taking steps to make that happen.

Henry N. Bowes: Captain Honhart asked if he could say a couple of words, too. Dave?

David C. Honhart: I just want to point out in a letter which we are preparing to send to our corporate members which I've simply titled, 'An Open Letter to the Aerospace Industry of the United States', dated 15 November 1987. A short paragraph to explain why corporations should be members of the AAS. It goes as follows: 'Last, but not least, being a corporate member also provides the opportunity to influence the youth of our country through participation in the Goddard Memorial Symposium. In the youth of the country lies the strength of our future, what better investment of time and money than to cultivate their young minds?' We made a decision -- a decision last year within the AAS, that is, by our national leadership, that in the Goddard Symposium which occurs in March, we have a session just for the students. It's the afternoon preceding the full symposium. We also made a decision that during the two days of this symposium we would select, on a competitive basis, papers written by students at the colleges and universities, graduate and undergraduate level. And we clearly see, as far as the AAS goes, that it's incumbent upon us to promote this education and enthusiasm of the youth of the country.

Henry N. Bowes: Thank you Dave. And I'd like to put my own two cents in. There is an organization in the Clear Lake area of Junior Achievement, Bill Hayes of McDonnell Douglas is the leader of that effort. Its purpose is to provide programs in space technology, at the high school level. So, there's a number of things already going on. And certainly, Dave, AAS could take a stronger position to encourage participation in the kinds of things that Mr. Briggs suggested. I have two more questions and then we'll take a break.

Question: "The plan that has been presented to deliver the Space Station to orbit requiring a total number of shuttle launches and flight rates we have not yet been able to achieve seems unreasonable to me (a member of the program). How can we expect the U.S. government and the American people to fund/buy such a plan?

Recommendation: Shouldn't we be presenting a plan that we in the space community believe in? And this is by Ray Barrington, Shuttle Flight Controls, Software Verification from Rockwell.

Willis Hawkins: I have listened for a long time to the people who have tried to figure out what is the rate that you can use for shuttle launching. We need, of course, the new orbiter, and we hope it gets here reasonably soon. I really am not close enough to the Space Station plans to know what rate they need to have in order to get all of the hardware into space, but this is an element of what I was talking about in making building blocks that fit. We certainly don't have to launch all of the pieces of the Space Station as rapidly as we would like. We don't have to launch them in quite the order, maybe, that we first planned and will eventually apply. I was trying to suggest that we do indeed need to do this planning so that we can fit both the budgets we know that are going to be available and the realities of life as we know they're going to be. There was a large study out of the National Research Council, that took a look at the rate of shuttle firings which really ought to be planned for and that was published some time ago. I don't know what NASA has done about that or whether they have changed their plans because of that report. But the rate at which you can launch shuttles is certainly going to depend on (a) the paperwork and (b) and how well we have fixed the hardware.

Henry N. Bowes: Thank you very much, Willis, and the others. I would like to ask that we acknowledge some pretty good fielding on some really tough questions, so thank you very much, panel.

HONORS AND AWARDS BANQUET - MASTER OF CEREMONIES PRESENTATION

Aaron Cohen*

Aaron Cohen: I should introduce myself, I'm Aaron Cohen, Director of the Johnson Space Center, and I'm the Executive Chairman of this symposium, and I also will be your master of ceremonies tonight. Many of you know that the AAS is a society that has the principal goal of strengthening the United States space program. The caliber of panelists we've had here at this symposium has been excellent and we truly had some outstanding discussions that will help this country and our nation resolve these issues and extend our international leadership in space. Your participation is providing an exchange of ideas and information that will serve policy-making organizations, both nationally and internationally.

The highlight of our program tonight is the presentation of awards to leading contributors to our space program. It is particularly fitting that we pause and honor a few of those very important people whose contributions are providing the foundation of our space work. There are so many people here tonight that I'm a little bit awed by the number of outstanding people we have in the space program, both from Government and industry, and I do appreciate your coming.

We will begin the ceremony with the presentation of the colors by the United States Coast Guard Detachment from Ellington Field. Commander Stephen Goldhammer, Commanding Officer of the Coast Guard Air Station at Ellington Field, will lead us in the pledge of allegiance.

Commander Stephen Goldhammer: Parade to Colors.**(PLEDGE OF ALLEGIANCE)** Post to Colors.

Aaron Cohen: Please remain standing for the invocation which will be delivered tonight by a gentleman who flew as the pilot on STS 61C in January of 1986 and is currently lead astronaut for the vehicle integration testing at both Johnson and Kennedy. Ladies and gentlemen, I'd like to present Colonel Charles Bolden.

Colonel Charles Bolden: Please bow your heads and join me in a word of prayer. Lord God, we thank You for the beauty of the Earth and our continuing search for knowledge of the Universe. Guard us in our frustrations as we find, continually, that

* Dr. Cohen is Executive Chairman, 34th AAS Annual Conference. He is Director, NASA Johnson Space Center, Mail Code AE, Houston, Texas 77058.

the more we learn, the more there is to know. We ask that You be with us this evening, that You bring us a message that we can understand from our speaker. We ask that You bless him and everything that he says and does. We ask that You bless this food that we are to receive for our use and we ask that You bless us to Your service. Amen.

Aaron Cohen: May I have your attention again please. At this time I'd like to introduce our guests at the head table, on my left, is my wife, Ruth. Beginning at the far end is Marilyn Goldhammer, and her husband, Commander Stephen Goldhammer, who gave the pledge of allegiance. Most of you already met Hank Bowes; he's Vice President and Program Manager for Lockheed's Houston Engineering and Science Programs and was really the muscle behind getting this symposium ready. In view of all the work he did, I think he deserves a big hand, and also for all the people who helped him; I thank you all very much. Next to him on his right is his wife, Betty. Next is Captain David Honhart, President and Chairman of the Executive Committee of the AAS; he has also served as Chairman of the Planning Committee of the 1986-1987 Military Space Symposia. Captain Honhart was actually involved in planning the utilization of a Space Shuttle for oceanography and also developed the concept for an oceanographic satellite project, known as the Remote Ocean Sensing System. He is presently serving as Deputy Director for Ocean Sciences and Technology and the Director of Space Oceanography on the Staff of the Chief of Naval Research. Tonight Captain Honhart, in his capacity as President and Chairman of the AAS, will present the honors and awards at this banquet. And sitting next to Capt. Honhart is his guest, Carolyn Fine. Carolyn is responsible for the Lockheed Leadership Program and is Secretary for the Symposium Host Committee. At the far left, is Colonel Bolden, and his wife, Jackie. Next to Jackie is Willis Hawkins, who gave the Keynote Address and is a very well known consultant in the aerospace industry and he will also be on the final closing panel tomorrow, which should be very interesting.

Now, I have the pleasure of introducing our Speaker for this evening, Dr. Thomas Paine. We all know him, and we are most fortunate, that he was able to find time to be with us tonight. We truly thank you. Some of you may know him as the former NASA Administrator, directing NASA during the first seven Apollo expeditions, including the first landing on the Moon. I heard him talking to a very good friend of ours, Dr. Chris Kraft, a few minutes ago, telling him he always knew that Chris would get the software there on time. And I guess you did, Chris. Others may know him from his outstanding work that he and his team did for the Presidential Appointed National Commission on Space, created to chart civilian space goals for the 21st Century. I will not say more about that, because I presume he will go into that. To illuminate a little his background, Dr. Paine has served three Presidents, directed several major corporation, such as RCA and NBC. And has been the recipient of many distinguished awards for his leadership and achievements, both national and international. Of particular interest to each of us is that Dr. Paine just returned a few weeks ago from Moscow, where he not only attended but delivered the closing remarks for the Space Future Forum which was sponsored by the Soviet Institute for Space Research. He truly is a great American, a great support of the space program. Please join me in welcoming Dr. Paine.

HONORS AND AWARDS BANQUET - KEYNOTE ADDRESS

Thomas O. Paine*

It's a great pleasure to be back in Houston tonight. I see a lot of friends here. Talking of Chris, I didn't say that I knew that Chris would always deliver the software on time. As a matter of fact, I had a lot of doubts. What I said was he always did. It wasn't always obvious.

We meet in the 30th year of the Space Age here tonight, and that was the allusion that Aaron made, that one month ago tonight, I was in Moscow on October 4, which of course, is the 30th anniversary of the launch of Sputnik which occurred on October 4th, 1957, an event which changed the lives of everybody in this room, and really changed the lives of people around the world. The Soviet Union in its new spirit of *Glasnost,* which we're all waiting to see how real it's going to be, invited about 900 people from many countries to this get-together in Moscow a month ago and used this symposium approach to take a look at the future.

Now, as the Chairman of the National Commission on Space, I had been charged by the Congress and the President to come up with a 20-year space program. We did that and published it in a book which you can buy for $14.50 from Bantam Books. I thought it was a pretty good job, but it wasn't supposed to be a book or literature;, it was supposed to be a program. Well, I'm delighted to tell you tonight, that it has now moved from a book to a program. The good news is that it has been accepted. The bad news is that it was Moscow that accepted it, not Washington. One of the events of the 30th anniversary, again in the spirit of *Glasnost,* was an effort on the part of the Soviet Joint Government and News Bureau -- another branch of the Government -- which resulted in a publication that laid out the next 15 years -- and they came pretty close to 20 -- of the Soviet space program. It was interesting to go through it and note that it was certainly very close to our National Commission on Space recommendations to the United States, which are still being debated.

However, about two weeks later, there was a meeting in Mexico City of cosmonauts and astronauts -- Rusty Schweickart's little group that gets together. Rusty had started this organization. It didn't go anywhere at first, but then the Soviets, wanting to have more international discourse, started sending cosmonauts to these meetings. So we had quite a crew in Mexico City, and it was interesting to talk to them about what they wanted to do with this astronaut-cosmonaut organization. One of the things they decided to do was to take a look at the future and to come up with

---

* Thomas Paine & Associates, 2401 Colorado Avenue, Suite 178, Santa Monica, California 90404.

recommendations from the viewpoint of people who had flown in space, which is not a bad idea. But the interesting thing was that on the very last day, in fact, as I was leaving the hotel, with my suitcase in my hand, when one of the Soviet cosmonauts came up to me -- it was Leonov who speaks pretty good English, better than my Russian -- and said, "You know, you will be interested to know that in our next year's meeting, where we're going to be looking at the future, we needed some kind of a common document that we could all use, and we've adopted your report as the international document." So maybe we've sold it twice. But we're still working on Washington.

I must confess to you that it was a new experience to me to be in Moscow and to have a group of Russians playing the role that I remembered so well as those of us in this room that worked together have played. The Soviet scientists were confident of their program. They were very sure of their goals. They were publishing them covering the next 20 years, so that other people can cooperate with them on their programs, and they were welcoming and inviting international cooperation and bragging about the number of people who were going to be working on their programs with them.

Next year, when they launch to Mars, on the Phobus Mission, for example, they're going to have participation by 14 different nations, not counting the U.S. support that is being given to that mission. So, it was kind of a turn around. We used to be the ones who had the programs that we were enormously confident in. We were the ones that used to announce our goals well in advance. We were the ones that wanted to work with people from many nations. A thought went through my mind, as I listened to the speeches of the Soviet leaders. It sounded like the 1958 Space Act that created NASA. They were all the same words. So I think you could say, that the United States and the Soviet Union have reached an agreement in principle to cooperate in space, since they are using the same words out of our Space Act and that's the law of the land here. It's just that we've got a lot of work to do, and I think the work is very clearly geographically pinpointed in this country. All of the problems of the civilian space program in the United States today are confined to a very small triangle. They are bounded by the White House, the Capitol and the Pentagon. If we can work that little area, and get that part of the equation right, everything else will click into place very quickly.

When you look at the actuality of the American civil space program, when we resume launching with the upgraded Shuttle next year, we will be launching the most advanced launch vehicle in the world and one that will remain so for many years. The cargoes that it will be taking up to orbit will demonstrate the most powerful spacecraft that have ever been built. The Space Telescope is going to create a revolution in astronomy; there is no question about it. The Galileo Mission is going to tell us a great deal about the formation of the Solar System in its early evolution. We have a magnificent program ready to go. But what we still lack is the decision, at the highest levels, as to what should be the goals and objective over a long-term period. And that is what we must get behind.

I had a thought, which I'll share with you. It occurred to me that if we took the people who are responsible for developing the goals and objectives and converting these to a United States 20-year space program and if we were do tie to each of their ankles a leg iron with a long chain that would be long enough to reach from the Washington Monument to the Capitol and then let them go anywhere they'd like, in that way they could still reach the White House, reach the Pentagon, we would keep them tethered to the Washington Monument until we had a program. That would be a satisfactory way of emphasizing the fact that we must get some high-level decisions. We've had enough committees, we've had enough studies, we've had enough reports, and I'm sure Sally Ride would agree with me, if she were here tonight. It's time for some decisions.

I think that when you look at some of the problems we have in the country that are very dramatic, such as the skirmishing in the Gulf where we sank a fishing vessel; we lost a trillion dollars on Wall Street -- well what's a trillion dollars to a bunch of space people like us, except a few years' business? -- you will note that many of the problems that are so exercising the country today are really problems of the moment. We're not spending enough time looking at the fundamentals. We're not looking at the long-range things. Now, I have no doubt that the goals and objectives in the space program that we spelled out in our commission, that Sally Ride put down in her study and we added this to the Cobb-Whitey(?) five-foot shelf of space studies, that these are all going to happen. Our problem is to translate them into a program with a time table with long-range budgets, so that we can get the needed support. Many people will say the problem is that we've got to get more support from the American public. I don't believe it. The public is with us. The public wants a strong competitive, far-out, really challenging American space program with America in the lead. Our problem is to get this public support translated into an acceptable program and we can't let short-range considerations rule out the very important long-range considerations that must be the basis for decisions.

I would like to indicate where I think we ought to be going, and in a sense, where I think the international space community will be going. In laying out a program like this, we thereby create the capability of international cooperation. I think that there are many people around the world waiting to announce their cooperation with the United States. Only first they want to know what it is we're going to be doing. I go back to the days of Apollo/Soyuz. We started that by sending to the Soviet Academy essentially a sort of a road map, as to where the United States will be going in space in the next ten to fifteen years and inviting them to take a look at it, and then asking the question, where they would like to mount some cooperative efforts. In raising specific questions, like coming up with common hatches, so that for docking under emergency conditions we could carry out a rescue mission, and so forth, we were able to establish common ground. It was on the basis of being able to lay out where we were going that we got others to go along with us.

When we announced the Shuttle Program, immediately we were able to get the Europeans to say, we'll do a Spacelab. We got the Canadians to say, we'll do the arm. But we must have a program before we can get the kind of international cooperation which is a natural forest in which everybody wins. It's not a zero-sum game. Everybody wins when we cooperate. But it must start with a decision on the part of the United States as to where we're going in space and what kind of a time table is there. And it's that kind of a proposal that I'd now like to discuss with you for the remainder of the talk.

We will of course, be back in space very shortly. We will be back in orbit. And we will be there not only with people but also with robotic spacecraft able to perform the many different things we've learned to do in orbit, like establishing the thermal distribution of the entire planet. We will be able to go back into the business of using the Shuttle the way it is supposed to be used, which is to permit us to carry out a wide variety of missions, involving both men, machines, carrying things to orbit, bringing them down, all of the things you know so well.

The Space Telescope will be only one part of the Great Observatories Program and we must keep these on track. One of the very strong parts of the Soviet program that was announced last month was an Observatories Program paralleling our own. The electromagnetic spectrum is wide open once you get out into orbit, and it can tell us a tremendous amount about the Universe in which we live. And it's very important that this be a cornerstone of our program. You ask the question, what is the most fundamental return we get from our space program, what should be the thing that we place right at the top? The scientific research return and the technological advances required to get this new knowledge are the principal values that we get, in my view.

There are three major transport systems that form the basis for the next 20 years in space: the cargo transport, the passenger transport and the transfer vehicles beyond orbit. The Soviets have very vigorous programs in all of these areas. We, of course, have in the past been very much in the leadership, not only with the ability to take higher payloads to orbit than anyone else, but we pioneered the hydrogen/oxygen engines, and other things. We must now take this great heritage from the past and, for cargo transport, move this out into the next generation, get away from the technologies of the 50s and 60s, move out to the technologies of the 90s with the major objective of increasing automation, robotics, the ability of reuse with the objective of getting the cost down. Of course, for passenger transport, we are about to go back into business with the upgraded Shuttle. But the Shuttle will require a replacement.

The Aerospace Plane is another one of the subjects of the conference. It's a terribly important and absolutely basic project for this country. But many of the things that are going to be accomplished in the Aerospace Plane Program in the areas of very advanced materials, structures, many of these things would apply equally well to rocket-propelled vehicles. I think we need to broaden not only the Aerospace Plane project, we need to make very sure that we are also taking a very creative view in other ways, of cutting the cost and increasing the reliability of getting back and forth to orbit.

Thirdly, we need to be working on the transfer vehicles that go beyond Earth orbit. It has been many years now since we have operated beyond Earth orbit, and we haven't even been sending planetary missions out for some years. It is time that we begin to move beyond Earth orbit. It's time that we begin to get into things like the aerobraking technologies and the modular types of constructions so that we can have sensible vehicles to go to geosynchronous orbit, to lunar orbit, to Mars, to the outer planets, and have enough power so that we can put together enough of these modules to accomplish landing missions on the moons of Saturn. These are the kinds of devices we need. And the basic point of all this is that we must advance the technology that will give us low-cost access to the Solar System.

I see that my friend, Willis Hawkins, is going to conclude the meeting tomorrow with a talk on that subject. This evening he was not only preparing his talk, but also completely redesigning the Shuttle system. And that got us a little nervous. But this is absolutely basic, and when you ask the question, what is it that is going to make the 21st Century, the extraterrestrial century? It is low-cost access to the Solar System. And NASA must work with industry, universities in the great team work that we've seen in the past to really tackle this problem and get those costs down. It's fundamental.

There are three critical space facilities that we need to be moving forward on: the Earth Space Port which will evolve from the Space Station or will more likely be a second generation of it. We need to start doing some variable gravity research and we need to begin to think of the lunar surface outpost. We, of course, have pioneered the Space Station business. One of the great disappointments to me was that the Skylab Program was not recognized as the magnificent success that it was. This program was done as an offshoot of Apollo. It demonstrated, with the Orange Curtain that Pete Conrad and his friends put up there, the tremendous importance of being able to operate with people in space, to combine that with your robotic systems. It was a great success and we learned an enormous amount from it.

It is time that we stop the study, study, study on the Space Station and go to work on a Space Station for the 90s that represents the new technology that has the automation, the robotics, all the other things that we need on it. The ability to move up and put permanent facilities in Earth orbit is long overdue. We have really not taken advantage of the capabilities and the reason is that we've gotten hung up on the transport, and the transport has been too expensive. We must solve that low-cost access to space and when we do, the Space Station programs will be greatly facilitated. They will grow. They will be able to accomplish many more things.

The type of space base that we need in the future is a sort of a Cape Kennedy in orbit. We need to be able to use the transport systems from the Earth to the Moon, Earth to Mars, Earth to all orbits beyond low-Earth orbit, through a facility that has the capability of providing the life support, the logistics support, the refueling capability, the reworking of experimental parts, where you don't know before you do the experiment all the things you're going to find out. You don't want to bring them back all the way to Earth and take them up again. You need to have very capable systems in this space base. And it is time we began to specify this as a system for the early part of the next century.

The variable G is a very controversial subject. We did variable G way back in the Gemini Program, in a sense. I think we got it down to .001. At the meeting in Moscow a month ago, the Soviet cosmonaut who at one time had held the record for the longest duration of any human being in zero G announced that he would certainly volunteer for a Mars crew. He knew he could stand it for that long, but on the other hand, if it was all the same to everyone else, he'd like to try it with variable G next time. He was sort of booed by the audience and it didn't go over. But we do need to begin to understand artificial gravity, the parameters, and if we do need this in the future, we need to have the engineering numbers available for designing it.

Of course, we have been to the Moon many times, right here from Houston and we are going to go back. The ability to traverse the surface to do useful work there has been all worked out and in my mind; it is almost getting shameful that the United States would have such capability and not put it to use. Again, the key is to develop the low-cost transport, so that the next time we go to the Moon, first with robots then with people, we can go to stay and support the program year after year in an economical way. The Soviets are talking about establishing a robotic base on the Moon about the year 2002 equipped with robot vehicles with perhaps a range of 1000 kilometers. The biggest difference between our past trips to the Moon and the future will be in the area of life support systems. Again, here is a technological capability that will be one of the limiting factors. We must learn how to live off the land on the Moon.

You are having sessions here, discussing the problems of closed ecology life support. They're a very big topic in the Soviet Union right now. And the Soviets have another interest in them and that is, many of their far northern installations could also utilize them right here on Earth. But in order to learn how to handle an entire food cycle from bottom all the way up to human, and then how to add automation to this, we need a real revolution. There is an experiment in Tucson, Biosphere II, that is now starting to do this. But we are low by one or two orders of magnitude in the level of investment in this technology that we should have underway in this country. The Soviets are putting a lot of emphasis on this; we must do the same. If we can indeed learn to do the cutting of the umbilical cord back to Earth for many of the things that we carry out on the Moon, there is no reason why we can't have a very substantial and growing colony on the Moon, largely self-sustained, except we'll have to bring in some water, carbon, other things, but far less than we would have to carry in the early days of the century.

In the early half of the next century, beyond the Moon, of course, lies Mars. If there was one topic last month in Moscow that led all the others, it was the topic of Mars. The Soviets announced that the principal focus of their planetary program was going to be the Red Planet. Of course, they would love to have the capability of taking a picture like the Viking picture that we took of Mons Olympus. That magnificent volcano lies there, as Arthur Clarke once said in one of his stories, it lies under the stars waiting for our grandchildren. We need a program of going to Mars and we need to follow up this program that the Soviets will be launching next year to put laser probes into the Martian orbit to rendezvous with Phobos, perhaps also with Deimos, although that was not clear, and find out what these moons are made of. If indeed they are carbonacious asteroids that have been captured by Mars, as we suspect, it should be possible to extract water from them, which would give us a refueling base right there in orbit around Mars. The Soviets will probably be the first to find this out if the two spacecraft they are launching next year are successful. The ability to go to Deimos is not quite clear. The Soviets could launch their second spacecraft next year to Deimos. But I don't believe they have the capability of doing that after the first vehicle arrives at Phobos. They can't reprogram as easily as we do; they don't have the software control that we have in our planetary program. So I think they have to make up their minds before the launch whether or not they are going to target Spacecraft II to Deimos. But in any event, the chances of the Soviets' finding out the composition of Phobos and Deimos are very good. They have also announced the very vigorous program to land surface rovers on the surface, and to effect sample return missions as, of course, we have in our own planning and recommendation.

The two programs, those of the United States and the Soviets, looking at Mars, have so many points of potential cooperation that it will be a great shame if we don't get our own program in order, and then discuss with the Soviets how we can work together.

The Soviet sample return mission will bring samples back to Earth where they can be studied in the laboratories of both countries -- an ideal international mission in many ways. If either country fails, the other country still has a pretty good mission. But if they're both successful, you multiply the effect 100 times. Whatever the timetable -- and that's open to a good deal of question -- there is no doubt that man will be landing on Mars in the next century. The Soviets assured us that they had no timetable, no specific program, and that they regarded the first manned mission to Mars as a magnificent opportunity for international cooperation. And their first preference would be to join, perhaps lead, but join in an internationally planned expedition to Mars. They were very sure that it would not be carried out in the next ten years. After that they weren't quite clear. A lot depended on the research that they are still doing and their very long duration flights aboard the present MIR space station, which will then be augmented to a more elaborate space station that they plan to build using the new and Energiya launch vehicle that will allow them to build a capability, perhaps by the turn of the century, to carry something like a hundred cosmonauts, in a very large orbiting space structure.

Lest you think that a hundred cosmonauts, in a big space station is excessive, I cast my own memory back. For three years of my life I lived in a steel tube that was 18 feet in diameter and about 250 feet long. And there were about 90 crew members in that World War II submarine with me. So it can be done. It isn't terribly comfortable, but a space station with 100 people aboard, sometime around the turn of the century, is probably going to happen. With that kind of capability, the ability to mount a long-term, large space vehicle capable of taking people back and forth from Earth orbit to Mars orbit seems very straightforward indeed.

I am not one who advocates that our landing on Mars and the establishment of an outpost, and then a base there, and then maybe eventually even a community, should be done as a joint venture. I like the idea of an American program and a Soviet program cooperating with each other and with other nations free to contribute to and join, either or both, of those two. In my view, our Martian eggs would be much safer in two baskets. But in any event, once we have cosmonauts, astronauts and people of other nations, I'm sure as you look at the long-range plans of the Japanese space program, that they will certainly be ready to join us by that time. The capability of establishing a live-off-the-land ability on Mars is very clear. I'm sure, with all the resources on Mars, particularly the water and the carbon, that we can indeed learn to live off the land very successfully there and we must do that.

Beyond Mars, of course, we are going to get our first look at the planet Neptune, when the old Voyager spacecraft gets out there. Again, there's some good news, and bad news. The good news is that that is absolutely magnificent -- the Voyager, the looks that we've had of Jupiter, Saturn, Uranus, and the Voyager is still healthy and able, in a somewhat creaky way, to give us some very good pictures of Neptune and its moon, Triton. That's a wonderful program. The bad news is, that with that capability the United States has not continued to send out spacecraft based on what we already know and what we could have done at a very modest cost. I remember telling President Nixon, when we were planning this program, "Mr. President, the last time the planets were lined up favorable to a mission like this was back when Thomas Jefferson was President and he blew it. It's up to you to authorize this." Well, after a lot of negotiation, one thing leading to another, we finally got the Voyager programs going.

The Soviets have very high on their agenda of desirable missions a landing of a probe into the atmosphere of Titan, the big organic moon that circulates around Saturn. We don't know a great deal about Titan except that it seems to have a great deal of methane in the atmosphere. It may have oceans of methane. It looks as if it has a sort of a preserved surface, very much like the early surface of the Earth. With all of those organic compounds there and with processes of a very intense electrical nature, it could well be that some of the conditions that led to the beginning of life on Earth are now going on on this cloud-shrouded moon. It would be very desirable to get some probes into that. The Soviets are determined to do it. It's also very high on our list of planetary objectives. Again, a wonderful opportunity for some cooperative programs.

Of course, Mars will not be the end of Man's adventures in the inner solar system. The Soviets are planning the first soft landing on asteroids before the end of the century. They are going to use some gravity assist off Venus and perhaps Mars. They may dump off some balloon probes into the atmospheres of those planets before they fly on out to the asteroids. They plan both docking and also a potential sample return mission, perhaps around 2000 or so and there is no question that, beyond that, the day will come when the cosmonauts and astronauts will want to rendezvous with some of these mysterious objects.

Now in closing, I'd just like to ask, are all of these pipe dreams that are too expensive to contemplate? The answer is, absolutely not. One of the things that we did in our commission study, was to take a look at the economic growth of the United States and then try to plug it in 20 to 50 years into the future.

In the right-hand column, the numbers show the growth in per capita income in constant 1985 dollars, where we are now up to something like 16,000 dollars per man, woman, and child in the United States. The announcements in the paper this week point to a very substantial increase in productivity for the first time in a long time in the United States industry.

There is no question that this productivity growth goes all the way back to the beginning of the republic. When we look at all of the things we have available now, with artificial intelligence, robotics, automation, to the biological engineering revolution coming on, we've only begun to get the maximum productivity out of our vigorous American society.

There's no question that the $16,000 will become $20,000 per capita. We projected that by the year 2035, probably the United States per capita income would be on the order of $40,000 of present day purchasing capacity. In a nation of that sort, the kind of space adventures we're talking about, the investments on the space frontier in science, technology, are really minimal -- a fraction of one per cent of the Gross National Product. And it was on that basis that we laid out our timetable but don't try to look at the detail of this! If anybody is interested, he can look in the book, but I think it is a completely realistic, a highly productive, and a program that would attract enormous international cooperation. We would almost have to fight our partners off; they would be so anxious to join us in these adventures. But we must start with a plan.

The United States has, of course, the greatest space accomplishments. We have opened the exploration by humans of other worlds and those in this room can take great pride in the part that Houston, and we, and our predecessors played in this. The time will come when the things we have done on the Moon, will be done on other bodies. It is inconceivable that the United States will not be in the leadership when this happens. We are the wealthiest nation that has ever existed. We are the most technologically advanced nation in history and we have the great pioneering tradition.

Five hundred years ago tonight there was a little caravel from Portugal approaching the Cape of Good Hope. And by December the Portuguese had discovered the tip of Africa and opened the routes to the Far East, bringing the Atlantic nations into the forefront. Of course, 500 years ago tonight, the seizure of Malaga was under way and Columbus was with the Royal Court of Spain, submitting his unsolicited proposal. It was unlike some other proposals; it was accepted and five years from then he would set out to discover the New World. We are the people who have inherited this tremendous past, and we are the beneficiaries. We must lead the world out onto the Space Frontier. Thank you very much for inviting me to be with you tonight.

Aaron Cohen: Dr. Paine, thank you very much for your interesting and stimulating talk. And here's a little momento. At this time I would like to invite Captain David Honhart, President of the American Astronautical Society, to present the 1987 AAS Awards.

AAS 87-657

HONORS AND AWARDS PRESENTATIONS

David C. Honhart*

The AAS is a relatively small organization in comparison with other professional groups. But it is a highly esteemed organization. And the Awards it gives out and its fellowships are well received. And well they should be because, in order to be elected to be a fellow or elected to receive one of the awards, it's done by your peers, people who are equally successful, who have achieved the highest levels of success in the Government and in private industry.

So with that let me make the announcement for the first award. It's called the AAS Space Flight Award for 1986, and it's given for outstanding contributions to the advancement of space flight and space exploration. This year's award goes to Eugene Krantz. He is the Director for Missions Operations at NASA Johnson Space Center. The citation reads "For your exceptional contributions to the development and execution of spacecraft operation concepts from Project Mercury through the flights of the Space Shuttle."

Next, the Victor A. Prather Award 1986, for significant contributions in scientific research. This year's award winner is Dr. Joseph P. Allen, who is Executive Vice President for space Industries, Inc. The citation reads, "For significant contributions in scientific research and explorations from the Apollo Missions to the Shuttle Program".

The next award is titled the Lloyd V. Berkner Award, which is given for outstanding contributions to the commercial utilization of space technology and this year's award is given to Dr. Maxine A. Faget, who is the President of Space Industries, Inc. The citation reads, "For your pioneering efforts in the development of the first privately developed man-tended space facility based on your productive career in the nation's manned space program".

The next award is titled the Melbourne W. Boynton Award, which is given for outstanding contributions to the biomedical aspects of space flight. This year's awardee is Dr. Margaret Rhea Seddon, who is a NASA astronaut. Her citation reads, "For your exceptional contributions to NASA, the space program, and the biomedical sciences as evidenced by the preparation for and participation as a crew member of the highly successful STS 51D Mission". Dr. Seddon is not here tonight, but she will receive her award at a later time.

---

* Captain Honhart is President of the American Astronautical Society (1986-1988). He is affiliated with the Office of the Chief of Naval Research, 800 N. Quincy Street, Code 13D, Arlington, Virginia 22217.

The next award is the Dirk Brouwer Award for 1986 for space flight mechanics and astrodynamics. The awardee is Dr. Andre Duprit, who is a senior fellow from the National Bureau of Standards. His citation reads, "For significant contributions in the use of computers and the development of perturbation methods and their applications to analytic orbit theories and to the furtherance of our knowledge and understanding of satellite orbits."

The next award is the Eugene M. Emme Astronautical Literature Award for 1986, and this year it goes for the book, *Pioneering the Space Frontier,* and the recipient is actually the National Commission on Space, and receiving the award is Dr. Thomas O. Paine.

Now, the nominations for the rank of Fellow which is granted to individuals who, over a long period of time, have continued to contribute to the space program. The other awards we give are for specific things. But this rank calls for a continuous contribution. Again I point out that the nomination and election to be a Fellow of the AAS is done by your peers, by people who are Fellows themselves, so it should be regarded as a very high award. The first Fellow who, unfortunately, could not be present is Dr. Michael I. Yarymovych. The second individual elected to be a Fellow of the American Astronautical Society and also, unfortunately, not present, is John J. Quann. The next Fellow of the American Astronautical Society, is Peter B. Teets. Peter is the President of Martin Marietta Denver Aerospace. For 25 years he has participated in key roles in American space programs, beginning with flight control analysis on the Titan Program in 1963, continuing with the program management of several launch vehicle programs, missile programs and space system programs and culminating with his present assignment as President of Martin Marietta Denver Aerospace. The next individual nominated for the rank of Fellow is Carl Henrikson, whom I met tonight for the first time. He is the Senior Program Manager for Ball Aerospace Systems Division in Boulder, Colorado. Carl has been a key engineer, designer, and technical manager of several outstandingly successful space programs in his 29 years in the aerospace industry. His work stretches from the Viking Lander, Voyager, Lunar Orbiter, and DinaSoar while at Boeing, and continued at Ball through the Mars and Venus orbital studies, the angular momentum control device and the small instrument deploying system to his most recent position as Program Manager at Ball Aerospace. (I have to tell you another thing about doing this; you have to take linguistics to learn how to pronounce all of these names.) Our last Fellow awardee tonight is Jer-Nan Juang who is a Senior Research Scientist for NASA Langley Research Center. He was the principal developer of a new system identification algorithm which extends accuracy, speed, and confidence in identification from model test results for large space structures. He was a principal developer of a closed form solution to feed back control with quadratic cost and terminal constraints with applications to slewing control of flexible spacecraft. He made the experimental demonstration of fast slewing control of a flexible solar panel, and development of continuing model identification methods. He is a developer of a unified approach using a system realization theory to derive and correlate model testing methods for space structures. Jer-Nan.

That completes our awards program for this evening.

I would like to remind everyone that to date this conference has really been outstanding. We've had some of the best panels that I've ever seen in a symposium. We have some more for tomorrow and the wrap-up session should really be outstanding. Again, I would like to thank Dr. Paine for joining us and that concludes the meeting. Thank you very much.

GOVERNING POLICIES AND ISSUES

PANEL ONE

Chairmen:	Aaron Cohen Center Director NASA Johnson Space Center
	Willis Hawkins Aerospace Consultant
Host Chairman:	William G. Straight Regional Manager Lockheed Missiles & Space Company

The following paper numbers were not assigned:

AAS 87-607

AAS 87-608

GOVERNING POLICIES AND ISSUES - INTRODUCTION

William G. Straight*

Good morning again, Ladies and Gentlemen. I'm William Straight, the Host Chairman for the first panel, entitled, 'Governing Policies and Issues'. It is my pleasure to introduce this distinguished panel. However, before I do so, I should say that the AAS would like to stress that we have a limited amount of time with these distinguished panel members who have influence on policy. We encourage your participation by filling out the audience participation form which was provided to you when you entered. They will be collected and distributed to the panel for discussion.

Now, for the introductions. Our Co-Chairmen for this panel are Aaron Cohen, the Director from Johnson Space Center and Mr. Willis Hawkins, our keynote speaker, both of whom were introduced before. To my right and your left is Courtney Stadd. Courtney Stadd was named to his present position as Director of the Office of Commercial Space Transporation for the Department of Transportation on December 22, 1986. Secretary Dole said, "Courtney Stadd brings nearly 10 years of experience in a wide range of commercial space endeavors. He has the kind of qualifications and leadership needed to help the department implement the President's commercial space transporation policies." Before joining the Department of Transporation, Mr. Stadd worked in the Department of Commerce, Office of General Council, advising the Office of the Secretary on commercial space policy issues. He has been active in a variety of national space forums. In addition, he has authored numerous articles including a collaboration on the Aerospace Industries Association report on space commercialization. He also belongs to many societies, one of which is the AAS. Mr. Stadd holds a bachelor of science degree in international economics from Georgetown University.

Next to Mr. Stadd is a replacement for Gregg Fawkes, Director of Commercial Space Programs, U.S. Department of Commerce. Jeffrey Manber is currently the Project Director for the Commerce Department's Office of Commercial Space, a position he assumed in September, 1987. He is particularly involved with the financial communities' efforts and interests in space-related investments, as well as the Office's outreach campaign toward the U.S. industry community. Prior to joining the Commerce Department, Mr. Manber served as a consultant for the Lockheed Corporation and advising NASA Earth Resources Laboratory on efforts to commercialize remote sensing technology. In 1986 Mr. Manber was employed by the New York firm of McCloud, Young, and Weir, to establish a space unit at the brokerage firm

* Regional Manager, Lockheed Missiles & Space Company, Inc., 1150 Gemini, Houston, Texas 77058.

for the creation of investment banking relationships with space-oriented ventures. He has authored numerous articles on commercial space. Mr. Manber also serves on the board of the New York Space Business Round Table.

Next on my right, is Dr. Joseph Allen. Dr. Joseph Allen joined Space Industries Incorporated as Executive Vice President on July 1, 1985, and is responsible for marketing and public relations, user development, and relations with Congress and NASA. From 1967 until his employment with SII, Dr. Allen was an astronaut with NASA. He served as a mission scientist for the Apollo 15, and was a mission specialist on two shuttle flights, the most recent of which was a successful satellite salvage mission in November, 1984. Dr. Allen also served as a NASA Assistant Administrator for Legislative Affairs from 1975 through 1978. He is the author of *Entering Space - An Astronaut's Odyssey* -- a personal account of his flight experience and he has published widely in the fields of physics and space research. Dr. Allen received his undergraduate degree in mathematics and physics from DePaul University and holds a master's and doctorate degree in physics from Yale University.

To my left, is Rear Admiral Richard H. Truly. Rear Admiral Richard H. Truly was named NASA Associate Administrator for Space Flight on 20 February 1986. Before assuming this position he was the first commander of the Naval Space Command, which was established on October 1, 1983. The Navy Space Command became the Navy component of the U.S. Space Command on October 1, 1985. During his career in the Navy he was a Naval aviator, and made more than 300 carrier landings. In 1965 he was among the first military astronauts selected for the Mole Program. He became an astronaut for NASA in August, 1969. Rear Admiral Truly has received many meritorious service awards both in the Navy and in the NASA. He received a B.S. degree from Georgia Institute of Technology and his ensign commission in 1959.

And last, but not least, Dr. Duncan, the Assistant Secretary of Defense, Research & Technology, and Director of the Defense Advanced Research Projects Agency. In this dual-headed capacity, he is the principal technical advisor to the Undersecretary of Defense, Acquisitions, on space-related matters and serves as a chief executive officer of an independent defense agency responsible for research and exploratory development concepts. Dr. Duncan also had a Navy career, and was a fighter pilot. In 1964 to 1968 Dr. Duncan was assigned to NASA. During this period, which will be of interest to Houston, he was Chief of the Guidance and Control Division at NASA's Manned Spacecraft Center in Houston. In 1968 Dr. Duncan joined the Polaroid Corporation and in 1975 was elected Vice President of Engineering. Interesting enough, during this same period he had the additional duty of program manager of the SX70 camera, its film and its equipment. Dr. Duncan received a bachelor's degree from the U.S. Naval Academy in 1945. He holds a bachelor of science degree in aeronautical engineering from the U.S. Naval Postgraduate School, a master of science, and doctorate of science degree in aeronautical engineering from MIT. He has authored one book, *Dynamics of Atmospheric Entry*, has co-authored three other books, and has published many technical papers. Dr. Duncan has received an exceptional number of awards for his outstanding contributions. I present the panel.

AAS 87-609

GOVERNING POLICIES AND ISSUES: DEPARTMENT OF TRANSPORTATION PERSPECTIVE

Courtney Stadd*

First of all I want to thank the organizers of the session for inviting me. I think that these AAS policy forums, which I have observed either as a part of the audience or in which I have had the privilege of participating in as a panelist during the last few years, are very important in bringing people from various disciplines together. Regardless of what your interest is, in the nation's space program, be it the space science or the manned part of the program, or the commercialization part of the program, you are all joined together here, I think, by a common desire to see the space program recapture the leadership and capabilities we know that this nation is capable of in space.

I'm not going to insult your collective intelligence and claim that I have an answer to the national quandary that Mr. Hawkins so rightly characterized as our current situation *vis-à-vis* this Government grappling with commercialization in space. I'm not going to claim that there are any easy answers to what the partnership ought to be between the public and private sectors. What I would like to do is spend a couple of moments to give you a sense of what one department in this Administration is trying to do to organize itself to meet its mandate, to help encourage one part of the program which is transportation.

In 1984 Congress passed the Commercial Space Launch Act. I believe that was a path-finding piece of legislation. I think that in years to come it will be looked back upon as a landmark piece of legislation. It provided this department very broad authority. It basically said, 'encourage, facilitate and promote this nascent industry while regulating to ensure that the public welfare is protected'. We have now spent the last two years working to implement this mandate. We're within a couple of months of issuing our final regulations which, if we stay on course, I believe will be equally pioneering in the sense that these regulations will be performance-based, not design-based, in an attempt to allow the licensee as much latitude as possible in terms of developing the technology without getting into a degree of micro-management that perhaps we've seen in some other transportation sectors.

The other key responsibility provided the department was the somewhat daunting task of establishing insurance levels for the launch industry. We are now in the process of developing what we call a "draft rule" which we hope to get out in the next

* Director of Commercial Space Transportation, Department of Transportation, Washington, D.C. 20546.

couple of months. It will be the first step in helping to rationalize our approach to establishing liability levels for this new industry. Let us not forget that we are basically talking about the transportation of explosives to orbit. This is a risky venture, but I think one of the benefits of having this regulatory department establish itself in these areas is that we are beginning to use some of the same expertise that we have applied to other hazardous sectors to space. One of the research efforts that we've had under way since I joined the department was to draw some analogues to transportation of hazardous material in other areas. And without prejudging the character of the results, let me say that, based on the preliminary results, we're finding that what the anecdotal evidence tells us about launch activity is being substantiated by our studies. In short, space transportation is no more hazardous than certain other activities. We're finally beginning to establish a basis for quantifying risk, which in turn becomes a basis for the ability to make prudent judgments regarding the levels of insurance coverage that are commercially reasonable and, at the same time, adequately protect the government.

One of the most fundamental developments in making this transportation industry a reality was the President's announcement in August 1986 in which he said, "The Shuttle will no longer be launching routine commercial payloads". That was the first time that the launch industry had a clear signal -- a clear sense of what the boundary was between Government and the marketplace. By making that announcement he basically turned over nearly a billion dollar industry for competition by the private sector. Subsequent to that, the Air Force, under Secretary Aldridge, as you're aware, announced two substantial contracts, one for the Titan program, one for the Delta, and those production bases have also been equally fundamental in helping to establish this industry.

Let me give you some indicators of progress. I estimate that about $400 million has been invested in start-up capital for commercial space transportation. The bulk of that is basically monies that the three primes, General Dynamics, Martin-Marietta, and McDonnell Douglas, have reallocated for commercial space endeavors. The remainder is what I'd call entrepreneurial money being invested by the new-start companies.

Considering where we were a year ago, we are doing very well. We have won competitions with Ariane in the last couple of months. Nearly a half billion dollars in satellite orders have been signed to date. In fact, I'm optimistic that there will be some very significant signings in the next couple of months. So, in terms of dynamics, I think that the industry is beginning to come into its own -- not to mention several entrepreneurial organizations that have emerged just within the past year or so in direct response to these policy announcements mentioned earlier. The challenge that we have is to ensure that the environment that we, NASA, Air Force, and Department of Commerce have the responsibility for fostering, is truly one that's commercially reasonable. And as Mr. Hawkins mentioned earlier, it's a tough challenge for a government that's been speaking to itself essentially for 30 years to make the transition. To give you one example, at the national ranges where the bulk of the launch activity will take place, at least until the next few years, the Government has been both the launch agent and the range operator. By leasing these facilities to private operators Government is suddenly confronted with having to deal with a commercial tenant. And that introduces all sorts of interesting problems to a local Air Force officer who has to deal with questions of scheduling, and capacity, and so forth. Those are real meat-and-potato issues that we have to struggle with at the department just about every day. Again I see some light at the end of that tunnel but it is a struggle.

In the case of liability, the Government has basically picked up the tab for that. However, in the commercial world you have to talk about allocating the risks and quite frankly we really don't have models that are very appropriate to dealing with those types of issues. And we at the department again are struggling to reach out to some of the other transportation sectors that we oversee to see what types of analogues would be appropriate to insuring space transportation. Again, we're hoping through our rule-making actions that we'll begin to rationalize that whole area.

Let me make a couple of concluding observations: James Webb was probably the most eloquent administrator, I think, in terms of expressing a broad world view of the Government's role in the space economy. He believed in a university/Government/industry complex that would have a pervasive impact on society. There's some merit to that. I certainly agree with the major premise which is that space is clearly a major driver in innovation in the economy. But I think we've inherited some of the consequences of that world view which possibly are not so good. You can't ask one organization to be all things to all people because, given the political sensibilities of Washington, you're unlikely to get the resources for any one purpose all the time. To wit: we've ended up with a rather disjointed space program for 30 years. The Saturn V lying outside Johnson Space Center is perhaps symbolic of that. The fact that we allowed Skylab to fall back without any follow-up is perhaps also symbolic.

The Russians às in the tortoise/hare story have taken a rather plodding classic approach to space and we find ourselves in 1987 faced with a dilemma perhaps more profound than Sputnik -- which was essentially a rather discreet event for which we certainly had the wherewithal to respond.

The challenge we face now, however, is far more complex. We have many more actors in the space race. We have governments that are very aggressive in ensuring that their so-called private sector, or at least how they characterize their private sector, will have their place in space. And we're struggling with how to re-landscape -- I suppose it's the best metaphor -- the Government's role in private sector. What you witness in this Administration has really been an attempt to struggle with it. *Aviation Week* and other publications have taken lots of shots at us for what they perceive to be protracted, endless debates, endless interagency squabbling,and clearly we deserve, I think, some of the blame for that. But I think that we do ourselves an injustice as a nation if we do not realize that we've been struggling with some very complex, very profound issues these last few years. The *Challenger* accident of January 28, 1986 was really the loose thread on the sweater. It brought to a head lots of issues and consequences of decisions made many years earlier that Administrations in the past did not for one reason or another grapple with, not the least of which was where the Government's role in developing technology begins and ends. I think we're now beginning to get our hands on it.

In the case of transportation, I think once we finish the task at hand we'll have a regulatory apparatus that will be a good basis for encouraging the space industry. I think we need to work some more with NASA so that they get the adequate funding they need to ensure that the future propulsion systems and so forth are being pioneered. You can't look to the private sector to do that alone. So we're grappling with it and I'm hoping perhaps in the discussion that will ensue, we can delve into these concerns in more detail. Thank you.

AAS 87-610

GOVERNING POLICIES AND ISSUES: AN OFFICE OF COMMERCIAL SPACE DEPARTMENT OF COMMERCE PERSPECTIVE

Jeffrey K. Manber*

The Office of Commercial Space was created within the Department of Commerce, as recognition of the growing need for the government to encourage the development of a genuine commercial space industry.

Despite the U.S. government's spending of over $25 billion per year on the U.S. space program, the private sector has been thwarted in its efforts to freely compete for space-based goods and services.

The Department of Commerce is today engaged in a policy exercise whose goal is to remove government impediments to private sector participation in space, and to instruct federal agencies to engage in commercial practises when entering the marketplace.

Given the fiscal realitites now facing all federal programs, it is imperative that we reward the imagination and cost-efficiency of the private sector in creating a true space-based infrastructure.

Introduction

The Commerce Department's Office of Commercial Space was created in April, 1987. The Office was created within the Office of the Secretary in recognition that the commercial space industry still requires the sort of active support best provided by a high-level, highly-visible office.

The Office has two primary objectives: to serve as an industry advocate within the federal government and to participate in ongoing policy exercises concerning commercial space objectives.

This participation in the emerging space industry is part of the historic role of the Commerce Department which, since its formation has served as the federal agency most concerned with ensuring legitimate markets are allowed to develop within each new commercial frontier.

* Project Director, Office of Commercial Space, U.S. Department of Commerce, Office of the Secretary, Washington, D.C. 20230.

Background

The clear need for an effective advocate with government for commercial space ventures became clear during last year's Cabinet review of space shuttle pricing and use policy.

Space technology is now sufficiently mature to allow a wide range of possible new commercial applications. The reasons this has not taken place to the degree often anticipated can be traced, in our opinion, to a number of factors, including government competition against the private sector for space services, as well as the inability of the government to serve as a reliable customer of these services. These two factors: competition from the government and inability to firmly plan multi-year demands, have distorted the commercial space market, preventing the full resources of the private sector from participation.

The imagination of the private sector is the strongest suit this nation holds for competing in the international marketplace and in the growth of commercial markets. There is no reason why this nation should swerve from what has been a highly successful model for economic development across all markets.

In the early days of aviation, for example, both the Post Office and the military ordered new airplanes, one for civilian use and for military use, but both agencies did so by entering the commercial marketplace and placing orders as would any other customer. Today, the same analogy holds true for launch transportation services and on-orbit hardware systems. It is inefficient for the government to operate an operational space transportation system or to build for itself commercial facilities in orbit. These are services best performed by the private sector.

Government as commercial customer is an essential component to the maturation of the commercial space industry. It would allow the private sector to develop its resources in a traditional manner, introducing competition in launch services, civilian remote sensing, space-based manufacturing and development and construction of on-orbit hardware. It is no surprise that space is one of the few technological markets where the cost has not dropped in the last twenty years. Unlike computers, telephones, calculators, or air travel, it is just as expensive today to fly a pound of cargo into space as during the Apollo program.

Space is just that--the latest commercial frontier. It is an economic frontier whose potential is well recognized by the Japanese and Europeans, both of whom have mounted long-term, integrated programs of national governments and industry to establish leadership. The pattern undertaken in this country since the Apollo program has thwarted full private sector exploitation of space. The United States has clearly lost its leadership in space and only through a well-diversified program will we enjoy the benefits of the sizeable capital already invested in space.

Current Activities

Policy

The Secretary of Commerce officially chairs the Commercial Space Working Group under the Economic Policy Council of the Cabinet. This Working Group is composed of the fourteen executive branch agencies interested in commercial space ventures. It is currently engaged in a a major policy review exercise to set the nation's economic goals in space, as well as to define the proper role between the federal government and the private sector, and recommend actions to be taken to remove any government-imposed barriers to U.S. commercial space ventures.

The Office has identified reform of federal government procurement procedures as one of its most important goals for the next year. The U.S. government is the largest customer in the free world for space goods and services. If it purchases space goods and services in a commercially reasonable manner, it will significantly expand the market and stimulate greater private investment and innovation. And, at the same time, <u>the government will benefit through reduced costs for space operations.</u>

The Department is an accredited member of the U.S. delegation negotiating a partnership agreement with the Japanese, Europeans and Canadians to build an international space station. The Department's primary concerns with the space station are that no U.S. technology be transferred to the partners in a manner inimical to U.S. industry interests, and, equally important, that the legal regime adopted for the space station not create obstacles to future commercial space development.

Facilitating Commercial Space Ventures

The Office acts as an advocate and facilitator as needed for any U.S. commercial space ventures interested in such assistance. Aside from space transportation (where the Department of Transportation has statutory responsibility for both encouraging and regulating the ELV industry), the Department is actively involved in working with commercial space ventures covering the entire range of potential applications: remote sensing, materials processing, on-orbit facilities, communications, geo-location services, commercial spaceports and return vehicles.

This facilitation role involves interagency activity concerning such matters as regulations and procurement. It also includes support of U.S. space companies in international marketing and joint venture formation.

Conclusion

The United States must confront the difficult task of changing how we conduct ourselves in the space arena. It is somewhat ironic that this nation has not unleashed the private sector in creating the infrastructure of space. We no longer have the luxury of a monopoly on space; indeed, unless we move quickly to utilize space more efficiently, this nation will become increasingly more dependent on other national space programs for space services.

The Department of Commerce recognizes the importance of space utilization across all sectors and is working to establish a U.S. space policy which reflects the true complexity of our dependency on the space frontier.

Table 1. MARS LANDING ACCURACY (DATA FROM JPL)

ENTRY-VEHICLE TYPE AND GUIDANCE CAPABILITY	LANDING ERROR, 3σ SEMI-MAJOR AXIS (KM)			
	DOWNRANGE	CROSSRANGE	EST. IMPROVEMENT WITH BEACON DR	CR
VIKING (70s), L/D = 0.18, UNGUIDED	138	53		
VIKING (90s), L/D = 0.18, UNGUIDED	76	26	73	25
LOW L/D = 0.3, APOLLO GUIDANCE	49	33	25	30
LOW L/D = 0.3, ADVANCED GUIDANCE	26	31	13	27
HIGH L/D = 1.4, APOLLO GUIDANCE	18	20	12	18
HIGH L/D = 1.4, ADVANCED GUIDANCE	8	8	5	5

AAS 87-611

GOVERNING POLICIES AND ISSUES: SPACE INDUSTRIES PERSPECTIVE

Joseph Allen*

Good morning. I'd like to follow in the sense of the theme that Courtney and Jeff began and say that I'm pleased to be here and have looked forward to being on the panel. I'll admit to the Chairman that I was flattered to be included and wondered why because I am the only non-government person here to talk about governing policies and issues. I, too, Jeff, wondered what I was going to say when I came over this morning and a few things were clear: I knew that, because of a commitment that I secretly made with you, Aaron, that I was not going to give a technical description of the industrial space facility and I think Richard will be pleased to hear that as well. I will ask the audience, though, to definitely attend sessions that Doug Lilly and Al Louviere will participate in later. I want attendance to be obligatory. I also have a secret pact with Dick that, if he does not show his two STS films, I promise not to show my 8,000 STS slides this morning. And I dare say, without even looking at him, that there are one or two stories that neither he nor I will tell concerning activities that we were sent on as young junior astronauts almost 20 years ago.

That leaves me with trying to decide what to talk about it. I mentioned that I am the only person on the panel that really is not in a policy setting responsibility of sorts. It is perhaps like being a citizen that wanders in out of the rain and finds himself on a panel of meteorologists. But as I sit here I will try to add something to the policies discussion. As Bill mentioned in the introduction, some years ago I was involved in a Government position that was related to policy, in fact, it was related to politics. It had to do with NASA legislation, working with the Congress. I had been in that position for a few years. I decided that I needed to get into a safer line of work, so I came back into the astronaut office. Even though there are some risks that we all recognize, believe me it's a lot safer line of work than being in politics and in many ways a lot more fun. I left that position a couple of years ago and have since been in the business world. But after the last two weeks, particularly starting with October 19, I'm thinking seriously of becoming an astronaut again. It's tough in the business world, particularly in trying to get commercial space endeavors funded when there are not as many funds around right now as there were two weeks ago. It's going to be a very, very tough business - a very challenging business.

---

* Executive Vice President, Space Industries, Inc., 711 West Bay Area Blvd., Webster, Texas 77598.

With that as a real short introduction let me indeed put in my two cents worth in regard to Government policies. It seems from this outside perspective that some things with regard to Government policies are essential. And I'll mention just three. It's essential that they be relatively simple. It's essential that they be consistent and we all hope that they reflect certain thought and considerable wisdom; that is simple, consistent and wise.

Courtney, I'm sure your report with regard to the encouragement of commercial space transportation services is true. At the same time I had the very sobering experience of sitting in a room here in Houston, Texas, not too long ago and hearing David Thompson, the President of Orbital Sciences Corporation, which must be one of the most mature commercial space undertakings and at the moment certainly the most successful, describe how his business was started. He said that in his entire business plan initially was based on a particular policy, which he believed would be consistent. It was a policy of commercial things going aboard the National Space Transportation System which indeed had been in place, as he put it, through eight Congresses and four presidential terms since the time when he had been in junior high school. So he assumed it would be consistent. And yet it was changed. Orbital Sciences Corporation, I think, has been able to respond to that. I'm hopeful they have. Nonetheless it's very difficult to cope with sudden square waves in policy changes like that.

I mentioned it should be wise. I am of the conviction, a conviction that increases as time goes on, that some of the procurement policies - and they're in fact laws, they go beyond policies - need some strong admixtures of wisdom put into them, even though they already extend over thousands and thousands of pages of laws. If I understand it correctly, in a Government procurement process lots of bonus points are given for low cost, estimating low cost, although I'm not certain whether penalties are added later if the estimates are wrong. I have never quite understood that. And those policies, or those laws, I should say, are in place to absolutely guarantee a totally level playing field - fair competition.

If you think about it that begins to rule out innovation - genius coming onto that playing field - a person or a small group of persons with really some brilliant ideas. They are leveled out by and neutralized by those very laws, procurement laws, of our land. And as far as I know, there are no bonus points given in procurement policies right now that have to do with timeliness, or how much time does something take?

I'm reminded of a truly traumatic experience that I underwent as a student, I survived it, thank goodness. I was about, I hoped, to receive a Ph.D. And as all of you know, one must go through a series of tests, including some oral examinations in front of a body of professors. I was in the process of going through this ordeal. I was trying to derive a certain long equation in electrodynamics on the blackboard. I was stuck. I came to a certain point, thought about it, finally turned and said to the group of professors, that I was unable to continue but I knew exactly where the derivation could be found if I ever needed it. There was a silence for a moment and then the oldest professor in the room looked up at me and said, 'Mr. Allen, you need it now'. There are lots of things that we in the space community and indeed in this nation need right now. We are not being helped by the procurement laws and policies that support those laws, that we all must wrestle with. I had the good fortune of serving on a panel with Dr. Max Faget and Dr. Joe Kerwin some years ago; in fact it was in 1974. This panel together with a number of other NASA workers tried to peer into the future and decide what a proper course of action for a civilian space program might be. A recommendation was given to the NASA administrators and others summarized in a short poem, which, if I can remember, goes more or less like this:

'Any study group with common sense
Must advise this nation,
When we've got our railroad into space,
Then we'll need a station.'

That was penned, not by an anonymous author, but by questioner Dr. Joe Kerwin and it seemed obvious a long time ago. We have now spent lots of money on that station and thought about it a long time. We have developed some designs - they may not be perfect - but they'll sure work and we should get on with it. That was in 1974. I hope that station is there by 1994. Sometimes I am somewhat discouraged. 1974 to 1994 is a long time! That is not very timely.

Those are just some random thoughts. Once again I'm the citizen that's come in out of the rain with this group of meteorologists and I'll implore you, in predicting the weather, make the weather simple, consistent, and with some wisdom. Now, I did not draw this analogy without some thought. Weather people, as everyone knows can't control the weather, and I dare say, even collectively, the panel, myself excluded, cannot control the policies of the United States with regard to the rate and the direction in which we, this nation, proceed on the space frontier. However, I tend to be optimistic and there are few weathermen, like Willard Scott, that are very optimistic. And I think through their optimism and talking about it with enthusiasm, that unlike the weather, some positive steps and some changes can be made. Thank you.

GOVERNING POLICIES AND ISSUES: U.S. NAVY/NASA PERSPECTIVE

Richard H. Truly*

All my life I've been following country boys like Joe Allen! First of all, let me say that I'm delighted to be here in Houston, to come home. I've spent a lot of my adult life here working on the space program. And I'm glad to be here whether it's the summer or the winter, but I am glad it's November.

Let me also say to Willis, when I was a young boy growing up and sitting in the pew, sometimes I didn't listen to the sermons. But I really enjoyed your remarks this morning, Willis. I agreed with almost everything you said and have forgotten the one or two things you said that I didn't agree with. But, seriously, I think you gave this group and this forum a lot of food for thought, and it was an excellent keynote to an area of national policy that deserves both the people on the panel, and in the media, and the public and the members of the industry and Government that are here to talk about and be a part of this open forum. You made mention that we are living in a new environment, and I can tell you that in the last year and a half or so since I've been in my present job in Washington at NASA headquarters, it is a new environment, and it is one at times that has been terrifying from the point of view of overcoming in order for us to get back to what our main love and dream and goal is, which is the space business.

I appreciate your comments on risk and particularly your comments about individual responsibility as opposed to committee decisions. Along that line, I must tell you that my experience over the last year and a half has been one where I've gotten a lot of help. The Presidential Commission, a full Congressional investigation, no less than four National Research Council oversight committees, the Europe Alumni group, the ASAP, and all my friends and neighbors that I've known in the space program throughout the years, all have given me a lot of advice. And, as a matter of fact, it's generally been good advice. And I have taken a great deal of it.

But to get back to your comments about individual responsibility, in listening to the questions that were posed this morning, it's a good thing that there are decision-makers in this country because you could take just the advice from the people sitting in this room and have license to do almost anything.

---

* Rear Admiral, U.S. Navy and Associate Administrator for Space Flight, NASA Headquarters, Washington, D.C. 20546.

In the next several months we're going to get a lot more help before we get the shuttle flying again. Frankly, I believe that we would be best served by people standing aside and letting the professionals, like Aaron Cohen, within the system do that job for the nation.

I will comment, or make a couple of observations on my personal views about policy on space. But first I would like to use this forum for a few minutes to give you an update on the realities as to where space transportation is in NASA and the civil space program. And I bring you this short update from a personal experience I think that is interesting because I do believe that I understand the national security requirements for the space business. I've been associated with the DOD programs since I first started in space, when some of you were starting the Manned Spacecraft Center in Houston. And as Commander of Naval Space Command, I was a component commander of the U.S. Space Command, so I appreciate the DOD and national security side of our policy discussion.

But I think we must remember that the number one job today is to get the shuttle system back flying. It is the backbone of our National Space Transportation System. It is not a viewgraph presentation of what we should do in space. It is a series of space planes that are today in hangars at the Cape and they are grounded. And our first and foremost priority is to get the system flying again. Where are we?

First of all, in the solid rocket motor we have selected a new design and we have tested it at full scale. A couple of months ago a full-scale development motor was tested, and we will test the next motor in the next several weeks. Post-test analysis of the solid rocket motor joints have borne out the adequacy of the new joint design - both the field joint and the case-to-nozzle joint. Frankly, the case-to-nozzle joint was a little more of a problem than the field joint design. In the main engine, we suffered a heat exchanger leak the other day down in Mississippi on our first flight engine, and we had to pull it out of the flow. We are back on plan. We have now fired our next flight engine. And we are back on a plan where we will have flight engines and a spare engine in place and back in the flow by the end of January.

We have a number of orbiter changes that we have decided to incorporate. One was mentioned this morning - the 17 inch disconnect valve. I was at the Johnson Space Center with Aaron last night and looked at a couple of other changes. The orbiter changes are in-flow to support the Cape schedules, and we are shooting for a June 1988 launch.* Within the program, we are shooting for the 2nd of June. My present commitment is the month of June, and we still have some slack but it is fast eroding with real problems, but I can assure you that these are finally the kinds of problems that we like to see in the program. Because they are real, they have to be fixed before we fly again.

We have a mountain of paper to climb between now and next June, including a design certification review of the shuttle system. But we will climb it, and I can guarantee that paper is not going to drive the shuttle flight schedule. The solid rocket motor redesign verification and qualification program is the pacing schedule item.

---

* Since, launch has been delayed to September.

In the people category, moral across the centers -- not just Johnson, but at Marshall and at the Cape -- is higher than I have seen it in a long, long while. Certainly, sky-high compared to the pits that we were experiencing last year after the accident. We have two flight crews in training simulations. We have KSC launch crews in training, and we have mission control crews here in Houston that are in training.

The shuttle recovery program fits into a larger picture of what is going on in NASA with space transportation. I would like to move then from the short-term projection, from a preparation for flight that is currently scheduled on the 2nd of June. But I can assure you that if it takes to the 3rd of June, or the 4th, or even to July, we will fly when we are ready and, when this system is flying again, it's going to be in far better shape than it was before the accident.

But there are other things going on that I think you ought to be aware of as we talk about policy. We are building today a new orbiter in Palmdale, California, fully funded and approved by the President. As a matter of fact, because of foresight of previous years in building structural spares, that orbiter is about 20-25% complete today. It will roll out in 1991 and join the other three in the winter of 1991-1992.

My first priority for a new start in the 1989 budget processes is an advanced solid rocket motor -- a full and open competition to obtain a new motor that would be available in 1993 or 1994, that would aid the industry and bring it to a more automated, higher quality process and also gain 8,000 to 12,000 pounds of shuttle performance. That performance gain was counted on from the early 70s and needed desperately by some of our larger payloads but not delivered. We have, in this same budget proposal, which includes the advanced solid rocket motor, adequate facilities and resources at the Cape and here at Houston to support a steady flight rate of 14 shuttle flights per year. There was a question earlier today of 14 shuttle flights per year. And my answer to the question about whether the 14 flights per year is achievable -- and I would be glad to later ask Aaron to comment -- is that it's a question of resources. It is achievable if we get the facilities and the resources to do it. Since I've been in this job I've made a point several times to say that it's my intention to balance our commitments for the future with the resources that we receive to meet those commitments.

Finally, we issued a new shuttle manifest that went out through flight 44, and it was worked out in cooperation with Secretary Aldridge of the Air Force. Because of some DOD commitments, we did not publish a shuttle manifest beyond the 44th flight. But in this same little blue book that lists our manifest is a commitment to a mixed fleet for the civil space program; in other words, a mixed fleet of ELVs and space shuttle transportation, and we desperately need it. If you look through this little blue book that includes this manifest, you will see NASA planning requirements for a range of ELVs -- from small Scout-class ELVs to the large Titan IV class.

The great majority of these ELVs will be achieved from the private sector through commercial launch services. I was recently the source selection official for the first contract for a commercial launch services for our Government procurement. We are not only talking about supporting this policy; we are doing something about it. I must say also that it is my view that, despite some of the things you might read in the paper, we have been working hand in hand with the Department of Defense in this total space recovery program. I fully support and have supported publicly and privately the procurement of additional Titan IVs and medium launch vehicles by the Air Force. We have had excellent cooperation with the Air Force and particularly with Secretary Aldridge. We have gone through the agony of not only recovering from two Titan 34D accidents, but also the Shuttle accident and planning for the future. We have plenty to debate between NASA and the DOD, but it's my view that that's going extremely well. An example of that is our work with the DOD on an advanced launch system -- plans for a heavy-lift capability for America. So, all is not black and all is not lost.

Frankly, the litany I just went through descibing NASA transportation planning is very healthy. We will get the shuttle flying again as quickly as we can because we desperately need to. It will not be perfectly risk-free. We have not lost our will to fly. We will get it up to a robust rate in the 1990s. It will be supported and matched with a mixed fleet of various classes of ELVs. And, frankly, the best thing that could happen, I believe, to the commercial launch vehicle industry is to fund NASA space science, research, and exploration payloads because that's how we intend to get many of them airborne.

In closing, let me just say a couple of words about Richard H. Truly's view -- not the Governments's view -- about our national policy. First of all, I think it should be just that -- a national policy. It should stand above NASA and the DOD and commerce and transportation. It should be America's policy and it should be bold. It should be a policy with a vision for the future, and it should be a policy that is enabling new programs in the national security arena, in the commercial arena, and in the civil space arena. It ought to be a balance between a policy that encourages you and our industry to invest in advanced technology. And it ought to be a policy that recognizes our tremendous heritage in humans flying in space and clearly does not back away from our successes, with our men and women who have flown before now, because they will be flying again permanently within a few years. I personally believe that we do have the leadership within the Government, within the industry, and within the Administration and the Congress together to come up with a national policy that is as I described -- bold, one that has a vision, one that is enabling, and one that stops the bickering between the manned and unmanned sides of our industry and recognizes that both of them have to be an important part of our national space policy. Thank you very much, Aaron, for having me here today.

AAS 87-613

GOVERNING POLICIES AND ISSUES: DEPARTMENT OF DEFENSE PERSPECTIVE

Robert C. Duncan*

Good Morning. It's a pleasure to be here. Dick talked about how difficult it was to follow a country boy. Think about how difficult it is to follow two astronauts who are highly articulate, who have been successful fighter pilots, one making over 300 carrier landings and I don't know what Joe did, but he probably did the same or more. That's a tough assignment. I'm very happy to be back in Houston myself.

As indicated earlier I spent much of the 1960s here when the Manned Spacecraft Center moved from Langley to Houston. I was a Navy detailee to be the Chief of Guidance and Control here in Houston and I believe the Guidance and Control Lab was the first lab opened up when the Center was built. We had a facility in downtown Houston and then moved to the Center with all the mud, the road building and all the other activities. Some of you probably remember. I came here for the Apollo program.

I was a Navy fighter pilot also and had advanced education in guidance and control, so I think somebody, somewhere, in the days when they had IBM cards, clanked them out and somebody with golf shoes stepped on mine and I ended up coming here. I've only been back once since then, and I'd like to tell you about that experience because it ties in with what Joe was talking about -- about entrepreneurial companies and what others have been talking about regarding the difficulty of working with the Government. I was at Polaroid where I spent close to 18 years as the Vice President of Engineering. We developed and announced the SX70 in 1972, -- a revolutionary development. It was a photographic system in which you no longer had garbage associated with photographic process and you no longer had the time problem in processing a picture. The timing was done automatically and chemically. There was an abrupt Ph change at the right point to freeze the photograph so it would not continue developing; it was equivalent to pulling the paper off.

This was attractive to NASA and the Skylab program as something that could be used in space. So somebody in NASA called Polaroid, when the announcement was made, to see if it might be possible to look into using the SX70 in space. Since I was the only ex-NASA employee in Polaroid, the telephone call was routed to me and in the discussion it turned out, yes, indeed, it might be possible to use the SX70 in

* Dr. Duncan is Assistant Secretary of Defense, Research & Technology/Director of Defense Advanced Research Projects Agency (DARPA), Washington, D.C. 20546.

space. So I talked to Dr. Land -- he's one of the entrepreneurs like Joe Allen who started a company from scratch. Dr. Land said, "That's great, but we don't have any contracts with the Government, and I don't want any contracts with the Government. So, Cliff Duncan, if you go down there and get a contract with the Government, you're fired!" Those were his exact words. You can imagine how enthusiastic I was coming down here. I went by my old house at Nassau Bay and I delayed as long as I could and then finally went before a group of people. (As usual in those days at the Manned Spaceflight Center you couldn't have a meeting with five people, there were 50 people -- Dick says it's the same now.) I had brought along a camera and film to demonstrate, and we talked about the problems of the space environment and there was a lot of interest. Finally I said, "If you want to use it, we can't have a procurement contract, we will give it to you." It took about a month before we could get a response. This is an actual story, a true story. Ultimately it turned out that NASA did the space qualification test, we provided cameras and films. It was used in Skylab and quite successfully. But there was no contract.

I don't recommend that, Joe, as a way of working with NASA. What I recommend is that we do something about that. Joe used words like 'simple', 'consistent', and 'wisdom'; and Dick used words like, 'bold' and 'vision' and 'stop bickering', and I think those are all great words about space policy and about working with the Government.

I'd like to take a few minutes to talk about two programs that we're doing in DARPA: one with NASA that has to do with space transportation for the future; and another program that's military, but it has some of the characteristics, I think, that Willis Hawkins talked about in his very enlightening keynote speech;that is, doing things smaller and lighter and cheaper.

The National Aerospace Plane is a joint program with the DOD and with NASA. Its partnership is four-fifths DOD funding and one-fifth NASA funding. It's a national program that is technology-driven rather than milestone-driven and I think that's one of the things that Willis was talking about earlier also. Now, it's very hard to have a technology-driven program in this Government because of the way we do business. That involves projections for five-year funding, the need to go to the Congress every year for funding,the need to show milestones and specific accomplishments, so I think we have an uphill battle to do this program successfully. But I hope we can be innovative as we try to take this major potential step forward and I think we can do it successfully.

The management of the program now includes a steering group with the Undersecretary of Defense for Acquisitions as Chairman and Ray Colladay, the Associate Administrator for Aeronautics and Space Technology as Vice Chairman. The Steering Group provides policy guidance and direction but the program is directed by a program management office in DARPA. That office has staff representatives from the Air Force, Navy, NASA and from SDI. The NASA representative is physically working in DARPA. He's going to give you a detailed description of the aerospace plane in the program of this conference later on in the schedule. Because of the joint nature of the NASP program, it does have a somewhat unusual management structure.

What's exciting about NASP, however, is its technological promise and challenge. The program undertakes to develop and demonstrate the technologies that could be applied to a potentially new generation of flight vehicles. These include launch vehicles capable of single stage to orbit, with horizontal take-off and landing like aircraft but capable of long-range hypersonic cruise in the atmosphere. Military applications include space launch and long-range interceptor vehicles. In addition to space launch the civilian applications include the Orient Express, that the President referred to in his 1986 State of the Union address -- the possibility of a two-hour flight, say, from Dallas to Tokyo.

We're now in Phase II in NASP. We have three airframe contractors, two engine contractors. In Phase II we're concentrating on maturing the critical enabling technologies and on ground-based demonstration of the major components required to produce an experimental flight vehicle. The major features of this phase include multiple technical approaches, vigorous Government, and private R&D activity and intense industry design and demonstration competition, as well as close integration of propulsion and airframe efforts.

NASP is an important national endeavor. We expect that these efforts will come together in a couple of years and that we can make a decision. It has the designation of the X-30 vehicle. We can make a decision as we enter into the next phase of the program, called Phase III, to actually start to put together a vehicle. But, as I said earlier it's technology-driven. Our time table calls for making the Phase III decision in 1990 and to be able to be flying in 1994, 1995. But this is something we'd like to keep flexible. As we develop the technologies and meet the milestones, we can firm up what we're going to accomplish. The X-30, like all X vehicles, is a technology demonstration -- it's not a prototype to do anything specific, except, we hope, to fly single stage to orbit, with a small payload of approximately 2500 pounds. From that can come a whole gaggle of vehicles -- military, non-military, civil transports, and we hope that this will be the bell wether of what transportation will be like as we enter the next century.

I'd like now to talk about the opposite end of the spectrum -- the other program I thought I would mention -- a satellite program, referred to by some as "Cheap Sat". We don't like that term. By others it is called "Light Sat". We don't really use that term. The official name is the Advanced Satellite Technology Program. This nation has developed an extraordinary technology base for satellites and with very high performance, outstanding reliability and, perhaps surprising to some people, outstanding economy, in light of the service they perform. Unfortunately, as Dick described earlier, our launch systems have placed the nation in a precarious situation. We have a very limited inventory of ELVs and not a very high production rate. Even when the shuttle resumes service, our capabilities will probably fall short of meeting our requirements. In addition, we are severely limited in the number of launch sites. Light Sat is attempting to address these sorts of problems in several ways.

First, we're supporting the development of a new technology base to significantly reduce the weight of critical satellite components. To get more bang for the kilogram. Areas of interest include optics, structural materials, microelectronic systems, cooling systems, attitude control and stabilization hardware, power storage and generation and others. These new technologies should be applicable to a broad range of satellites. During our program we plan to launch some small technology demonstration satellites on the way to space qualification. We intend to concentrate on the high-risk, high-payoff ideas which are the requirement for the DARPA type of program. In the same spirit we'll be looking at new distributed satellite systems.

Second, we will explore new booster technology and Willis referred to this earlier this morning. Unconventional launch ideas and advanced manufacturing methods that can be applied to boosters.

Third, as infrastructure for our research program we're looking at the use of rockets like modified Pershings to launch these technology satellites. This is surely not the least expensive way to launch a satellite if you talk about dollars per kilogram. Probably the least expensive way would be to collect all of our satellites together and then go out and contract with the Russian Proton. But what we're trying to do is to develop satellite systems in contrast to so many that we use today in the military. Without being very complex and multi-purpose, they are single-purpose systems and as simple as we can design them. Instead of having 99.99% reliability the concept here is if you have an inexpensive enough system, you can perhaps accept nine-tenths reliability, then if one out of 10 satellite fails, you add one more to it. This may not be a very viable system, but this is the kind of R&D approach we're taking.

We're currently looking at a small number of possible prototype satellites for restricted and specific missions. Each of these will be carefully selected and we'll do these in partnership with one of the military services or with one of the defense agencies. In doing this work we're particularly interested in seeing if it makes sense to use some commercial parts and commercial practices since current civilian technology in such areas as micro-electronics is very much better than it was in the 60s. We think the Advanced Technology Satellite Program is one of extraordinary promise. And we are hopeful that we can continue the program into the next phase.

In conclusion, both militarily and commercially, space is vitally important to this nation's well being. The policies that we at DARPA develop, and those of NASA, the DOD, the Commerce Department, the DOT, and all those involved are critical to our success in achieving and maintaining our preeminence in space. Thank you.

AAS 87-606

GOVERNING POLICIES AND ISSUES: PANEL DISCUSSION

Panel Moderator: Aaron Cohen
Panelists: Willis Hawkins
Courtney Stadd
Jeffrey K. Manber
Joseph Allen
Richard H. Truly
Robert C. Duncan

Aaron Cohen: I'd like to thank the panel very much for an outstanding discussion. And now they're ready for the questions. Do we have any questions?

This is a question to all of the panel members. And that's a good way to start.

Question: What elements should there be in a national space policy to be promulgated by the next, or this President and the Congress?

By: Pat Jackson, freelance space writer and videomaker.

Who would like to start that? Let me ask Dr. Allen to start off.

Dr. Joseph Allen: What I don't know, Richard truly knows. It's a tough assignment and in answering that, I'd have to come back to some of the thoughts I began with, certainly amplified by Richard's comments with regard to what the policy must be. I think Dick pointed out the fact that a national space policy must be precisely the statement of the nation. And how it's implemented and what parts of it are assigned to various establishments to implement must be contained in that policy. It's a matter of how one administers a relatively complicated program. Interestingly I have always believed that space may indeed be amongst the most simple programs that this nation has to undertake. It's essentially technical. Everybody gains from it. I think the polls show that by and large it's not even controversial to the voters -- to the man and the woman in the street -- it's not controversial, the way many other programs and policies are in this complicated world of ours. It's an addressable problem -- perhaps one of the easiest, which adds to the frustration that, for some reason, it seems not to have been addressed broadly at the very national level, as Admiral Truly mentioned earlier.

Aaron Cohen: Would anybody else like to take a shot at that? Willis, would you like to take a shot at that? Anybody else?

Dr. Robert C. Duncan: I would make a comment about it -- not directly to the question, but I would hope that the policies that we develop would make it so that those with the entrepreneurial spirit are inspired to invest their talents and to take the risk with their money and with their family and with their future to go into this field. I think of my own experience where a photographic company was created that way (Dr. Land and The Polaroid Corporation). I talked to David Packard about how he started in a garage and developed his amazing company which has had a great influence on the world of technology. I would hope that the policies of this country can nurture that in space and, if that's possible, I think there's nobody in the world that can come close to us.

Aaron Cohen: Anybody else want to take a shot at that before we go on.

Courtney Stadd: I have to be careful. I'm a lonely political appointee, starting a family so I don't want to prejudge my President. But speaking as a citizen, the commitment by the President to U.S. leadership across-the-board is necessary even though Sally Ride is correct in saying you can't have leadership perhaps in all areas. But I believe that this nation must commit itself to having the potential capability to exert its leadership in all areas.

Secondly, the United States should commit itself to an architecture in orbit that is American, the elements being the Space Station, the Space Shuttle, and the other elements in the mixed fleet that the Admiral referred to. Third, I think we ought to triple or quadruple the funding for advanced propulsion systems. We've got to move much more vigorously in that area. Fourth, I think that we must make a very concerted effort to address what Joe Allen was referring to earlier in the area of procurement; for example, there are other similar areas where we have to be very clear, very wise, very consistent in determining the transition from a government prototype to private sector. When something works, such as LANDSAT, perhaps we have to be smart about how to move that out into the private sector and move on to another state-of-the-art project. We're pretty dumb about that in this country and we've got to begin to grapple with that. We've been trying, as in these policy debates, but we've got to move much faster. We've got to be pretty sure that there's adequate room for the commercial projects such as Joe Allen's.

Aaron Cohen: Very good. Let me go to another question now. Dick, let me ask you to handle this one.

Question: Could you outline the chief requirements for an Advanced Launch System and discuss possible options? How do NASA's and DOD's interests compare in this regard?

Richard H. Truly: The long-range goal is a heavy-lift launch system that offers a dramatic decrease in cost. The system would be available, we believe, or could be available, no earlier than the late 1990s or just prior to the turn of the century. Such a system would be based on new technology. It probably would require propulsion system developments for both a core vehicle and what's called a booster vehicle. The booster vehicle might be one that Willis alluded to which would be a fly-back booster; it would return to the launch site. The concepts are being studied now by a number of contractors. There is also discussion on whether or not the nation requires an interim capability at an earlier date. And there has been quite a bit of discussion on this issue between the Department of Defense and NASA. My personal view and NASA's view is that for that interim vehicle a configuration that should be studied would be what we refer to as the Shuttle C or a cargo version of the shuttle which would incorporate our external tanks and solid rocket motors and main engines right off our production lines with no change in part number. Replacing the Orbiter with a payload canister and with a two-engine configuration, it would provide just over 100,000 pounds to low-earth orbit. It could be available for development costs of a little over a billion dollars with the first flight in 1994 or so. This system, however, would not compete with the end-goal that I just described which would be a new-technology, lower-cost system. A cargo version of the shuttle system would not be a dramatic cost decreaser; however, it would be, I believe, the way to get the earliest, cheapest and best -- which is very important -- and most reliable system in the mid 90s. That's going to be a national decision as to whether the nation believes that it needs to have a heavy-lift capability that soon. There's no question in my mind that, in the long run, for certain payloads and long-range human exploration of the Moon or Mars or even a large mission to Planet Earth type of initiative, a heavy-lift capability is necessary. It's not at all clear what the flight-rate of that capability would have to be. Certainly if SDI were deployed, there would be a very large requirement.

Aaron Cohen: Thank you, Dick. Courtney, let me address this to you.

Question: When will DOT licensing regulations for launch services be issued? What do you estimate the incremental cost and time required to complete licensing requirement for a launch?

Why Important and of Interest: Will impact cost of launch service and international competitiveness.

Recommendation: Minimize regulations and review process.

By: Henry Minami, Project Development Manager, Rocketdyne, Rockwell International, ELV Propulsion Systems.

Courtney Stadd: Let me clear up an impression I must have left earlier in my original remarks. We have an interim version of final regulations that are the basis of our current licensing action. What I referred to was the final issuance of those regulations. I expect those to be out within a month and a half to two months. We're currently negotiating some outstanding issues with a couple of agencies. The statute says that we're obligated to turn a license around in six months. We've already issued two licenses in the last two months. We did so in about 100 days. Admiral Truly will sympathize with my sensitivity in saying that I'm not pressured by outside events. We are not being paced by an outside number set by the Congress. They've made it clear that if we need more time, so be it. But so far we've stayed well within the time envelope. I'm within a few weeks of announcing a full-up license for a company, that is to say, both a launch safety review, as well as a mission review. We will have processed this application within the 160-day period. In terms of cost, one entrepreneur has put it on the order of a few tens of thousands of dollars to respond. I don't think that even with the primes it will be much greater than that. One of the things that we're doing, and have been mandated to do, is to try to absorb some of the opportunity costs. For example, in environmental assessments, the Government, having been the launch range operator these many years, basically handles its own environment assessment. But when you are a commercial system, entering the state of Florida, for example, the local EPA is very interested in your environmental assessment. The department has spent the past year taking the data from the ranges and organizing it in such a way that makes it understandable to the both the Federal and State level EPAs. In this particular area, we've probably saved, in my estimate, a few hundred thousand dollars for the launch companies. In the case of space entrepreneurs, I think, if they had to pay that, it would have been a show stopper.

Aaron Cohen: Thank you. Let me address this one to both Jeff and Joe.

Question: Can commercial space really hope to operate in a totally independent fashion from Government?

Why Important and of Interest: The commercial space industry would probably best function in an environment like a deregulated FAA.

By: William Bates, NASA - JSC

Jeffrey K. Manber: I'll take a first crack at that and then Joe can follow and tell me, if I express it properly from his vantage point. I think there was no effort here to have a commercial space market to be -- to use the term here -- independent of Government. The Government, by and by, is the largest customer. I think that the figures I've seen are that DOD and NASA spend over $25 billion a year. That's a sizable amount. That's *the* major amount spent in the free world. And, I think, to say "independent" misses the point. The point is that we're trying to have the Government recognize that it's a customer, and we believe that there will be cost savings and efficiencies in the economic scale if it would come in and purchase things as a customer. And that could be remote sensing data, that could be launch services, that could be on anything but hardware. But from our vantage point, there is no effort to try to make it "independent".

Dr. Joseph Allen: Jeff, let me just follow your thoughts and perhaps even back up a bit. Aaron, I'd say I'm not certain I know of a business in the United States of America today that operates independent of the Government and, by and large, we are economically a very successful country. Indeed, we're amongst the leaders of the world and much of our economic success has to do with the way our Government and our private businesses have always worked together. One can think of the aviation industry, as an example of this. The extraordinarily successful aviation industry really has never worked independently of the Government. We needed NACA laboratories to get it started. We needed, downstream, an FAA, as regulators to keep it going. Every pilot in the room knows for certain that we need air traffic control and feels a lot better knowing it's there. That's all provided by the Government. I might say one other thing, as a footnote. Our definition of commerce in space and commercial space is perhaps a bit different from the original concepts in the minds of those thinking about the policies. Our definition at Space Industries is that an undertaking puts some of its own money, or a lot of its own money, at risk and takes responsibility for either the service or equipment or both that is provided. If it doesn't work, the user doesn't have to pay; much like the business that goes on day to day right here in downtown Houston. It has some good examples back in the aviation industry. When a user, who might be Government, needs a test bed -- let me use the Air Force as an example -- for weapons development or for that matter even scientific lasers that happen to be airborne, it need not procure a new airplane. It can buy a commercial airplane, and often does. In that commercial airplane is very little advanced super state-of-the-art technology. It's proven, commercial airplane type technology and it is just bought by the Government just as the Government would buy filing cabinets or pencils. There is no DDT and E and all the phases of the RFP and the response to the proposal and so on in order for the Government to go out and purchase available, appropriate technology. And that is our view of what the distinction might be between a commercial endeavor and the more traditional acquisition of things by, for example, NASA, to use in space.

Aaron Cohen: Very good. Thank you. This question is addressed to all, but especially to Cliff Duncan.

Question: How reliable should a booster be? Can we get material to space with large numbers of moderately reliable, small, relatively inexpensive, mass-produced boosters built to commercial (automobile, VCR) standards, instead of military/aerospace standards?

By: Rodger Cliff, Sr. Engineer, Advanced Technology and Research.

Do you want to take a crack at that one?

Dr. Robert C. Duncan: When you talk about putting man in the system, obviously the reliability requirements are and should be extraordinarily high and I think we have to be prepared to pay whatever it requires to provide that sort of reliability. However, if you are launching a very inexpensive payload, with a small booster or with a booster that's inexpensive also, then I think you could take a different view toward reliability. Rather than build parallelism into the system at the subsystem level or the component level, you build simplicity there and your parallelism comes by just launching one more satellite. This is particularly true if you have a distributed satellite system, where there are going to be a quite a few that you're going to put up, then you'll add one more to that distributed system. I think that could possibly lead to considerable cost savings for unmanned satellite systems which require many individual satellites.

Aaron Cohen: Willis, this is left over from Panel 1 and it's from an Associate Professor of History, Ohio State.

Question: You mentioned the *Washington Post* as a large part of the problem in my judgment, correctly. Though I would add the *New York Times* and the major TV networks, news networks. What can we do to minimize the negative impact of their distortion of the issues?

Recommendation: We are finally developing a strong anti-establishment media - CNN, *Wall Street Journal*, *U.S. News & World Report*.

By: Joe Guilmartin, Associate Professor of History, Ohio State.

Willis Hawkins: I was afraid somebody would get to me. I was sitting very quietly. And, of course, I get the easy one.

Aaron Cohen: Willis, perhaps you should ask if the *Washington Post* is present. We do have press here. I wonder if they're represented here.

Willis Hawkins: I just happen to know a past President of the *Washington Post*. He graduated from the United States Army. He was once an Assistant Secretary in the Army, and after a while he sort of shook his head and said he wished he was the editor and not the President, because the President wasn't in charge. Anyway, I think I used the *Washington Post* as an example because they seemed to have the ear of the Government process somehow and they seem to be enjoying it. I think it has biased their judgment a tad. The *New York Times* is gradually catching up, I agree, and of course the *Los Angeles Times* isn't doing very badly. Anyway, the response I would give to this is that if you've done your homework, you'll have to talk to your boss and you'll have to explain what kind of homework you've done. Furthermore, I think your boss has to know that you've got the homework, because, in spite of everything, the bosses at NASA and the bosses of the Air Force, for that matter, of DOD are going to be exposed, no matter what. They're out there all the time and if your homework hasn't been made apparent to them or if you haven't put together a logical program, they're going to have a terrible time defending it. So I think the answer is strictly homework. Homework means more than just coming to the point where you think you've satisfied your judgment. You have to decide in your programs to satisfy somebody else's judgment and that includes the Washington media. This puts an additional burden on everything we do. But unless the folks that are actually doing the work understand that, they're not going to do enough homework.

Jeffrey K. Manber: I just want to add one thing as one who spent much of the last few years trying to make a living writing for the *Times* and some of the McGraw Hill and other papers and publications, so let me provide a perspective, and that is, that many of the editors of the major papers are very milestone-oriented. And I think in 1979 the *New York Times* ran an article and they mentioned Russ Ramson of Micro gravity and some of the entrepreneurs who would be taking advantage of the upcoming shuttle programs. Well, you can't go back to the editors at the *New York Times* four years later and say, "Let's take a look at where it is today." Their answer to you would be, "Well, it hasn't changed. Russ is still trying to do this. And this is still trying to take place." And the same is true with the Space Station. These people have been trained so that when news happens, they'll cover it at that time. And many of the papers are not really skilled at this point in covering things that we might almost call background. We have the newsletters in this society that handle the day to day events and we have the magazines that maybe take a broader look. But the newspapers are really in an uncertain position as to how to handle the space program. And it is something that all of us who are trying to write about it come up against. The newspapers were very happy to cover the *Challenger* unfortunately. It was a milestone. And so one of the problems that we always face is how to make them recognize that certain milestones may not be the sorts of milestones that take place in other industries, such as biotechnology where you might have a patent. And so that's one of the problems, in that space is such a broad area and has so many different applications. And it moves so slowly compared to the other industries they're used to dealing with. Their feeling is, we'll cover it when something new happens.

Aaron Cohen: We'll change to a little different subject here. Joe I'd like you to think about this one since you've had some experience and Dick has too.

Question: Robotics promise an opportunity to reduce crew exposure risk. Proposals include smaller orbital cruise, astronaut time and orbitals of values of resource. Is there a way to trade exposure risk, the payoff for man in orbit by using robotics?

Recommendation: Shuttles should be launched with full complement and robotics used to enhance productivity.

By: Lyle Jenkins, Project Engineer, NASA Johnson Space Center

Dr. Joseph Allen: Absolutely.

Aaron Cohen: Excuse me. If I had to define a robot and write the equations for you while you held that payload over your head for an orbit, it would be pretty hard to do. That would be a pretty smart robot.

Dr. Joseph Allen: Well I'm not so certain. Smarts were not required of that but you had to be versatile. The question is about trading risk of human exposure and the relation of this factor in the use of automation, robotics, automatic intelligence. There was a panel held out at NASA Ames last week. Some of you perchance may have been there. This was a symposium on technology that had been developed within the aerospace industry and how it might be accessed or shared or moved out into other industries. Because it was at Ames, the industries represented were, by and large, from the Silicon Valley, very high tech industries, that were there looking at aerospace developed technologies. One of the subjects discussed was, "How do people interface with increasingly intelligent robots?" Now, I was bothered by that. I wanted it to read, "How do increasingly intelligent people interface with increasingly intelligent robots?" But that was not the subject. However, I think, it contained a message. And that is, "Our lives really have been governed, and will continue to be governed by the way we humans relate to better machines. Our generation has been confronted with this situation. Believe me, our children's generation is really going to be confronted with it. The machines do get better. Thank goodness for that. I think it is reason for much optimism. The amount of human labor and sweat in factories today is considerably less than it was several generations ago because of these machines. The exposure of humans to danger, particularly, potentially dangerous industrial situations, is considerably less today, because of these machines. And there is nothing about what people are doing, have been doing, or will do in space that should be any different from lessons that we've already learned or are in the process of learning here on the ground. I would point, as a good example, to the nuclear industry. There was a time when people worked for a few minutes inside reactor vessels clad as though they were astronauts, in very bulky garments to protect them for just those few minutes from radiation levels. Less and less of that is being done right now. More and more of it is being done by robots being controlled by those same people. But those people are a safe distance away. Aaron, when all of us space workers design and ultimately operate the Space Station, there should be lots of robot-like activity going on there to relieve our versatile crew members to do in space what our versatile people do best on Earth. And that is, use our hands occasionally to fix things and, most importantly, our brains to accept still greater challenges.

Aaron Cohen: Courtney, let me ask you this one. Given the potential for commercial space, especially for launch services, shouldn't the Government provide some reasonable limitations on third-party liability for launch service providers, as a way to encouraging industry development?

Courtney Stadd: My bias is yes. We've been grappling with this for many months. There are several schools of thought on it. One is that the Government basically agree over a certain level to pay whatever the damages. We've rejected that. A second school of thought is a cap. We've been pushing that in other areas, such as product liability. Congress, more specifically, the American Association of Trial Attorneys has problems with caps. But I think there is a lot of potential in looking at the capping area. There was a third area. To try to come up with some commercially reasonable definition. Our dilemma is that the agreement that has been worked out to date with the Air Force for access to the natural ranges sets a rather untenable proposition for getting insurance. It says that you, the company, will obtain as much insurance as is available on the market. That's just commercially unreasonable. I mean, you can get five trillion dollars worth of insurance if you are willing to pay four trillion in premium, which is an untenable situation. We are currently exploring the potential of caps. Concurrent with that we are trying to speak to the best minds we can find in the insurance industry to help characterize this whole business. So I do agree there must be at least some limits. Part of this is driven by the fact that foreign governments, for example, the French government, are now willing to pay all excess liability. Likewise with the Chinese. And we have to confront that competitively. So going back to my earlier comments, we're going to be putting out a rule in which we will characterize what we think is an approach to help limit the third party area. And then we'll give the industry sixty days to comment on it. It's my feeling we have about a calendar year to get this liability issue sorted out.

Aaron Cohen: Dick, let me give you two of them together. What should be the relationship between manned and unmanned space flight and what should we try to accomplish on each? And this is a different question, but I think maybe you can answer at the same time. Should we aggressively investigate liquid propulsion as a replacement for solid rockets?

Richard H. Truly: Well, the answer to the last one about liquid propulsion -- I should have mentioned that, I did not. In parallel with the advanced solid rocket motor studies, we have a study that is specifically looking at the feasibility of replacing the solid rocket motors on the shuttle with the liquid fuel system either pump-fed or pressure-fed. We are also looking at whether such a system could be as a standalone ELV.

I'm not sure about the first question, "What should the relationship be between manned and unmanned space flight?" I have always been disillusioned, over the last 25 years I have worked on the space business. Going from unmanned systems to manned, and back to unmanned, and then back to manned again. Why should we argue this point? I think there are clearly missions that should be flown on an ELV and an unmanned system that have no relationship to the presence of man, get nothing from it, can be made cheaper and operated more easily. And when such systems come along, such as Cliff was talking about earlier, I think it would be crazy to fly them on a manned system and pay the premium that goes along with that. On the other hand, the greatest dream of man has been to fly, I think. And to fly in space is a challenge that we have stepped up to and have a heritage in, as have the Soviets, and almost every country in the world has that dream and ambition. And for us, at this point in this national policy to debate, to even consider turning our back on that heritage is something that I can't image. So, my answer is that I think there should be a balance. We should fly unmanned or ELVs when it's appropriate and fly manned missions when that meets our national goals.

Dr. Joseph Allen: I couldn't agree more with Richard. I think in the years that I've been associated with the space program -- in fact Dick and I had about the same length of time in both aspects of the space business -- the most unimaginative argument that I have heard is this manned versus unmanned. And there are those that argue it; in fact, Professor Van Allen is one of the strongest proponents of solely unmanned. He was a physics professor when I was a student. He was doing physics experiments on the ground that did not involve people at all, except for analyzing the data. That was his background, and it certainly is his right to argue it. But he bases his argument on a very narrow view of what people do in other arenas -- other than the very simple basic physics measurements arena. He is not, for example, a crystalographer. You do not hear crystalographers making that argument. Unfortunately, a lot of his arguments are picked up by the very papers that we've heard about. And those editors don't understand it either. And it goes on, and on. I am of the personal opinion that there will be a time when there is no reason to have people in space and that will occur at exactly the same time when there is no reason to have people here on Earth.

Dr. Robert C. Duncan: I'd like to add, to reinforce what Dick and Joe have said, one brief little story. When I was at the Johnson Space Center in the '60s working on Apollo, we were up to our ears in problems with relays, power supplies, bearings, gyros, and valves, and those were the things we were worried about. You and I were worried about, and Ken Cox, and others. At that time the Institute of Navigation sent me a book to review. The book was on the *Meaning of Man Going Into Space*, by a so-called space philosopher. I read this book and I wrote a review on it. It opened my eyes. This is the subject for the philosophers. The basic theme of the book was: man doesn't realize it, but his going into space is part of his survival. At some point in time, and it may be a million years, this Earth will require that it be expanded. That was the basic subject of the book. I did a terrible review on it because I thought that the important things were the relays and the power supplies and the bearings.

Aaron Cohen: We have time for about two more questions. I'm not going to get to them all. Let me try one which you said, Jeff, you understood. I'm not quite sure I understand it but let me read it and maybe you can answer it and explain it. Will Government agencies consider easing NASA/Air Force requirements to encourage commercialization of launch services?

Jeffrey K. Manber: It goes back to one of the major goals that we've been talking about here. And it's of interest to the Commerce and Transportation Department and all the other agencies. And that is, if we have a mature technology, do we allow the Government to purchase some of these pieces of equipment, even though they are space based pieces of equipment, as a commercial customer? One of the problems is that the Government is handicapped in the way that it buys certain pieces of equipment or contracts for certain jobs. One of the problems, however, is that ELV manufacturers are fearful that, if we are successful in changing some of the flaws, that they will make a commitment to produce what is probably a more economically efficient launch vehicle only to discover that six-eight years from now regulations will have changed, policies will have changed, and suddenly they may be at a disadvantage in bidding on Government contracts. So the answer to the question is yes, I think it is certainly agreed in the Commerce Department that we do want to change this and try and make it more efficient. But on the flip side of that, there is a skepticism on the part of private industry and the private sector as to whether we will be successful.

Aaron Cohen: Let me end on this final question which will be a round-robin. Can six or seven agencies be sponsors and accountable for delivering space benefits to DOD industry and Government agencies at economical cost? What mechanisms allow all these agencies to address coordinated issues? Would somebody like to take a crack at answering that question?

Willis Hawkins: If all those people are in charge, we'll never get there.

Courtney Stadd: I know that scientists and geniuses have pored over the political process, for good reasons I think. But in any organization the person who controls the file cabinets, controls the money, controls the life blood of the organization. Space priorities are set, for better or for worse, in Washington. And I think that it is important that this Society, which has so much talent, so much sophistication along with other space societies, such as the AIAA, face the challenge and hold the candidates who are out there, who presume to be our leaders, accountable. I think that in order to make what we were doing the last few years a reality, we have to have continuity. To insure the wisdom, the consistency that Joe Allen is talking about we've got to look to the next Administration, Democratic/Republican. So I really would challenge the leaders of AAS to work with your counterparts around the country and drag the candidates, drag their chiefs of staffs in and educate them. Because there is only one way to get the program going and that's top down. Thank you.

Aaron Cohen: Very good. That completes the first session here. We do have about eight questions we didn't get to answer and I'm going to try to figure out some way to get an answer to those questions. As you show your appreciation to the panel, I'm going to hand a little plaque to them, so I would like you to show your appreciation to the panel today.

SPACE UTILIZATION/APPLICATION

PANEL TWO

Chairwoman:	Elizabeth Young Vice President Maritime Services Comsat Space Communications
Host Chairman:	Douglas S. Lilley Vice President Space Industries, Inc.

AAS 87-614

SPACE UTILIZATION/APPLICATION - INTRODUCTION

Elizabeth Young* and Douglas S. Lilly†

Henry N. Bowes: We're now ready to begin if you would take your places please. I'd like to introduce Doug Lilly, who is with Space Industries. He is the Host Chairman who will get this panel started.

Douglas S. Lilly: Thank you, Hank .... It is my pleasure today to introduce Session 2 on Space Utilization and Application. I think we have a very good forum and a very good set of speakers for you today. First, I would like to introduce the chairman of this group, Dr. Elizabeth Young, who is Vice President, INMARSAT Policy and Representation, COMSAT Maritime Services. Also, she is a board member of the AAS. Earlier she was in public broadcasting for 12 years and has been President of the Public Service Satellite Consortium. She was formerly with Comsat General. Now I'm sure that there will be a lot of questions prepared for our panelists. Dr. Young.

Dr. Elizabeth Young: Thank you very much, Doug. We have a full agenda this afternoon. But the most important part is hearing from you. And we do want to give you ample opportunity to send forward your questions and we'll leave plenty of time for our discussion. In discussing with my panelists this afternoon how we would format this session, we came to the conclusion that a very good topic would be generally titled, International Cooperation and Competition in Areas of Space Applications. It has become a truism really to say that space exploration ignores national boundaries. And it may be equally true that some day space exploration will even ignore planetary boundaries as we seek to find others in our galaxy with whom we could work cooperatively. But certainly, we are rapidly reaching a time when we must take an ever closer look at the complex role the United States must play as it advances its space exploration policies and missions vis-à-vis those of other countries. We use terms like cooperation and competition rather generally. But operationally they have specific meanings and those meanings shift over time. I think no company has been more aware of that than Comsat, the Communications Satellite Corporation, for which I work and which was founded in 1962 specifically to enable the United States to participate in internationally related space communications, first, via INTELSAT and more recently for the Maritime System through INMARSAT.

---

* Dr. Young is Vice President, Maritime Services, Comsat, 950 L'Enfant Plaza, S.W., Washington, D.C. 20024.

† Vice President, Space Industries, Inc., 711 Bay Area Blvd., Suite 320, Webster, Texas 77598.

Today, we will hear from four people who are vitally concerned with certain aspects of the United States' future role in space applications and utilizations. I will introduce each of our speakers.

INTRODUCTIONS OF SPEAKERS

First on our panel today will be Robert Berry. Bob is currently Vice President for Space Systems Operation for Ford Aerospace in Palo Alto, California. He directs the prime contracts for satellite systems. Recent contracts for Ford have included the Intelsat 5 and 5A Series, the next generation of GOES Weather Satellites, the Japanese SuperBird large commercial satellites. And Bob has in participated and directed several teams working on packages for the Space Station. Before this position he was Deputy Director in Defense Research and Engineering, and he was Founder and First Chief Executive of the IRT Corporation in San Diego. His academic degrees include a bachelors of science from Manhattan College and a masters from the University of Pennsylvania.

Our second speaker is a friend of mine, and a colleague, and I'm very happy to have her on the panel. Dr. Heather Hudson, who is currently the Director of Telecommunications Management and Policy Program at the McLaren College of Business, University of San Francisco, is, I think, one of the foremost experts on the uses of new technology, in particular, satellites in developing countries. She is currently on leave from the University of Texas, where she's an Associate Professor in the College of Communication and Director of the new technology areas. I first knew her when she was Director of Telecommunications Applications, for the Academy for Educational Development in Washington, D.C. She has consulted for a number of telecommunications companies, as well as foreign governments, the U.S. Agency for International Development, UNESCO and the World Bank. Dr. Hudson received an honors B.A. in English from the University of British Columbia and the M.A. and Ph.D. in communications research from Stanford. She is going to focus her remarks on an area where the U.S. undoubtedly can regain and maintain preeminence, notably in technology transfer.

Dr. Robert Filep, our third speaker, is currently involved with program development for Advanced Communications Spacecraft at TRW Space and Technology Group in Redondo Beach, California. This work includes follow-on tracking and data relay satellites. Bob and I have worked together on a number of projects in the past but particularly for the last 12 months, when he has headed a committee of the National Research Council Space Applications Board in Space Communications Research Development and Applications. This committee was asked by NASA to assess policy issues affecting research development and application programs in space communications and to define past and future benefits resulting from such investments, and, if appropriate, to specify new requirements and the role for research in space communications. Dr. Filep has been a government official and an associate administrator responsible, among other things, for the ATS6 Satellite Program, which you'll hear about from Dr. Hudson, and, interestingly enough, for "Sesame Street". He has also been a university professor at the University of Southern California, a member of U.S. delegations to the World Administrative Radio Conference, a Fulbright lecturer to India, and Founder and Chairman of Communications 21 Corporation. Now you can tell one of Bob's biases, when he writes in his biography, he holds a bachelor's degree from Rutgers University (a Scarlet Knight) with a masters from Columbia (a Lion) and not last, and certainly not least, a doctorate from the University of Southern California (the Trojan). I'm not going to try to put all three of those together, Bob, but he also notes that he's a veteran of the last six years of the Space Station scrimmages. It gives you an idea of which sport he's in favor of.

Our first three speakers will talk today about a combination of uses of space, but focus on communication uses. Our last speaker will turn our attention to a very different area of commercialization and the possibilities of international cooperation. And that's industrial applications in the micro-gravity environment. To share these thoughts this afternoon, is an extremely well qualified speaker and another colleague of ours from the Space Applications Board and a participant on another SAB study which has just focused on this issue, Bud Evans. Bud, whom many of you know I'm sure, is currently President for the Center for Space and Advanced Technology, Inc., in Arlington, Virginia. Just prior to that he was Deputy Assistant Administrator in the Office of Commercial Programs at NASA. In addition to that, he has served in other positions in the Government, including Director of the Regulatory Improvements Office at the Nuclear Regulatory Commission and holds senior positions at the White House. Bud has his degrees in bachelor of economics from Colby College and his juris doctor from Cornell Law School. And it gives me a great deal of pleasure to welcome him as the final speaker on our panel this afternoon.

AAS 87-616

RETURN TO EMINENCE: ISSUES FOR U.S. COMMUNICATIONS RESEARCH, DEVELOPMENT, AND APPLICATIONS

Robert Filep*

Thank you very much, Dr. Young. Liz, I will certainly allude to your good work on the SAB committee as I share my comments with the group.

Houston certainly was, is, and will continue to be an historic site in the annals of space achievement. My last, very pleasant, visit here was the day the decision was announced to switch from a single to a dual keel on the Space Station and yes, Aaron Cohen, there were a lot of people in the room at the time. It is good to be back!

I'd like to share some thoughts in three areas this afternoon. First, on the commercialization of space, just very briefly; then I'll look to the future in space communication research development and applications areas that the SAB committee has been examining over the past few months. Finally, I'll share with you some issues that need to be addressed in order to get to the future and to put that train on the tracks, using the analogy that Joe Allen made this morning, that will allow us to get to the station -- a higher plateau and a greater capability during the next 25 years in space.

The first area. The concept of satellite communications was invented only 42 years ago at the dawn of the Space Age when modern rocketry was coming into being. In that short time, communications has become the outstanding success of America's civil space program. It has been thus far the only commercial success among the various U.S. endeavors in space, setting benchmarks for the others that I know will follow. There have been more than $10 billion in public funds expended to date on satellite communications, RD&A and the technology demonstrations about which you've been hearing. The return on the investments has been pretty good. If you would check with the people who have been involved, it ranges from 10 to 15 to 20 percent ROI (Return On Investment). It has produced international long distance, rapid response, news gathering, greatly expanded television reception and, more importantly, it has spawned an entirely new industry. The civil satellite communications industry comprises several sectors: manufacturing and marketing of satellites; spacecraft, components and assemblies (transponders, switches, etc.), the manufacturing, installation and operation of ground systems, direct provision of com-

* Dr. Robert Filep is Chairman, SAB Communications Task Group, TRW Space & Technology Group, One Space Park, Redondo Beach, California 92078.

munication services by satellite owners and operators and the commercial activities of end users, such as those who offer various information and business services via satellite circuit.

It is difficult to quantify the actual value of these markets because of their diverse and international character and also because comparable industrial financial data are difficult to collect and compare. However, the global market for civil satellite communications hardware and devices is thought to be on the order of about $10 billion per year and growing. Now, the context of space communications has changed in 25 years. This June, we celebrated the 25th anniversary of Telstar, which was an AT&T accomplishment but it should be noted that a team arrangement between NASA and the AT&T laboratories at that time produced what was then our first operating telecommunications satellite. Over 25 years there have been new players and partners, and many became protagonists for the future, be it the organzation of space communications, RD&A, or the role of the Federal Government, or the technologies that are most worthy of investment and development, or the means of achieving full-scale flight test. To paraphrase Joseph Conrad, our reach should exceed our grasp.

Our challenge today, is to look 25 years ahead and envision the form communications will take, stretching our imagination and lifting our sights in time and space. The issues cover near space, the Moon and environs, and deep space. They cover geostationary, low Earth orbits as well as areas where new communication channels need to be identified and explored. As has been suggested in *Star Trek IV*, we have many frontiers and frequencies to conquer. The benefits to date, as well as the prospects for the future, need to be reevaluated. We have only just begun, to use the parlance of the modern song, to identify the potential benefits of space communication research development and applications. If we plan to colonize the Moon for exploratory and commercial purposes, then we must also examine the role of real-time, broad-band communications from the Moon to Earth and back. What will be the characteristics of the broad band fiber optic link with its great capacity to transmit voice, video and data streams? And what about holographics? How large will moon communications stations be? Will there be a fiber optic light tunnel to the Moon? All challenges!

Recognizing the high cost of putting man in space, what RD&As are needed to use video images, for example, as a surrogate for humankind in a hostile and high-cost travel and maintenance arena? We assume that we are adequately prepared to handle the communication requirements of the Space Station. Are we? Are we proceeding with the RD&A required to provide real-time broad-band multi-color video and audio to minimize the high cost of man's presence on the Moon? That's not to discourage man presence on the Moon, but to expand his capability. Are we planning missions to explore the backside of the Moon, which may shield against interference, providing a superb locale for deep-space tracking and relay? Are we thinking of the possible benefits of Moon orbiting communications satellites to support many missions on the Moon as well as here on Earth?

Identifying our objectives for space will be a futile exercise if we do not resume a regular launch schedule that will include a sizeable expendable launch vehicle capability. I was heartened, as I hoped you were, by Admiral Truly's comments this morning and also by the release of the mixed flight manifest that we saw a few weeks ago from NASA. However, there's a lot of work involved and it remains a critical variable. It is extremely important that the capability be reestablished and that there be options, particularly if we're going to make gains in the commercial and civil use of communication satellites.

Some of the major questions that our committee study included were: What has the United States invested in space communications RD&A? How are these activities structured? What communication systems and technologies are appropriate for investigations? What economic and social benefits have been derived from space communications? What are the key issues of international competition in space? And what is the regulatory and policy environment for space communications research development and applications? The initial charge of the committee was to look at all this in the current climate for space communications -- one of deregulation and international competition.

Here today with us are two members of this committee who worked very hard, and continue to work very hard: Dr. Elizabeth Young and Dr. Kenneth Rose, who has come for this panel. Ken is now very much involved with fiber optics that provides major competition to the communication satellites. His prior experience was with Ford Aerospace Company in satellite programs. He's a man for all seasons and we respect and admire that. He has constantly been helping us to identify the synergism that can take place between the communication satellites and fiber optics -- the network questions, those issues as to where optics can be best utilized and where the communication satellites can be best utilized to provide the highest quality and lowest cost service to all our citizens. We applaud your being here. Thank you on behalf of the committee.

How about the current situation? We're aware that a single modern communication satellite can carry as many as 30,000 telephone calls and 12 television channels at once. I mentioned Telstar. In 1963 when AT&T began offering services to the South Pacific, the calls went through a radio operator in Oakland, California; that is when one was available, and it cost $33 for 10 minutes. Today a 10 minute call to the same location costs less than $17 and can be dialed directly. Other consumer prices, as we all know, have nearly quadrupled in the same quarter century. Satellite communication industries and operators have been in good health, and they continue to be.

We've heard about some of the applications. Some of you in this room today, along with John Pierce, who serves on this committee, were there at the beginning, although some might suggest that the first space communications took place in the Garden of Eden, but whether a receiver served as an apple is debatable. I'll leave you to ponder that throughout the afternoon.

As the nation celebrates, the 200th anniversary of the Constitution, the similarities between the Constitutional model and the RD&A structure become very apparent. A compromise between the central government and the states rights advocates was reached. Compare this with NASA Headquarters and the NASA Centers. There are checks and balances throughout, as well as redundant organizations and manpower. The structure of RD&A is a subject of the committee .... We all agree that science, space technologies, and engineering have become a pervasive part of the human experience in the 20th Century and our key to our future well-being, and that changes are occurring at a head-spinning pace. Therein lies part of our challenge.

Now, I'd like to move briefly to the third part of the presentation and share with you some of the issues discussed by the committee. One challenge was to conduct a symposium in June, where about a hundred members of the space communications community were brought together: Government, business, industry, NASA, the private sector, (international was excluded). We looked at some of the issues pertaining to space communications RD&A. This laid the platform for the committee's work and its report, which is in progress right now, and hopefully will clear the channels of the National Academy of Science sometime next year.

I will now cover some of the issues in this symposium that the participants saw as areas requiring examination in order to allow us to move to the high plateau in space communications RD&A .... There is a concern, and this reverts to Bob Berry. Bob, I'm grateful to you for setting that theme about the need for goals, that there is a need for coordination in establishing the goals when we establish our priorities. You'll see a little later the technology road maps for space communications RD&A. Again these are observations .... There was a concern that NASA has expertise which historically has not been utilized. The slots are there, the people are there as effective as they might be .... There are still communication satellites being sold in the civil and commercial market and the rate of growth is there. How fast the growth is clearly related directly or indirectly with the launch vehicle availability. There was a concern that fiber optics are not going to drive out communications satellites. In fact, there is a greater challenge here in planning and decision making and examination of the technology as to what technology can do what in best bringing about a synergistic relationship .... Bob Berry stated it very eloquently, "If we can set the goals, this issue may not be as great as we perceive it. But there are clear realities in terms of the levels of investment. We've been trying to identify those as a percentage of the gross national product invested in space in various kinds of space activities. Clearly we are not the leaders in that investment. Part of the problem is the definition of roles. The symposium group felt that, if the category means were available to plan a better mechanism for the establishment of the national goals, there would also be a clearer definition of the various responsibilities. This is getting a little closer to a technology road map which would undergird the goals and the agreement with them, and also shows concern that we need better planning, both in terms of the applications and the technology requirements.

I think we would all support multi-year funding, and there was a lot of concern that it's very difficult for the organizations as well as the agencies to plan any sensible and cost-effective research development and application with single-year funding and with the associated vagaries. We have enough difficulties even when we do get a multi-year funding program. Clear recognition of the fact that incentives for industry, as it is now structured, including tax structure, are not necessarily there for long-term payoff investment is critical. Some attention needs to be paid to the fact that to help facilitate and encourage investment -- not short-term payoff, some reenforcement for investing in high-risk, long-term, potential payoff areas is required. This was a perception by the group that this was an issue that needed to be examined. This involves the definition of user requirements to arrive at the amortization of the level of resources that are currently available in Government and industry programs. Thus, if we worked together on needs, it might focus a little more closely on the actual amortization of whatever funds were available. There was a realizaton that we are facing budget limitations, certainly Government budgets as well as industry budgets where the resources for research, development and applications are getting less, not greater. As a consequence, we all need to pull together. There was also a concern that some collaborative mechanisms are needed to increase U.S. commercial access to space, reduce cost and share risks.

The symposium proceedings are now available from the Academy, where those issues are discussed at length. They contain a whole group of papers, which I think you'll find very informative. Some of the best people in the field were involved, and I'm hopeful that our report will come out later this year.

In conclusion, I feel we have some wonderful opportunities. We've reached a high plateau. We've got a long way to go. But if we all pull together, in twenty years from now we'll be amazed at what we'll be doing in the way of space communications. Thank you.

AAS 87-617

TRANSFER OF SATELLITE APPLICATIONS AND TECHNOLOGY: THE NEED FOR A U.S. INITIATIVE

Heather E. Hudson*

In the brief history of satellite communications, the United States has passed through three major eras: the Era of Conjecture, the Era of Experiments and the Era of Services. NASA took the lead in the experimental era to demonstrate both technology and applications -- and to ensure their transfer for commercial use. The developing world has also entered the Era of Services, but without the benefit of an experimental phase. Several developing countries now have their own domestic systems; others share regional systems or lease domestic capacity from INTELSAT. However, the record of developmental applications of these satellites has been disappointing to date. Much capacity sits idle. The U.S. has a great deal to share with the developing world to assist in the effective utilization of this technology. A U.S. Satellite Applications and Technology Transfer (SATT) program is proposed.

INTRODUCTION

In the brief history of satellite communications, the United States has passed through three major eras: the Era of Conjecture, the Era of Experiments and the Era of Services. We are now in the advanced stages of the Era of Services, although experimentation continues on use of higher frequencies, greater power, more efficient use of frequency spectrum, etc. NASA took the lead in the experimental era to demonstrate both technology and applications -- and to ensure their transfer for commercial use.

The developing world has also entered the Era of Services, without the benefit of an experimental phase. The exception of course is India, which borrowed ATS-6 for a year in 1975 to test delivery of educational television to villages. Yet nearly a decade passed in India before the INSAT system was available for use. Indonesia, Brazil, Mexico, and the Arab

* Dr. Hudson is Director of the Telecommunications Management and Policy Program in the McLaren College of Business at the University of San Francisco, Ignatian Heights, San Francisco, California 94117-1080.

nations now have their own satellites, and more than 20 developing countries lease domestic capacity from INTELSAT.

The record of developmental applications of these satellites has been disappointing to date. Much capacity sits idle. The U.S. has a great deal to share with the developing world to assist in the effective utilization of this technology. This paper proposes the a new U.S. initiative dubbed the Satellite Applications and Technology Transfer (SATT) Program to transfer experience and expertise in the application of satellite technology for social and economic development and to transfer technologies appropriate for developing country environments.

THE THREE ERAS

The Three Eras of Satellite Applications

The potential benefits of satellite communications for development have been recognized since Arthur Clarke proposed a geostationary "space platform" in 1945. (Ref. 1) We may consider the history of the application of satellite technology for developmental purposes in three phases. The first, from Arthur Clarke's proposal in 1945 until about 1970, may be called the Era of Conjecture. The second phase, from approximately 1945 until 1977, may be called the Era of Experiments. We are now in what may be considered the Era of Services when satellite services for developmental purposes are being offered on commercial satellites.

The selection of these dates is based not on the history of the technology (e.g. launch dates and early technical experiments), but on the history of satellite applications for experimental purposes. Developmental applications for the purpose of this analysis include uses of satellite technology for education, health care and other social services, agricultural and rural development, and cultural enrichment. This definition excludes international satellite communications and domestic commercial broadcasting. It includes applications of thin route telephony, teleconferencing, and educational/public radio and television. (Ref. 2)

The Experimental Era in the U.S.

The era of experimental social service applications of satellites began in 1971, when NASA made its ATS-1 satellite available for biomedical and educational experiments in Alaska. ATS-1 demonstrated that satellites offered a reliable means of providing basic two way communication for villages that had previously relied on HF radio. The ATS-1 satellite was also used in the Pacific by the University of the South Pacific (USP) to reach its extension centers scattered over the vast South Pacific and by PEACESAT, an organization based at the University of Hawaii that linked groups in the South Pacific, Micronesia, New Zealand, and Australia.

NASA's ATS-3 satellite was also made available for basic voice and data communications to support social experiments following its technical demonstrations. The ATS-6 satellite was actually designed to accommodate social experiments, and was used extensively for projects in Alaska, the Rocky Mountain States, and Appalachia. ATS-6 was loaned to India for a year to provide educational television to Indian villages (see below).

The Communications Technology Satellite (CTS or Hermes) was a joint US-Canadian project that provided experimental opportunities in both countries in 1976-77. By that time, commercial satellites were available in Canada (the ANIK series) and the first RCA SATCOM and WESTAR satellites were opening up the U.S. commercial satellite market.2

The Era of Services in the U.S.

The Era of Services in the U.S. dates from the FCC's Open Skies Policy of 1972 which established the principle of competition rather than monopoly in domestic satellite services. However, the first commercial services did not become available until the second half of the decade, with distribution of cable TV programming (pioneered by Home Box Office) and telephone service to Alaskan villages being among the earliest services.

By the late 1980's satellite services have extended to a diverse set of applications including:

TV distribution: distribution of network and cable television signals for local redistribution; delivery of instructional programming directly to the workplace

Voice: provision of telephone service to remote areas (primarily Alaska) where costs of terrestrial facilities would be prohibitive

Data: computer networks for data transmission and electronic mail; data broadcasting of weather, news, financial information;

Teleconferencing: interaction among individuals or groups at several locations using voice, graphics, and/or video.

---

2 NASA has now re-entered the experimental satellite field, with the construction now underway of its Applied Communications Technology Satellite (ACTS).

By the end of the decade, mobile services for communication directly with vehicles are also likely to be commercially available.

STATUS OF SATELLITE USE IN THE DEVELOPING WORLD

Developing Countries with Satellites

Satellite technology can be particularly appropriate for developing countries, where terrestrial infrastructure simply does not exist. Satellites are especially useful for regions with large land masses and scattered populations (e.g. Brazil, the Australian Outback), populations scattered on numerous islands (e.g. Indonesia, the South Pacific), and difficult terrain (mountains, jungle, desert, tundra, etc.) Satellite earth stations can be installed where needed without the necessity of stringing wire or building remote repeaters for delivery of telephone and broadcasting services. Satellite networks are robust in that an outage at one location affects only that location, whereas terrestrial networks are vulnerable if one repeater or cable is damaged. Additional capacity can also easily be added as usage increases.

Developing countries that have invested in their own satellite systems include:

-- Indonesia: PALAPA (1976)
-- India: INSAT (1983)
-- Mexico: MORELOS (1985)
-- Brazil: BRASILSAT (1985)
-- 22 Arab nations: ARABSAT (1985)

In addition, approximately twenty developing countries lease domestic capacity from INTELSAT.

However, these countries have not had the opportunity to experiment with satellite technology before investing in an operational system. The exception, as noted above, is India, which carried out a year long experiment using NASA's ATS-6 to deliver educational television to more than 2500 villages. However, nearly a decade passed before India had its own domestic satellite.

Developing Country Problems

Despite the demonstrated potential of satellite technology to support development activities, the performance has not lived up to the promise. Domestic satellite systems are generally underutilized, with a high percentage of transponder capacity left idle. There appear to be many reasons for this underutilization. Generally, the satellites and earth stations are owned by the government and operated by a ministry of telecommunications or government telecommunications authority. Government efforts to procure and install earth stations to operate with the satellite may take years. Tariffs designed to recoup investment quickly may actually be prohibitively expensive for some potential users.

In addition, there may be a lack of understanding of how the technology could serve development goals. Ministries of telecommunications are staffed by engineers who are unlikely to have experience in communicating with non-technical potential users. Such potential users in government ministries of health, education, agriculture, and rural development, etc. may not understand how satellite technology could be used in their programs, or may think that the technology would be too expensive and complex to meet their needs.

Even if officials in social service ministries and other development agencies make a commitment to use the satellite, they may need several years to plan, obtain funding, train personnel, and implement the projects. The timeline may be especially long if projects involve production of new materials and programs (as opposed to extending existing services to more remote areas). Meanwhile, the satellite has moved inexorably from construction to launch and into its finite ten-year (or less) expected life.

The result is a waste of an expensive investment. Yet the impact may be far greater in developing countries because the investment in satellite technology has used limited resources including foreign exchange and skilled personnel that are usually extremely scarce. If squandered, these resources cannot easily be replaced.

THE NEED FOR ASSISTANCE

Thus simply obtaining satellite technology will not guarantee developmental benefits. In order for developing countries to obtain maximum benefits from their investment, they are likely to need assistance in:

Planning of a nationwide system, particularly to take into consideration the nation's development plans and priorities;

Management of a complex satellite system that must interface with terrestrial facilities, reach regions not previously served, and deliver new services;

Understanding of developmental applications and benefits so that government departments, development agencies, and other potential beneficiaries can become participants in planning and use of the technology;

Development of software including techniques for needs assessment, training, production, pilot testing, and evaluation;

Technology transfer so that the most appropriate technology available is used, consistent with development goals of developing national expertise and indigenous industries. (Ref. 3)

A U.S. INITIATIVE: THE SATT PROGRAM

The SATT Program

The U.S. should recognize that it also benefits from developing countries' use of satellite technology. A visible commitment by the U.S. to developing country utilization of satellite communications could take the form of a satellite applications and technology transfer program similar to the Applied Technology program administered by NASA in the U.S. in the 1970s, which gave numerous organizations the opportunity to try out satellite technology. This initiative would also build on the experience of the AID Rural Satellite Program which sponsored development projects in Indonesia, Peru, and the West Indies. This Satellite Applications and Technology Transfer (SATT) program would also provide a vehicle for U.S. industry to enter developing country markets.

To help developing countries to obtain the greatest possible benefits from their investment in satellite technology, the SATT program would provide support for:

-- <u>applications development</u>: training, internships, planning assistance for developing countries that have or intend to obtain access to domestic or regional satellite capacity;

-- <u>technology transfer</u>: an active trade initiative to promote U.S. technology including VSATs, transportable terminals, teleconferencing equipment, low power transmitters, cellular radio, etc.

-- <u>demonstrations and pilot projects</u>: technical and/or financial support to implement new services on a pilot basis.

The SATT Program could be administered by AID and the Department of Commerce, working in conjunction with U.S. industry, satellite users, and universities to provide training and expertise so that satellites can be effectively used to contribute to Third World development.

Initiatives to Date

The U.S. has already undertaken initiatives that have developed and demonstrated techniques for assisting developing countries to derive greater benefit from new communications technologies.

SITE and AIDSAT: As noted above, NASA's ATS-6 satellite was loaned to India in 1975 for the Satellite Instructional Television Experiment (SITE). As ATS-6 was moved back to an orbital location in the western hemisphere, NASA and the U.S. Agency for International Development (AID) collaborated in the AIDSAT project to demonstrate satellite applications in several developing countries.

The AID Rural Satellite Program: AID established the Rural Satellite Program in 1980 to investigate potential uses of satellite communications in rural development. The Rural Satellite Program sponsored three pilot projects using teleconferencing networks to link universities, development agencies, and national centers. Pilot projects were conducted in Indonesia using the PALAPA satellite, in Peru using INTELSAT, and in six countries of the West Indies using a combination of satellite and terrestrial technologies. (In each project terrestrial technologies including land lines, radio links, and/or microwave were used for the "last mile" to reach participants and to extend the communications links from the satellite stations to other sites.) (Ref. 4)

USTTI: The U.S. Telecommunications Training Institute (USTTI) is an initiative supported by the private sector to provide short term telecommunications training to developing country professionals. Established in 1982, USTTI had provided training to more than one thousand participants from more than 90 developing countries by the end of 1987. Training is both technical and managerial, ranging from equipment maintenance to frequency monitoring to system management. The U.S. government through AID contributes funding for travel and living expenses of participants. (Ref. 5)

Project SHARE: From 1985 to 1987, INTELSAT operated a program known as Project SHARE (Satellites for Health and Rural Education) which provided free satellite time to developing countries for tests and demonstrations of satellite applications for health and education. INTELSAT has announced an extension of the program to be called Project Access. (Ref. 6)

U.S. Participants in SATT

The initiative should build on the experience outlined above, with major involvement by AID and NASA. The Commerce Department should also be involved because of its mandate for stimulating foreign trade and its expertise in telecommunications in the National Telecommunications and Information Administration (NTIA). The program should also draw on the experience of USTTI in involving the private sector and in running training programs.

Other organizations that could make important contributions by demonstrating effective applications of satellite techology and providing management and production training include:

-- Learn/Alaska
-- The National Technological University (NTU)
-- The Learning Channel
-- university instructional TV (ITFS) networks
-- health education networks
-- agricultural extension media and computer services
-- teleconferencing organizations

Organizations and companies that could assist in transfer of technology and technical management skills include:

-- satellite equipment suppliers
-- satellite carriers and system operators
-- universities and colleges.

LINKS WITH U.S. POLICY AGENDAS

The U.S. can tie the SATT initiative to one or more other near term policy agendas, for example:

Space WARC (ORB 88): The second session of the World Administrative Radio Conference on the Use of the Geostationary Satellite Orbit and on the Planning of the Space Services Utilizing It (ORB 88) will be held in August-September 1988. U.S. proposals to ensure that all nations have access to the orbit could include an announcment of the SATT program. This initiative would provide a means for the U.S. to demonstrate commitment to share expertise on developmental use of the technology rather than simply to sharing of the orbit/spectrum.

ITU Plenipotentiary (1989): Again, the U.S. could use this forum to demonstrate a commitment to assisting developing countries to benefit from new communication technologies. The SATT program could be announced as a follow-on or new initiative of the U.S. Telecommunications Training Institute (USTTI -- see above) which itself was announced at the 1982 ITU Plenipotentiary in Nairobi.

The U.S. could thus further its own policy goals while demonstrating a commitment to helping developing nations not only to obtain satellite technology but to use it to support their own development.

References

1. Clarke, Arthur C. "Extraterrestrial Relays." **Wireless World**, October 1945.

2. Hudson, Heather E. "Satellite Communications for Developing Countries: From Conjecture to Reality." **Space Communication and Broadcasting** 3, 1985.

3. Hudson, Heather E. **When Telephones Reach the Village**. Norwood, NJ: Ablex, 1984.

4. Agency for International Development. **The AID Rural Satellite Program: An Overview**. Washington, D.C.: Academy for Educational Development, January 1987.

5. United States Telecommunications Training Institute. **1988 Course Catalog/Annual Report**. Washington, D.C.: USTTI, 1987.

6. Pelton, Joseph N. "Trends in International Telecommunications in the Third World." Paper presented at the Fifteenth Annual Telecommunications Policy Research Conference, Airlie, VA. September 28, 1987.

INTERNATIONAL SPACE APPLICATIONS: WHERE COMPETITION AND COOPERATION CO-EXIST

Robert E. Berry*

All who come here have to have a story about Houston and what Houston means to them. Mine is about Johnson Space Center, and what Johnson Space Center means to me. That's the place where the characteristics were put into our INSAT satellite. Now there's a Ford Aerospace INSAT satellite over the nation of India providing for TV and weather and telecommunications. And it was at Johnson Space Center where the characteristics were put into that mission to make it so successful. It's the place where our Indian customers went and after the satellite had separated from the shuttle and had gone into its transfer orbit, that's where they broke the coconut. And that's where they said the words when they broke the coconut that made it so successful. I couldn't understand the words, but that's where it happened. And that's why I'm glad to be here in Houston. Now we live our lives out there in India at the facilities of this particular space contractor, involved with international projects in space. And there will inevitably be more applications projects.

This is a panel that deals with space applications. It has become common practice for people to talk about these other nations that are doing wonderful things in space when we're not and we're losing out in this competition. I thought I'd just give you some views on what competition is and what cooperation is and how they work. Among the points that I want to make are that there exist commercial projects in space and companies do compete for commercial projects, but that nations and space agencies don't compete; they work toward achieving their national goals that are set by their national policies. So I'm going to have something to say about national goals because national policy reflects a nation's goals. Another thing I think that we all have to realize is that you can't talk about goals without talking about resources. Somehow each nation has to come to grips with the resources they're going to allocate to a particular goal. And when we talk about space goals we've got to come to grips with the resources that are going to be allocated to space.

Some space projects are made for competition. They're commercial; they take place within the bounds of free markets under world trade agreements. They're the same as any other commercial activity in the trade field. These projects are pursued in space by commercial companies from different nations and they are competitive. Here are some examples: There have been five generations of INTELSAT satellites,

* Vice President, Ford Aerospace & Communications Corporation, WDL Division, 3939 Fabian Way, Palo Alto, California 94303.

INTELSAT 3, 4, 4A, 5 and 6, and now INTELSAT 7. The UK and Luxembourg recently bought -- and there was commercial competition for satellites manufactured around the world -- American satellites. INSAT, that I mentioned before, for the nation of India, was a free project, and there was competition. Indonesia, Australia, Brazil, Mexico have all bought satellites in the free world market. Colombia and China were nations that wanted to buy and then cancelled their projects, but the competition was held. In Japan, there are two American companies selling commercial satellites and there was competition with European companies and Japanese companies as well. There are regional systems, such as UTELSAT and INMARSAT, which are subject to free trade and free competition. That's where competition exists, where it must exist. That's where we want it to exist and it's healthy that it should exist.

As in any other sphere of trade activity any commercial company, including ourselves, makes a case for and receives more or less assistance from national governments. So when we see a Japanese company receive assistance from their government, or when a French company does likewise, it doesn't bother us, because we receive assistance from our government. The assistance involves government agencies who may get into negotiations with each other. That sets the ground rules and the basis for this competition. In the United States we use the Export/Import Bank to our advantage, or in some Administrations, it may be to our disadvantage. But it's one of the ways we get assistance from our government. And we observe that our competitors overseas get assistance from their governments. So, in the competitive world of space, both private companies and government agencies are involved.

We have observed that most of these projects up to now have been associated with communications because that's where we first were able to affect the pocketbook of men and women on the ground and affect the quality of their lives through communications satellites. There's nobody that exists on the face of the Earth that hasn't received some impact from telecommunications satellites. You can't watch an Olympics, or the Pope, or Summit meetings, or place a telephone call without being impacted by the reductions in telecommunications cost that have been brought about in space. But on the other hand, nations do have space policies that do not offer an open field to international competition. National space programs are estabished for purposes which are related to unique national goals. Now these goals are often quite specific and result from processes of decision-making that are appropriate to each particular nation. These nations have space programs and within the framework of those space programs there are activities going on which are not open to competition.

Regarding some of the applications satellites, I gave you the ones that were open to competition; now lets talk about the ones that are not. In Europe there was the ECS, European Communications Satellite. There was a project called L-SAT, or known as Olympus. SPOT was not open to competition. METEOSAT, the DFS in Germany is not open to competition. ITALSAT in Italy, the TDF and TV satellite in Germany, the TDF in France, the Earth Resource Satellite of ESA and the French TELECOM, those are national projects. They are applications projects, not really open to competition. There are some others. In Japan, ETS-5, ETS-6, MOS-1, the

Ocean Survelliance Satellite, and Earth Resource Satellite 1. There are some other satellites not open to competition. They are called ATS-6, ACTS, GOES, NOAA's Polar Orbitors, LANDSAT and TDRSS. So it's quite natural that nations, when they have projects, even applications projects within the framework of their own national space programs, reserve them for themselves ....

Why do they have these national space programs? Is there competition among nations when they develop their space programs. I've often heard people say, "Look at that Marine Observation Satellite in Japan. The Japanese are ahead of us. They're winning the competition." What does that mean when you say that somebody else's space program is ahead of ours. I think you have to look into the space programs and realize what they are there for. When I look at Japan and Europe, they seem to have a space program because they understand it's part of moving into the technological world of the future. It seems that they feel they're going to create a useful data base, a technology data base, that's going to be useful as a tool for future generations. But I can also see them, in Europe and Japan, viewing this as a societal matter -- something that stimulates young people to get a technical education. And it stimulates their elders, the mature people, to recognize the wisdom of paying for this knowledge and technology base.

For example, in Japan, they are very specific about their space goals. They have a vision, that they have published and agreed upon, calling for the expenditure of approximately $48 billion dollars in space over the next 15 years .... This vision of the future includes a base of applications and science projects, a Space Station module, even a shuttle and space plane.

The European Space Agency is the instrument for combined activity which serves as a collection of space goals because each nation has its own set of goals. As they integrate these goals working through the European Space Agency, we can now see something other than competition. It's called cooperation. And they can achieve more through cooperation than through any single member's individual efforts. Now, in Europe, more than in Japan, they seem to put a relatively heavy emphasis on science in addition to applications.

However, in Europe and in Japan, you can't talk about space applications without recognizing that they're doing something we're not doing. They have a conscience and active applications program underway in materials research leading toward materials processing in space. And they are using not only their government agencies but their commercial companies. They have somehow made the commercial companies feel it's worth their while to get involved in materials processing. In the UK, they seem to be going the other way. They appear to be repeating what they did in the 1960s, in taking a step backwards from space in terms of their own resources.

There are other national programs. India happens to have a space policy that specifically focuses on applications. They've decided that they should use space to increase and enhance the development of what is really a giant nation. And so their focus is on applications in remote sensing, in meteorology, in communications and in educational television. Consider the USSR. They use space, obviously, for economical development. You see that in a lot of their programs. Again, it's a country that needs a lot of development. But here's a country which obviously places military matters at a very high rank of national priorities, so you see them with an extensive military programs as well. But as a super-power, we ought to pay attention to them because they seem to have the largest space program of all. They have, as they set their goals, somehow decided that space is a good way to enhance the values that they hold important as a nation. As far as applications are concerned, they have programs for communications, remote sensing, positioning, navigation, TV, broadcasting, meteorology and materials research. We all know about their manned program. It's extensive and it's a world leader in the sense of providing information on what happens to man when in space for a long time. Their planetary programs set the agenda for the rest of the world. I've already talked about their military programs. They even have SDI .... Included in their military missions are the same types of goals we are striving to achieve in SDI.

Let's just take a look at ourselves in the U.S. There's a big debate in progress. What are our goals in space? What are our policies in space? They're not clearly stated. They're there. You can always look at them and define them in retrospect, which is probably now the best way to establish goals and objectives by looking back and saying this is why we did what we did. When you do look back at the United States, you see that we've had lots of starts, lots of stops. One thing that we haven't had is continuity. I mentioned that Japan was going to spend $48 billion in putting into effect its vision in space. Over that same period of time, 15 years, the U.S. will spend on the order of $300 billion. And my point is that without a set of goals, without a vision that's similar to that of Japan, we run the danger of winding up the next 15 years with less accomplishment than Japan would get for one-sixth of the investment. In other words, we would have a lot of words and we would have a lot of paper, but we won't have a lot of accomplishment.

I just want to point out that competition occurs between companies and it is right, proper and fitting, and not surprising that those companies are aided by their governments. It occurs in the field of space applications which deal with commerce that takes place under the rules of free trade. In a sense, when a nation sets goals, those goals may overlap and they may have competition. But if a nation has its own goals, and it's working to achieve those goals, then when somebody else has a success it's a cause for celebration, because our celebration is going to come when we achieve our goals. If we can't set our own goals, then when we see somebody else achieve something, we feel that they win and we lose. But if we can find a way to set our own goals, then we can celebrate when they win, when they achieve. And we can celebrate when we make our own achievements. If we have our own space goals, there isn't such a thing as competition between space programs, it's a win situation. Everybody wins. That's how competition and cooperation can co-exist. Thank you.

Dr. Elizabeth Young: Thank you very much, Bob. I know that those remarks will provoke some thoughtful questions and responses from our audience. Also your reference to what the Russians are doing in space reminds me that there are two arenas that tend not to be very well known, whether it's excellent international cooperation and certainly, foremost, between the Russians and the U.S., and that is first of all, in INMARSAT, which to date has handled maritime communications and is now moving into aeronautical and perhaps landmobile. We have six eastern block countries in the INMARSAT consortiom, the Russians, Bulgarians, Poles and East Germans, among others, who are very active users of the system. Russia has been quite an influential and cooperative partner in the INMARSAT endeavor and takes a great interest in the affairs of the Consortium. I was just at an INMARSAT assembly recently, which means a meeting of the delegates .... Our government was represented by our several agencies, as were other governments. The senior representative from Russia is retiring and several of us had a little informal party for him. We asked him what he was going to do upon retirement, and he stated that he was going to be able to continue to work full time. But then he brightened up and said that he would be sailing his yacht, and then brightened up even more and said that perhaps he could finally get around to buying a Standard A Terminal, which is the main means of communication in the INMARSAT systems. So we had a lot of fun working in that environment.

Equally, cooperative and even less well-known, I'm afraid, has been the COSPAS/SARSAT endeavor, which of course was founded principally to aid in rescues both on land and at sea with the EPIRBs by the Russians, the United States, as well as Canada and France. And now other countries have become equal participants in that. Indeed, for a short period of time, somewhat embarrasingly, as I'm sure Bob will remember, we had no beacons flying on U.S. satellites, because of unfortunate policies here that had cut back on the number of satellites that we were going to launch. That's no longer the case. There are two U.S. satellites and three Russian satellites, I believe, that currently fly the COSPAS/SARSAT beacons. They make search and rescue operations available worldwide and the system has been responsible for saving over 300 lives in countries throughout the world.

AAS 87-619

INDUSTRIAL APPLICATIONS OF THE MICRO-GRAVITY ENVIRONMENT

L. J. Evans, Jr.*

In the previous session this morning, the panel addressed how government and industry must work together to re-establish access to space. Restoring U.S. access to space is an essential step in order to enable a broad spectrum of business opportunities to develop in space. In addition to restoring U.S. access to space, two other fundamentals are needed. A strong and growing, independent U.S. enterprise of risk-building investments in space requires that the private sector also have confidence in (1) our ability to live and work in space, and (2) the ability of government to perform as a reliable partner.

LIVING AND WORKING IN SPACE

The first essential fundamental, to be able to live and work in space, requires that the U.S. have a fully capable U.S. Space Station. The Space Station will be a major culmination of all that the U.S. has learned in bringing man to space, and will provide the essential scale of activity and capabilities to permit the full spectrum of benefits of U.S. space activities to be realized. Like the early settlement of the U.S., there will be examples where private initiative can establish limited outposts and activities that are noteworthy and important, but the widespread economic development of space cannot occur until the features to be found in any human community can be established in space.

U.S. private efforts will help provide the capital and initiative as space activities grow. But they cannot be a substitute for the Station. Government will always have a major and complementary role in space, in conducting the basic research which is too broad for the benefits to be captured adequately by individual firms, in providing necessary oversight and regulation, and in providing some of the essential elements of fundamental infrastructure comparable to that which governments routinely provide here on Earth.

To adequately fulfill those roles, the nation's civil space budget needs to be increased by 50%. Let's look at what needs to be done:

* President, Center for Space and Advanced Technology, 4001 North Ninth Street, Suite 230, Arlington, Virginia 22203.

- Research on Space Station utilization
- Technology development for U.S. leadership into the twenty-first century, especially in composites and propulsion
- Unmanned heavy-lift vehicles
- Reinvigorated planetary exploration program
- Manned alternative to Shuttle
- Space tugs and rescue vehicles
- Get ready for Lunar Base/Mars visit
- Preparation for full operations and utilization of the Shuttle.

Large-scale U.S. private space enterprise cannot exist without a continuing strong and complementary governmental effort. Also, the government will be one major early consumer of many of the products and services of space-based enterprises to permit government to meet its essential missions. And it is, in part, this business base that will help private firms to compete with other nations and to develop products and services for non-government users. Hence, the second fundamental is that the government be steadfast in its commitments and perform reliably as a partner with private industry.

GOVERNMENT PARTNERSHIP

The Hesitancy of the Private Sector

Unfortunately, business leaders are receiving mixed signals as to the U.S. commitment to the Station and to the establishment of the permanent U.S. manned presence in space within about ten years. Recently, Senator Proxmire chairman of the HUD-IA subcommittee threatened to zero out Space Station funding. If his threat had not been met with a clear and resounding "No" from the space community, it would have spelled the death knell not only for the program, but for any prospect of regaining U.S. leadership in space and for large-scale private investment and involvement in space. However, in the last several years, the Space Station budget has been repeatedly cut and the Initial Operational Capability date repeatedly delayed. These key uncertainties, together with the unavoidable delays in access to space resulting from the tragic *Challenger* accident, are creating major obstacles to near-term private space investment and involvement and are seriously increasing the risks of these budding enterprises. Given these circumstances, it is remarkable that U.S. industry has been willing to invest in exploring new space technologies and business opportunities to the extent that it has.

Successful business men abhor uncertainty, and perhaps more than any other factor, what business leaders require from government, in order to justify the investment of their stockholders' funds in space investments, is steadfastness in our government's commitment to leadership in space and the programs and capabilities that this commitment represents. Above all, a company planning to perform work in space needs to know that it will be able to perform according to its plans. Most of the new high-potential enterprises that we see for the future require, as a central feature, the existence of a fully capable U.S. Space Station and Shuttle on the schedule now envisioned by NASA. Company executives must be assured of their availability. Without this continued commitment, these promising new enterprises cannot survive and even the existing investment levels will wither.

What is required for the U.S. to remain competitive? In addition to a Space Station and Shuttle which must be accessible to a company's research and technology endeavors, a number of policy initiatives on both the NASA and national level are, in my view, required:

- Implement the critical elements of the President's and NASA's commercial space policies
- Undertake a concerted research and technology development program to reduce the cost of space transportation, utilities, and hardware
- Re-emphasize NASA's research and cutting-edge technology roles and maximize private sector responsibility for space operations
- Develop a strong working partnership between government, industry, and academia. Establish more and stronger centers of technical expertise where government, industry, and academia can pool their resources
- Provide a national commitment to steady, consistent space policies and adequate resources to maintain America's leadership.

The Congress is now at a critical juncture. With the current deficit crisis, funds for all programs are necessarily limited, and painful choices must be made. But in making these choices, it is essential that this nation make the investments and policies required to protect its future, as well as to meet its current needs.

Civil space research and technology development is just such a critical investment in this nation's future. Building upon this critical core activity, I can foresee a strong future for commercial enterprise which will bring growing benefits to all Americans. Without them, solid development of this new enterprise cannot occur and such benefits as may develop will be largely due to the efforts of other nations, which will be first to reap the benefits of their leadership for their own citizens.

The essential truth that we must recognize is that, like the legs of the proverbial three-cornered stool, America's success in maintaining its historical leadership in space technology depends critically upon the ability of business, government, and the academic sectors to function efficiently and reliably together.

CONCLUSION

We at CSAT are perhaps involved with a wider spectrum of commercial space businesses than any other firm. If there is one message that stands out from the work that we have been doing every day, it is that this nation must not fail to get the shuttle flying safely, routinely, and quickly; it must not fail to develop a fully capable manned Space Station by the mid-1990's; and it must not fail to undertake the necessary research and technology development to keep itself competitive with international developments. To do otherwise would mortgage our future and sacrifice America's historic leadership position in the face of intense and increasing world competition.

AAS 87-615

SPACE UTILIZATION/APPLICATION - PANEL DISCUSSION

Panel Moderator: Elizabeth Young
Panalists: Robert E. Berry
L. J. Evans, Jr.
Robert Filep
Heather Hudson

Dr. Elizabeth Young: Thank you very much, Bud. I think we're right at the point where we can collect questions .... I do want to add -- because Bud's last thoughts triggered my recollection -- we did try very hard to get someone on the panel who could talk to us directly about what is clearly an expanding interest on the part of the Russians in working cooperatively. And as you know, there is continuing interest, not only on the sale and use of their Proton launch vehicle around the world but on many other aspects of cooperation in their space program. If I'm still around next year, we'll pledge to you to have at least one speaker on that subject in a panel like this .... The first question is addressed to Bob Berry ....

Question: What you say about national goals is correct, unless a nation's goals include supremacy in space (competition) or explicitly cooperation with other nations for diplomatic purposes and these kinds of goals are the rule rather than the exception. I would postulate that either course risks inhibiting the country's aerospace industry -- competition by restricting technology transfer and therefore international contracting, and cooperation by giving business to foreign nations, and in the long run by creating additional competition. How would you proceed in establishing U.S. goals to avoid these risks?

Robert E. Berry: That's obviously not a question to which I can answer yes or no. The thing to keep in mind is that we want to avoid having poor people. Is that right? We all know we want to avoid having poor people. That means we have to create wealth. We have to create well-being. And well-being in a material sense. Because you can't expect everybody to derive their well-being strictly from a spiritual sense. So it's absolutely fundamental that we must have a continually growing economy. If we have a growing economy, don't anybody believe it that you can do it without world trade, because you can't. It's absolutely essential that there be trade between nations. Once we say there's going to be trade, you don't have to think a lot about it to decide that, only when you have world trade, do you have constantly a growing economy here, as well as in other places. We need other people to trade with. Heather talked about the advantages of building telecommunications capabilities in underdeveloped nations. Think about it! That's a fundamental building block. That builds economies. We want those people to have money. We want them to have economies. We want them to be rich because we want to buy what they make. And we want to sell them what we make. That's the only way we're ever going to grow the economy. Once you get into space commerce you're going to have to have free trade. We're going to have to face the fact that other people can make satellites and sell them in competition with us. And we're going to have to be in competition with them. Once it's commercial, it is competitive. Ford cars are sold in a lot of countries around the world. But there are a lot of other people selling their cars here. It would be rather stupid of me to tell you that we ought to lock up America so that they could only buy Ford cars. Once you do that, you buy less cars. This is how you drive everything to be competitive, to be efficient, and to build a growing economy. So I want to say, our companies' success is based upon success in the commercial market place around the world. I don't want to cut out the competition. I say, let's have that competition. It's absolutely essential, because once we start cutting it out, we're cutting out ourselves. I want to be competitive but I want to play by the same rules. If taxpayers in France feel good about paying taxes so that their co-workers can succeed with high-technology products involving communications satellites, I'd like to have the same kind of advantages in this country. We ought to think about it. I'm not going to tell people how to spend their tax money. But, obviously if it does work toward a growing economy, then that's good for us. I'd like to have the same ground rules as they have in France or Japan or ask them to abide by the same ground rules we think are fair. That is necessary in international competition, and we can't be trying to step on the other runner's foot so that we can win the race. We have to pay attention to our own speed. That's how you win a race. So we need the advantages. We need to think through, in this country, where we want to put the money, where we want to put our resources, what we can be good at, and then go and live in a world of international trade, recognizing that that trade is going to involve competition that is more or less free.

Dr. Elizabeth Young: Thank you, Bob. I'm going to take these more or less in the order in which they came. Our next questioner writes:

Question: We tend to lose considerable business on the world scene, I believe, because of an obsession with limiting export of technology to the Eastern Block nations. Timely marketing will reduce our national deficit. I feel we are overly obsessed with fear of compromising national security. We need to speed up the declassification and approval for the export process. I request comment.

Dr. Elizabeth Young: I'm going to give each panelist a minute, because I know all four of them have some unique experience in this area. Heather, we'll start with you.

Dr. Heather Hudson: I just would like to say briefly, that I generally tend to agree. I think you see in developing countries, or the markets of developing countries, some of this concern, and there is obviously legitimate concern on behalf of the U.S. Government about transferred technology. On the other hand, if we are slow to make that technology available to other countries for civilian applications -- I think of spread spectrum technology and other advanced electronics systems for satellite communication and remote sensing -- which is another very hot topic and important issue these days -- we will lose the opportunity, I think, to see those countries benefit from our technology, and they will get it from other sources, although what they get may be inferior or more expensive. So I would generally support that comment.

Dr. Robert Filep: Yes. However, I think there's another challenge. And that's to turn it around the other direction and get the technology organized and ready for marketing in the international market in the commercial context, instead of being reactive to a challenge or some competing technology. I think that the technology is going to be obtained, one way or another, no matter how hard we try to prevent it. So I think we ought to be pro-active, instead of re-active.

Dr. Elizabeth Young: Bob, I'm going to take the privilege of making a comment of my own there, because you reminded me of an instance of a company which shall remain anonymous, and a foreign government, which shall remain anonymous, but it was in the Eastern Block. They wanted some help, but the company they approached was trying to sell a very particular product and service, and there was a complete mismatch. Eastern Block people were too polite to say so. And the company that was doing the selling, obviously, didn't want to acknowledge that fact. The result was no sale from the U.S. They picked up what they needed elsewhere. Now, I don't know if that logically takes a leap to some kind of multi-national commission in the U.S. I'd hate to think we have to solve all our problems that way. But it does point out a difficulty. Here was a ready customer and there was U.S. technology that was appropriate. But somehow we didn't match it up correctly. Maybe all we're talking about here is a little bit more cooperation. In the U.S. probably this was a case where a simple phone call to another colleague in an appropriate company would have solved the problem.

L.J. (Bud) Evans, Jr.: This question really is a tough one and it brings up two experiences that I've handled over the last few years, just to give you an idea of how it cuts two ways. When I was heading up the White House Commercial Space Group, Fred Smith, of FEDX, asked for some help in getting some DOD technology that he felt would enable FEDX to win some world markets. I thought that was a good idea. So we went after it, and after a lot of uphill battles, we finally succeeded. I still hope that FEDX will be able to use that in a positive way. As a part of the question that talks about getting classified information out faster, I think that's an important thing to do. Now let us talk about the other side of the issue. Recently several people who have seen the Russian shuttle described it. It just happens to have the same dimensions, is the same size, the same shape as our shuttle. Now that just drives some people up the wall, and probably, somewhat rightly so. So when you look at these very legitimate counter-balancing kinds of factors, I think you really end up with trying to arrive at a balance that produces the greatest good for our nation and for the world. And that's not always an easy call.

Dr. Elizabeth Young: Thanks Bud. I have another one that I think is directed primarily to you. Bob Berry might want to make a comment.

Question: To date, commercial success in space has been in the field of communication. Once the satellite is in place, it is relatively cheap to transport protons to and fro. Not so for materials. How and when will this change? This is what Space Station is all about.

Dr. Elizabeth Young: I'll give Bud another crack at that and Bob Berry and anyone else who want's to comment.

L.J. (Bud) Evans: I think you're quite right. I've been saying, "When everybody talks about commercial space, in my opinion outside of communications, there are no commercial space businesses." Nobody's made a plug nickel to my knowledge in anything outside of communications. And therefore it is my contention that, when people come along from the Government and say, "We can't subsidize commercial space businesses," I say, "You have that reversed." Because what's really happening today in the materials area, and any area outside of communications is that private companies are investing in basic research, which, as far as I know, both conservatives and liberals believe to be a legitimate government function. Therefore, any money that private industry is putting into materials research and space today is helping to subsidize a government function, not the other way around. So we ought to be applauding the companies that are willing to step up and look ten years down the path and invest that kind of money. By the way, that turns out to be over 100 companies and, very conservatively, over $100 million. So I think that where we are today in the materials area is basic research. That is the thrust of our report with the Academy, and what we have to do is to get the basic data together on the phenomena in the research arena. Then it can be used by companies to create applications and eventually commercial businesses.

Dr. Elizabeth Young: I know Ford has got involved in a space station. Bob, do you want to give us a prognosis?

Robert E. Berry: Oh, there are a lot of things that can be done with a space station. It's a transportation node, it's a place for doing, taking advantage of micro-gravity, which is an environment, that has to have its impact on man's activities. We're going to find that you can do things better and cheaper and some things you're going to be able to do in space that you couldn't do at all. But just because the Government is involved and industry or commerce will profit, that's never stopped America from finding a reason to make a Government investment. We certainly made it worthwhile for people to put their money into railroads. We certainly set up land grant universities and caused this country, because of 100 years of history in R&D in agriculture, to totally lead the world into a new economic era. Now the Government is putting up a facility, a place in space, where new kinds of research can be undertaken, leading to new kinds of economic activity leading to the wealth of its citizens. That's what government policy is all about.

L.J. (Bud) Evans, Jr.: Liz, if I could, I'd like to add something. We took a look at the technologies that NASA and industry will be developing as a result of the Space Station effort. And when you look at them, it turns out that practically every one of them is critical to the future competitiveness of the U.S. in world markets. And you're talking about areas and sectors of the economy that are literally worth hundreds of billions of dollars. Now that returns real economic growth and wealth to this country. Bob, I think you're right on. It's a very, very important point.

Dr. Elizabeth Young: We have three more questions. I think we're only going to have time for two. And one of them is an overall policy question. So let me get to the more specific one first. Our next questioner says:

Question: What structure or resources are set up for the businessman to obtain funding (venture capital) in the commercial space area? Also, how much do you think a space plane would cost? To purchase or to build one, once the technology is matured which would be necessary for the growth of the space industry?

Dr. Elizabeth Young: I'll take 30 seconds on the first part of that question. You know, fundamentally to get into any business or to get one started, you have to have an idea somebody believes in. There are well over, by last count, 400 major venture capital companies in this country, most of them interested in investing in hardware-related projects. It's relatively difficult to get venture capital for such things as writing and producing films or starting up service companies although that can be done. So in theory the dollars are there. There are other ways of doing it. Public stock offerings which aren't going to be terribly popular, I suspect, for the next few months given the state of the market. But there are ways of doing it, debt financing obviously, or partnerships with larger companies, which is something that isn't always recognized. If anyone on the panel would like to comment specifically on obtaining funding, he can do so. I just want to add, that there was mention made this morning, by one of our speakers of Orbital Sciences Corporation, which is a small company that has grown. I know, that it started literally with a good idea, and one major investor. It has now been able to expand its product line. But often the start-up activities, are particularly discreet, and frankly, very risky and very scary. Bob, any comments from your experience?

Dr. Robert Filep: I think we ought to give credit to the aerospace companies. Because in their various plans and discretionary funds, there are possibilities for funding. I don't want to guess on the Space Plane. Maybe somebody is working the numbers while we talk. I think the investment community has been very receptive to briefs. Not to mention any by name, but there are some from New York, who have been particularly active in the communications satellite area. In the international market they've been willing to consider opportunities and now you see the insurance companies joining hands with those investment companies to support $200-$300 million ventures.

Dr. Elizabeth Young: I know in the case of almost all the mobile satellite applicants, with the notable exception, of course, of Hughes and a couple of others, all of the other companies have a considerable amount of venture capital. They've simply obtained it by knocking on doors. Anyone else want to comment on that very briefly. What about the Space Plane? Heather.

Dr. Heather Hudson: I would just like to say, in terms of the overseas market, the U.S. industry has complained that it does not get government support in concessionary financing and packages that some other countries offer. But the Export/Import Bank has programs available to support U.S. industry that is trying to market new products and services overseas. And those programs really are more accessible than we may think, although not too many, especially the smaller companies, know about them.

Dr. Elizabeth Young: Thank you. Does anybody want to talk about the Space Plane? Bob, are you going to build one for us?

Robert E. Berry: No, I'll tell you what it will cost. It will cost more than a jet plane. Once it's in production, its cost will probably be less than 10 times as much. The 707 certainly cost more than a DC-7. Everytime you go to a new generation it costs more. But the payoff is more. That's the other message for people who are looking for funding a new venture. Make sure you have a plan to make money. If you have a plan that makes money, the more obvious it is, the easier it is to get the money. The less obvious it is, the harder it is to get money. A very simple rule. But if you don't have a plan to make money, then don't ask.

Dr. Elizabeth Young: Good advice. Doug, do I have time for one more if I limit the answers? All right, this is a general question to the panel.

Question: Dr. Filep named some elements of the national space policy. What additional element should be included and do you agree with the ones that he selected.

Heather, if I may start with you and I'll even give Bob a crack at that, just in case he doesn't agree with the committee's conclusions.

Dr. Heather Hudson: I very much agree with what the panelists have said about the need for national policy. I think in Washington these days we see a lot of strategy without much policy or what was described by Bob Berry as allocating funds without any sense of where we hope to end up. I would say that the technology transfer and applications transfer mechanisms, not only in communication, which is what I tend to focus on as the areas I know best, but in other forms of satellite applications should be very much a part of that program.

Dr. Robert Filep: We've grown used to having redundancy in space. It's a natural outgrowth of what we do and how we plan and how we work. As we've been examining the various structures in Government for defining Space Communications Research, Development, and Application, the redundancy issue has been very high on the list. It begins to get in the way when you start having a period of limited resources. And it clearly underscores the need to minimize the redundancy, in order to maximize the output of whatever is going to happen in RD&A. And that means a lot of people joining hands who have interest and commitments and are willing to bring some level of resources to bear. We have even toyed with the idea of the Space Communications Council, (SCC), which would provide the forum for the industrial and commercial sector to get together with all the aspects of the Government and community to begin to do the planning, at least to define what the goals and requirements are and then also look at the available resources. An interesting phenomenon happened on the Hill this Spring. Two major committees -- I won't mention either of them, one happened to be in the defense community, the other in the civil community -- suddenly discovered that they were both losing funds in the same budget line category, which happened to deal with communications. Lo and behold, they finally got together and sat down and talked to figure out a way to resolve that problem. There had to be some way of minimizing those losses so that they'd still reach their goals, because they had the same common set of goals and also some unique ones. It all worked out. They went back and dealt with that in the legislation. People have to start talking with each other and sharing their requirements. And let me say something else that we learned about planning policy. It was that one way to get your ideas out quickly to another group that's working on a similar related areas, is to have the same rapporteur. For instance, in a symposium we had a rapporteur, who then subsequently, went and sat with another committee at Woodshole. They wrote their report and we both were thinking about the same goals.

Robert E. Berry: Goals are a function of resources. It wouldn't make sense for Upper Volta to put a man on Mars. In our country the resources come from the people. The problem in space is that people in our country spend the money where they want to spend it. And until the people want to spend it in space, they're not going to do so. The message is: What is it about space that is going to cause our fellow citizens to want to spend their money there?

L.J. (Bud) Evans, Jr.: That is so true. If you look at the Ride Report, she set out four potential goals. They ranged from mission to Earth, which is basically remote sensing. (Let's see what has been happening to our environment type of mission,) to a lunar mission, a Mars mission and beyond that. I would buttress what Bob said. I have a dream. I think it should be all of our dreams, and that is that we use space not just to go to Mars or to the Moon, but to create a better life here on Earth. If we do that -- and there are many ways that space can be used to do that -- if we do that, and if we sell the people on the fact that that is what our goal is, I guarantee that the people will support more funding in the space program. I've recently had an opportunity to be with a number of leaders in the Administration and on the Hill. I've shown them a tape that asks people off the street a question: Would they support additional space funding? And the answer is almost unanimously "No." And then we showed them the benefits that had come out of this space program such as an intensive care unit in a hospital showing all the high-tech equipment that is saving lives of people in that unit and eliminating every piece of equipment which was a spin-off of the space program. When you get done, there is almost no equipment left. In any event, we showed these people those kinds of benefits, and when we were finished, we asked the same question. Would they support more funding for the space program? And the answer was unanimously "Yes." Now, don't think that doesn't have an impact on the political leaders of this country who are going to make the decision as a result of what money is spent here. So, Bob, I think you're a 100% on the money and I think that we need to get ourselves organized to have a goal for creating a better life here on Earth in terms of what we do in space.

Dr. Elizabeth Young: With that rather eloquent summary, I'm going conclude this panel. I have plaques that I'd like to present to the panelists. And thanks on the part of the AAS. I'd like to express on my part personal thanks to all of them and to you for being a very participative audience.

Henry N. Bowes: Thank you, Dr. Young and your panel. In fact, it sounds to me, as though they just wrote a prescription for what the American Astronautical Society ought to take on as a challenge, and more particularly, what the Houston section might take on as a challenge. So, thank you again, Dr. Young.

FLIGHT MECHANICS/GUIDANCE AND CONTROL:

EXPENDABLE LAUNCH VEHICLE AND UPPER STAGE MODERNIZATION

PANEL THREE -- SESSION I

Chairman:	Robert L. Gates Program Manager Rapid Retargeting Program Martin Marietta Denver Aerospace
Host Chairman:	Kenneth Cox Chief Avionics Space Systems Division NASA Johnson Space Center

The following paper number was not assigned:

AAS 87-622

AAS 87-621

EXPENDABLE LAUNCH VEHICLE AND UPPER STAGE MODERNIZATION - INTRODUCTION

Robert L. Gates*

The Guidance and Control Session deals with the modernization and update of attitude control systems for spacecraft. It's pretty safe to say that most of these systems could withstand a single-point failure and continue to function. But in looking at the expendable launch vehicle mechanizations that we have today I'd say that's probably not true. That's why we are addressing the subjects of where we are in ELV and upper stage avionics and what the issues are that ought to be discussed about that situation.

We'll have two speakers who will present what is available today: What you can get out of the cupboard today in the way of avionics and guidance and control systems. Then we'll follow with a requirements paper, in terms of what we would like to have for the future expendable launch vehicle in avionics....

The first speaker/presenter is Mike Davis from Martin Marietta. He's the Deputy Program Director of our Small ICBM Program. Mike has been in the business here since 1963 when we were together at Kearfott. He's a graduate of Cambridge University.

Dr. Jim Allder, The Aerospace Corporation, will be the second presenter. He's the Principal Director of a guidance and navigation subdivision at Aerospace and has worked the requirements issue quite recently. He has a bachelor's degree from Cal Tech and a Ph.D. from UCLA.

Dr. Cox, one of our local gentlemen, of course, of Johnson Space Center is Chief of the Avionics Systems Division. He's responsible for the Shuttle and Space Station avionics. He'll serve as the moderator to referee the discussion. His highest degree is a Ph.D. from Rice.

John Kares, is with General Dynamics. He is Chief of the Space Systems Avionics Organization and is a graduate of Georgia Tech.

John Hardtla from Boeing Aerospace is their Manager of Guidance Technology. He was quite active in the IUS program for which he developed adaptive guidance algorithms. John has a master's degree from the University of Southern Mississippi.

---

* Program Manager - SAVI, Martin Marietta Aerospace, Rapid Retargeting Program, P.O. Box 179, Denver, Colorado 80201.

Finally, Lloyd Thayne, Martin Marietta, is Martin's Manager of Electronics Engineering for their Space Launch Systems Company. He's a graduate of Brigham Young University.

We'll now proceed with the presentations.

AAS 87-625

AVIONICS FOR TODAY'S EXPENDABLE LAUNCH VEHICLES

Michael J. Davis* and Donald A. Willis†

Today's Expendable Launch Vehicles were developed from early ICBM and their Avionic Systems are largely derived from the ICBM Guidance and Control Systems. This paper surveys the status of ELV Avionic System, assess their availability and performance and concludes we should be replacing them in the near future. The basis for replacing obsolescent system should be those systems either in production or being developed for aircraft avionic systems. This will make available a far greater production base than the ELV market by itself and will use standardized interfaces, architecture, etc. Simple approaches to elimination of single point failures are recommended.

INTRODUCTION

Today's Avionics for Expendable Launch Vehicles (ELVs) go back to the 1960s and in some cases, even further beyond. The majority of the ELVs are derived from early ICBMs and today's avionics are also derived from those ICBM guidance and control systems. The Titan Avionics system was replaced by an aircraft navigation system in the early 1960s but nothing further has been done since then. They have served us well. By examining their record and assessing their design features, we conclude that we should be seriously looking at replacing them.

The purpose of this paper is to give a brief survey of what we are flying today. Assess what are potential existing systems available to us from the industry and what are the major issues facing us as we make this decision.

We will discuss the pros and cons of staying with the systems we have flying today and make some recommendations of where we believe we should be going in the future.

---

* Deputy Program Director, Small ICBM AT&SS Program, Martin Mareitta Astronautics, P.O. Box 179, Denver, Colorado 80201.

† Manager, Program Improvement, Titan IV Program, Martin Marietta Astronautics, P.O. Box 179, Denver, Colorado 80201.

STATUS OF EXISTING ELV AVIONICS

The current United States large Expendable Launch Vehicle Programs were developed directly from strategic ICBM or MRBM with the exception of the Space Transportation System (Shuttle) (Figure 1). The original avionic systems were, for the most part, the avionic system developed for the weapon systems, which were used during the early years of these programs. On Titan and Atlas, the computers were rotating drum machines for some of the configurations, and on other configurations of both of these programs, radio guidance is still used for missions from the Western Test Range.

The avionics systems developed for these programs during 1960s and early 1970s were designed using discrete components, floated single-degree of freedom gyros and large accelerometers (pendulous integrating gyros for instance) resulting in large heavy gimballed platforms. The digital computers of the 1960s and early 1970s used core memories, each had unique instruction set with around 100K ops thru put. The supporting avionics system like rate gyro system and other electronic hardware were of the same technology, all resulting in very expensive systems to build. The electronic piece parts are approaching obsolescence, which is the state of most of the 1960 vintage ELV programs.

There were no redundant systems in the 1960s and early 1970s except for the manned missions, such as Gemini, Apollo and Shuttle. Redundancy was not affordable for unmanned mission during these years. The inertial measurement units (IMU) and the computers weighed over a 100 pound each and required 400-600 watts of power--to install redundant versions of these such systems during these years was not practical.

During the early 1970s, the aircraft programs, including the commercial carriers, were developing and using redundant inertial guidance systems in large quantities. The 2-degree freedom dry tuned rotor gyros development made platform affordable for redundant applications for aircraft. The Delco Carousel system using a single degree of freedom gyros was incorporated on Titan II and Titan III in the early 1970s. About this same period, the system derived from the moon launcher strap down "Delta Inertial Guidance System" was incorporated on the Delta system.

In the 1970s, ELV programs were directly affected by the development of the shuttle program. The nation's priorities were assigned to this program and all the other ELV programs began a natural reduction in upgrading those Avionic systems. Only those programs directly associated with shuttle incurred any new development. The Inertial Upper Stage (IUS) was a new stage development in the mid 1970s and included the development of a dual redundant avionics system based on a Hamilton Standard airbearing single degree of freedom gyro which is the only totally redundant avionics system in use on "ELV" programs today, except for the STS program.

In the same time period, the laser gyro development was proceeding and reached a maturity in the 1980s. A number of aircraft programs are using this gyro in strapdown IMUs.

The aircraft programs again are the source of this development, this same period has seen the development of the new high speed computer systems with a standard instruction set (MIL-STD-1750A). It is now possible to find both IMU and computer systems that will easily satisfy the ELV requirements and are now affordable in cost, weight and power for redundant configurations applications. The IMU and computer will weigh like 30-40 pounds each and require 100-150V watts total power depending on the I/O requirements of the ELV.

Of course the shuttle accident in February 1986 caused the nation to reassess the United States access to space and the liabilities associated with having only a single program assigned to this important task. As the result of this reassessment, both military and commercial ELV programs are now well underway in developing or modifying existing programs to meet this need. Fortunately, there is an abundant amount of avionic hardware available that has the capability to meet the required upgrade of the 1960-1970 development system to meet the 1980-1990 time period requirements.

The ICBM avionic hardware has also progressed but the accuracy requirements are so much greater than that required for the ELV missions, with the associated cost and complexities. The aircraft derived system with large production bases are a preferred source of hardware (Figure 2). The aircraft production base is very large and will make the upgrade of the ELV program avionics low risk and low cost, well into the 1990s. As we move into this next decade, the development of other gyros (hemispherical resonator or fiber optic gyro) and VISIC computer chip development show real promise for size, weight and cost reductions.

Upgrading today's avionic systems is a major development task since there is no standardized interfaces and architecture. The aircraft programs have developed a standard computer architecture (1750A) and a standard data bus (1553) to handle communications. If we develop our new systems using this standard approach, we will be able to be more flexible in the future as we face avionic obsolescence and develop upgraded systems.

MAJOR ISSUES

The major issues we are faced with in assessing the status of ELV Avionics are susceptibility of the systems to single point failures, the availability of these old and almost obsolete systems, and cost.

Figure 1

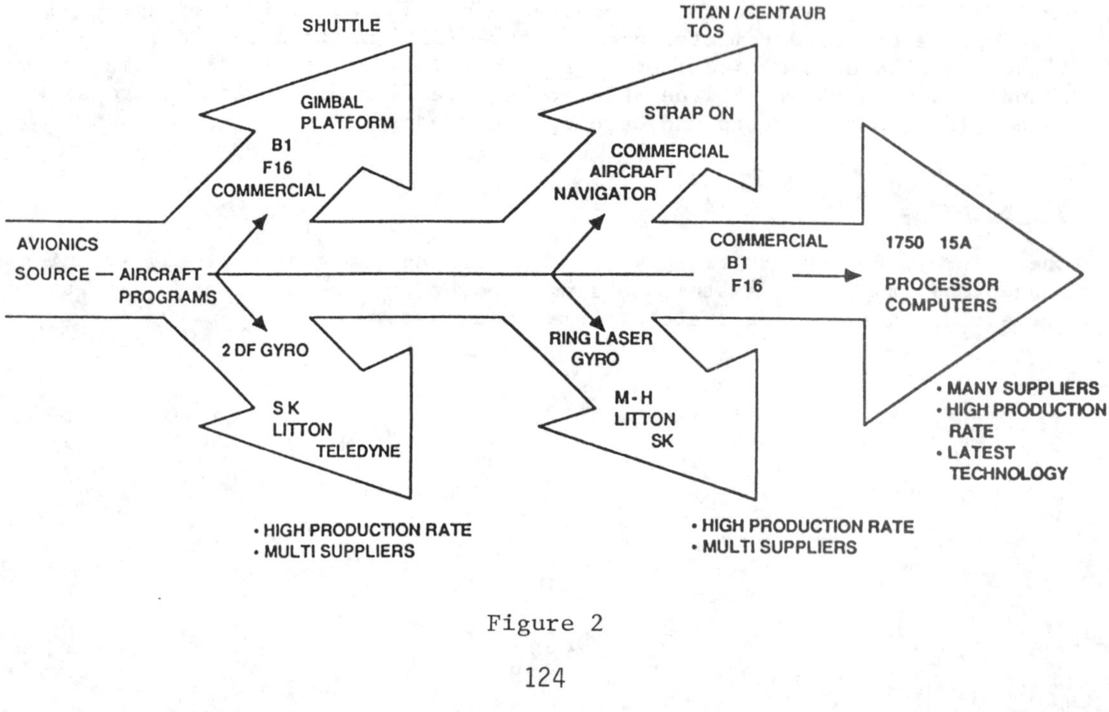

Figure 2

All of the ELV avionics systems flying today are single string configurations. That means that if any of the components should fail, the mission would be a failure. When you consider the cost of one of the payloads for an ELV mission, this appears to be an unnecessary economy. We do not believe that we should embrace complex redundancy schemes such as those used by the IUS upper stage but rather opt for simple majority voting schemes and use triple strings of hardware.

All of today's ELV avionic systems date from the 1960s or earlier. This means that their piece parts are obsolete and the production lines are being kept open only to provide for the very small base the ELV market requires. The aircraft industry has been developing several new avionic systems over the last 20 years and a great deal of progress has been made in accurate, low cost and large production base systems. These include advanced computers using ADA software and standardized 1750A architecture; two-degree-of-freedom gyros or ring-laser gyros used in simple strapdown configurations.

The cost of today's systems is large compared to their performance. In fact, their cost is approaching the costs of the far more accurate ICBM and SLBM guidance systems. Moving to the larger production base of today's aircraft, avionic systems should allow us to obtain redundant systems at no increase in cost from where we are today. Even if we were to see slightly increased costs, it would be cost effective to reduce the susceptibility of single string failures with there attendant losses of very costly payloads.

SUMMARY AND RECOMMENDATIONS

Today's ELV Avionics are:

Obsolete.
Expensive.
Single String.
Not Standardized.

A set of requirements for new avionics should be developed. Economics and availability dictate that they should incorporate the following features:

A/C production base.

Simple redundant approaches to eliminate single point failures.

Standardized interfaces - 1750A and 1533 and ADA.

AAS 87-626

AVIONICS FOR FUTURE EXPENDABLE LAUNCH VEHICLES

J.R. Allder*

We have heard an excellent talk which summarized where we have been and where we are in the area of avionics for expendable launch vehicles. I would like to turn the discussion to the critical issues for future avionics systems. While the talk specifically addresses avionics for expendable launch vehicles, we will see that most of the ideas apply equally well to avionics for non-expendable systems.

The key issue is one of cost. In the past, the cost has been higher than desirable due to (1) the extremely high cost of failure for highly complex and technologically advanced missions, (2) rapidly changing technology which renders avionics systems obsolete and additional production and maintenance difficult, (3) the number of distinctly different avionics systems developed (one for each different launch vehicle), and (4) the relatively small number of units bought at a time with the attendant small production base. The opportunity we now have is to develop more cost-effective systems for the new expendable launch vehicles which are now being developed (Chart 1).

One of the most significant costs has been that associated with avionics system unreliability, which can result in a flight failure. The chart shows what these costs are for various levels of avionics system probability of success for a combined launch vehicle-payload cost of $1 Billion. The unreliability costs are statistical and might be thought of as insurance premium costs to protect against possible loss. With this view, the unreliability costs are fixed and recur with each flight and therefore result in linear growth with the number of launches.

Experience to date with avionics systems has been in the range of 0.98 to 0.99 for probability of success - in marked contrast with that computed during the design phase. Therefore, unreliability costs in the hundreds of millions of dollars for a modest launch program of 20 launches are realistically to be expected, and it is clear that a probability of success in the area of 0.999 is highly desirable to limit these costs. One of the major challenges, then, is to develop new avionics systems with a <u>real</u> reliability of 0.999. I would like to emphasize the word "real" in that this reliability has much more to it than the typical reliability computation based on parts count and flight

* Principal Director, Guidance and Navigation Subdivision, The Aerospace Corporation, P.O. Box 92957, M4/953, Los Angeles, California 90009.

stress factors. This computation for past avionics systems usually turns out to be something like 0.999 but is seldom realized. The discrepancy between computed and experienced reliability results from extraneous factors such as unrecognized design problems, manufacturing procedural errors, parts contamination, etc.

It should also be added that the costs discussed here are only those associated with loss of the hardware and do not include the very real but less quantifiable costs associated with launch slips and delays, failure review and assessment activities, and costs of redesign and testing. These costs quite likely exceed the direct costs associated with unreliability (Chart 2).

Since improved reliability is the key to the reduction of unreliability cost, the technical challenge is how to achieve true reliability. The author believes advances in technology have been such that redundancy provides a practical means for achieving the requisite reliability. Analysis has shown that this reliability improvement is possible for a properly designed redundant system because first order failure probability terms are removed from the system failure probability equations.

Two basic contending redundancy configurations have emerged from past studies. These configurations are dual and triple in nature. Either will work and either can be used for the whole system or within a system at various subsystem levels. Thus hybrid systems are possible. The dual configuration employs less hardware at the expense of more complicated implementation for failure detection and isolation - for both hardware and software. Design of the software algorithms has been both technically challenging, interesting, and difficult. The triple configuration is simpler in that failure detection, per se, is not required - only the guarantee that a non-failed system is utilized so that the effect of a failure is isolated. From an algorithmic point-of-view, this is very simple in concept and involves "center selecting" from the three signals provided by the triple system. This approach avoids the use of thresholds and the potential for failure false alarms which are possibilities of the dual configuration. The question is then which configuration can be designed, built, and operated in the most cost-effective manner (Chart 3).

Rapidly advancing technology clearly provides advantages but, as has already been mentioned, there are practical disadvantages. The disadvantage associated with obsolescence can be mitigated by designing the system so that subsystem improvements can be introduced in a manner which minimizes the impact on the rest of the system. Thus modularity with standardized interfaces are highly desirable together with the use of data buses for communication among the various subsystem.

Special mention should be made in the computer area since this is an area where technology is still rapidly advancing with no signs of slowing. In order to exploit future technology, we should use hardware with a standardized instruction set (e.g., that of the 1750A). With regard to the greatly increased computer performance capability, these improvements should be utilized very carefully and in a manner focussed specifically on reducing software development, testing, maintenance, and operations cost. Toward this end, we should emphasize simplicity, use a high order language for the software, and finally consider distributing computational functions among the subsystems. This distribution approach is new for launch vehicles and will now be discussed further (Chart 4).

It is believed that the distribution of computing can result in a lowering of avionics system costs. Before discussing the reasons, the distributed computing concept will be discussed. Shown is a schematic of the partitioning of the computation into the major avionics functional areas of navigation, guidance, and control. The navigation computer processes the inertial measurement data to obtain vehicle attitude and navigation (position and velocity) information. The guidance processor handles the mission related functions associated with guidance and sequencing. Finally the controls processor is responsible for carrying out the guidance and sequencing processor generated commands and for stabilizing the vehicle (Chart 5).

One of the main advantages accruing from distribution of the computing is in the software area. Software becomes more modular and subject to separate and independent development and testing. In some cases, hardware and software testing can be integrated to advantage. Finally, from a contractual point-of-view, the software job is more readily carried out by the most knowledgeable and therefore natural contractor (Chart 6).

Distributed computing has further advantages in enhancement of system utilization and growth options. The avionics system can more readily be adapted to new and unforeseen missions. This results from the fact that the major changes required to support new missions are mainly in the guidance and sequencing area. Thus a new mission requires software development and testing in a relatively small part of the system only. Similarly, should it be desired to employ the avionics system on a new vehicle, the majority of changes should occur in the control system area with some modest changes in guidance and navigation. Finally the overall system has benefited by becoming more modular and therefore can more easily accommodate introduction of a complete new subsystem (e.g., IMU) (Chart 7).

The major technical issues which we have discussed are summarized here. Of prime importance is the matter of reliability where we have seen that we need a 0.999 avionics system reliability. I have suggested that redundancy is the way to proceed, however if someone can really achieve the reliability without the added complexity of redundancy, let's hear about it. Standardization of interfaces and the use of a standard data bus appears to facilitate the system's ability to accommodate future changes. Finally, in the computer area, if we allow low hardware cost and greatly increased capabilities of computers to result in excessive complexity, the technological advantages are nullified (Chart 8).

Since technical issues are not the only ones, as the final part of the discussion I would like to bring up two programmatic issues for consideration. First, in the past we always seem to develop a different avionics system for each launch vehicle. Can we not move towards an across the board use of a single avionics configuration for more launch vehicles thereby achieving more commonality and a larger production base?

And secondly, I think it is appropriate to give the last challenge to the software developers. I am sure we all have had the experience of an involvement with a new software procurement. Very few of these, in my experience, are carried out as originally intended. Computer memory size and timing becomes inadequate. Much time and money is spent trying to "shoe horn" the software into the available space. The computer configuration is ultimately changed and compromises made. So the question is "How can we effectively manage software development so as to stay within the resources allocated?" We have heard the many ready and logical answers but they don't seem to have worked (Chart 9).

Key Issue - Cost

WHAT DIRECTION SHOULD FUTURE AVIONICS DEVELOPMENT TAKE TO BE MORE COST-EFFECTIVE IN VIEW OF:

- VARIETY OF HIGH COST MISSIONS
- RAPIDLY ADVANCING TECHNOLOGY
- NUMBER OF DIFFERENT LAUNCH VEHICLES
- RELATIVELY SMALL BUYS

87-4165

CHART 1

Avionics Reliability Improvement is Required

AVIONICS FLIGHT FAILURE COSTS

LAUNCH VEHICLE + PAYLOAD COST = $1 BILLION

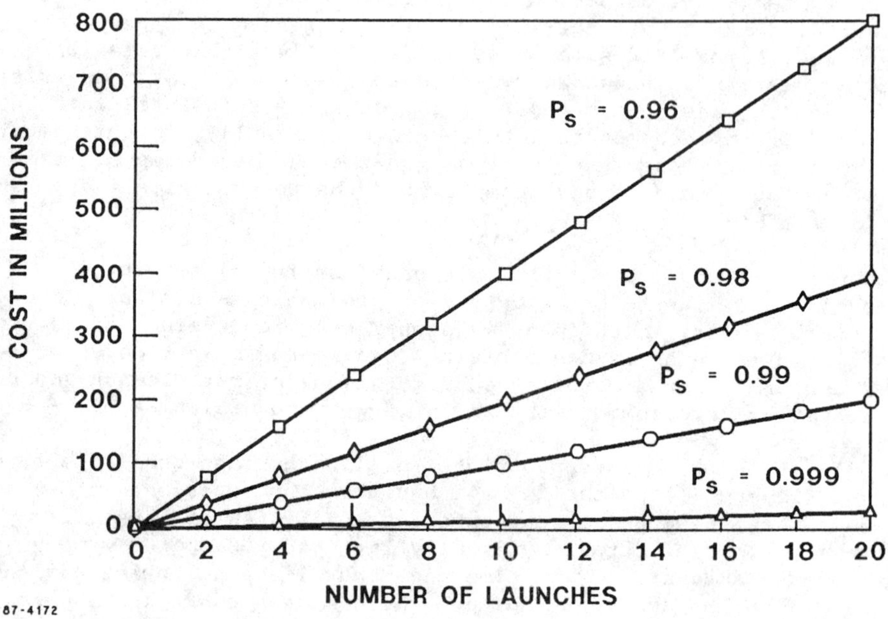

87-4172

CHART 2

Reliability

WHAT IS THE BEST WAY TO ACHIEVE RELIABILITY?

- REDUNDANCY IS A POWERFUL MEANS FOR ACHIEVING RELIABILITY
  - ELIMINATES FIRST ORDER TERM IN FAILURE PROBABILITY
  - WEIGHT AND POWER OF REDUNDANT SYSTEM IS NOW ACCEPTABLE

- THERE ARE TWO CONTENDING CONFIGURATIONS
  - DUAL REDUNDANCY
  - TRIPLE MODULAR REDUNDANCY

  } WHICH IS BEST?

87-4186

CHART 3

Rapidly Advancing Technology

HOW CAN WE BEST (1) AVOID THE OBSOLESCENCE PROBLEM AND (2) EXPLOIT NEW CAPABILITIES?

- PLAN SYSTEM CONFIGURATION TO FACILITATE IMPROVEMENTS AT THE SUBSYSTEM LEVEL
  - STANDARDIZE SUBSYSTEM (e.g., IMU) INTERFACES
  - USE DATA BUS FOR SUBSYSTEM COMMUNICATION

- PROPERLY EXPLOIT NEW COMPUTER TECHNOLOGY
  - USE STANDARDIZED HARDWARE
  - UTILIZE INCREASED PERFORMANCE TO REDUCE SOFTWARE DEVELOPMENT, TESTING, MAINTENANCE AND OPERATIONS COST
    - DRIVE TOWARDS SIMPLICITY
    - USE HIGH ORDER LANGUAGE
    - DISTRIBUTE COMPUTING INTO SUBSYSTEMS

87-4187

CHART 4

Distribution of Computing has Potential for Reducing Cost Associated with Software Complexity

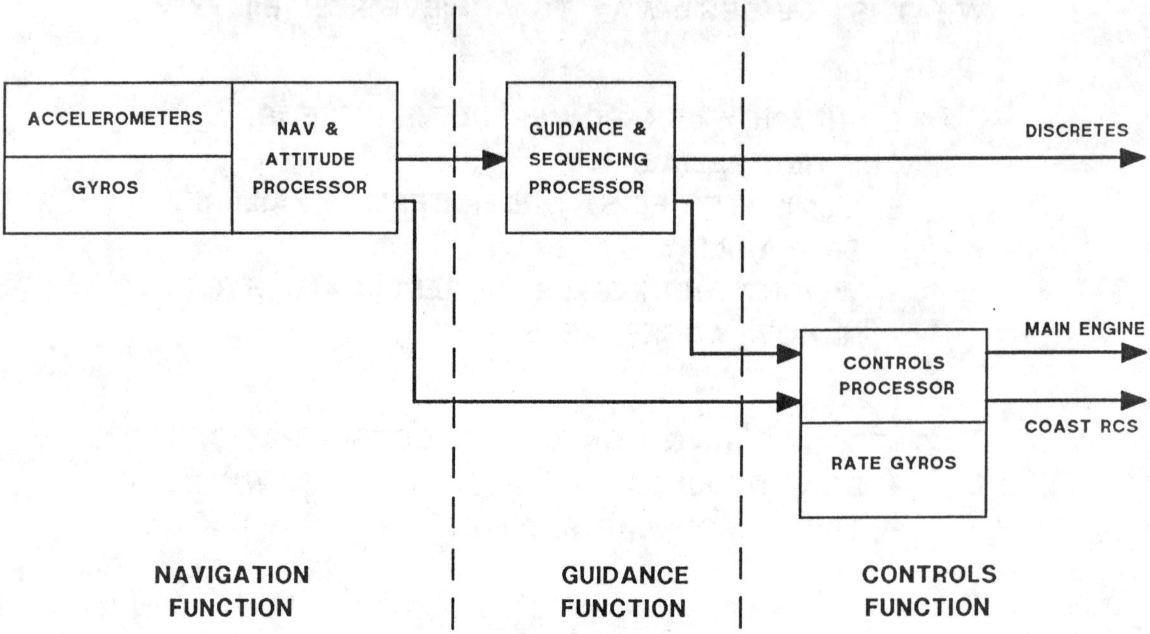

87-4174

CHART 5

Distributed Computing

ADVANTAGES - FLIGHT SOFTWARE DEVELOPMENT

- SOFTWARE PARTITIONED INTO SMALLER INDEPENDENT UNITS
- EACH UNIT CAN BE DEVELOPED AND TESTED INDEPENDENTLY
- FOR SOME UNITS (e.g., IMU), HARDWARE AND SOFTWARE TESTING CAN BE INTEGRATED
- UNITS CAN BE DEVELOPED BY A "NATURAL" ORGANIZATION
  - NAVIGATION REQUIRES DETAILED KNOWLEDGE OF INSTRUMENTS TO SUPPORT ALIGNMENT, CALIBRATION, COMPENSATION & FAILURE DETECTION
  - CONTROLS REQUIRES DETAILED KNOWLEDGE OF VEHICLE/SATELLITE DYNAMIC MODELS
  - GUIDANCE/SEQUENCING REQUIRES DETAILED KNOWLEDGE OF MISSION AND MISSION REQUIREMENTS

87-4188

CHART 6

Distributed Computing

SYSTEM GROWTH ADVANTAGES

- MORE READILY ADAPTABLE TO NEW MISSIONS
  - CHANGES CONFINED TO GUIDANCE/SEQUENCING
  - SIMPLIFIES SOFTWARE MODIFICATION & TESTING
- MORE EASILY ADAPTED TO NEW VEHICLES
  - MAJOR CHANGES CONFINED TO CONTROLS
- SYSTEM HAS BECOME MORE MODULAR
  - CAN ACCOMODATE NEW TECHNOLOGY DEVELOPMENTS
  - CAN ACCOMODATE ALTERNATIVE SOURCE PROCUREMENTS

87-4169

CHART 7

Summary of Avionics Issues

TECHNICAL

- WHAT IS THE MOST COST-EFFECTIVE MEANS FOR ACHIEVING A REAL AVIONICS SYSTEM WITH $P_S = 0.999$?
  - REDUNDANCY - YES OR NO
  - FORM OF REDUNDANCY (e.g., TMR vs DUAL)
  - AT WHAT LEVEL (IMU, computer or both)
- CAN SYSTEM BE DESIGNED TO READILY ACCOMODATE FUTURE TECHNOLOGY DEVELOPMENTS?
  - USE OF STANDARDIZED INTERFACES
  - USE OF STANDARDIZED DATA BUSES FOR COMMUNICATION
- WHAT IS THE RIGHT EXPLOITATION FOR NEW COMPUTER CAPABILITIES?
  - MOVE TOWARDS SOFTWARE SIMPLICITY
  - USE HIGH ORDER LANGUAGE
  - DISTRIBUTE COMPUTING INTO SUBSYSTEMS

87-4170

CHART 8

Summary of Issues (Cont'd)

PROGRAMMATIC

- CAN AVIONICS COST REDUCTION BE ACHIEVED BY BETTER UTILIZATION ACROSS PROGRAMS?
  - MOVE TOWARDS MORE COMMONALITY
  - ACHIEVES LARGER PRODUCTION BASE
- HOW CAN WE EFFECTIVELY MANAGE SOFTWARE DEVELOPMENT SO AS TO STAY WITHIN THE RESOURCES ALLOCATED?

87-4171

CHART 9

AAS 87-627

EXPENDABLE LAUNCH VEHICLE AND UPPER STAGE MODERNIZATION - DUAL VERSUS THREE-STRING TRADE-OFFS

John C. Karas*

Kenneth Cox: The process we'd like to follow is to have our three panel members discuss a particular issue that I have selected. We will take them one at a time and in addition to discussing the specific issue, we will allow them a little time for some of their own items that they want to bring up. Of the three issues that I chose, two were technical, one was programmatic.

Questions: What's the best way to achieve reliability, as you've just heard? What's the best utilization of some of the advancing technology where a lot of emphasis is on the digital capability? And then the programmatic item: How can one provide a better utilization of cross programs?

With that as a background, I'd like to introduce first John Karas, from General Dynamics, and he's going to address the best way to achieve redundancy and then add some additional comments.

John C. Karas: I don't know if I'm going to talk about the best way to achieve redundancy. I am going to talk about the dual versus three-string trade-offs. In general, I want to make three points with some subpoints.

First, generally in the architecture we talk about dual systems or three-string systems. It really is highly dependent on the launch vehicle itself, the launch vehicle requirements, the subsystem interfaces that you have to deal with. And that greatly depends on the heritage of the vehicle, heritage being where you are today with an Atlas or Titan vehicle or maybe even an IUS vehicle that exist today, and where you'd like to go and how you want to evolve that vehicle. Those are some things you have to consider when you think about a dual system or three-string system for implementation on a launch vehicle.

The other thing about three-string and dual-string systems is that that really is a sweeping generalization. Often you just pick the IUS system; for example, it has three-string or multi-string characteristics, it has a multiple clocks that determine the synchronization of data, or its real time interrupt. Or with the three-string system that we've been working with and advocating, we've cut down to dual-string inter-

* Chief, Avionic Systems, General Dynamics Corp., Space Systems Division, Mail Station 24-8710, P.O. Box 85990, San Diego, California 92138.

faces with certain subsystems. So there's no pure interface. As Jim Allder mentioned, there might be some combinations where you have a skewed self-testing IMU and a three-string computation system. Really, the purity of dual versus the three-string aspects really doesn't exist. It's not very pure and each system exhibits characteristics of the other....

The second point concerns implementation. Some thoughts on that. The trade-off is pretty classical. People can make their whole career trying to trade-off that one point, I know .... I've done a lot of that. The basic trade-off is the potential weight and cost savings of a dual-string system versus three strings. You only have two strings, essentially, versus the unique aspects of that hardware because of the built-in test or fault detection required of a dual string and the uniqueness of the software to implement the fault detection isolation. That's the trade-off. Now, let's look at it from two examples, one being an existing system and a brand new system, a white-sheet-of-paper approach.

For an existing system like the Centaur, we are upgrading our computer and guidance system today and that system is going to fly the first quarter of 1991, on both Atlas Centaur, Commercial Atlas Centaur, and Titan Centaur. So we're upgrading to a newer technology. We're also maintaining much of the software the way it is, as far as requirements are concerned. A lot of the software and analysis tools exist. So, using the Centaur vehicle where we have a new suite of avionics with the weight and cost savings that Mike talked about, it might be easier -- at least it seems that way -- if you wanted to be failure tolerant, to proceed with a three-string system, a TMR, Tri-Modular Redundant system.

For a brand new system, with a brand new set of requirements, maybe longer mission durations, you might want to think of something different. If you look at the technology available today such as Custom VLSI, our new system uses a standard off-the-shelf, VLSI, one-board computer. Well, we're doing a 180° turn, where the Custom VLSIs now are enabling greater levels of built-in tests. And that's really a necessity in view of the number of gates they can cram onto one square millimeter of silicon nowadays. So eventually, if you consider the situation three or four years from now, there probably will be standardized built-in test capability and Custom VLSIs, so you might gravitate back toward a self-testing dual-string system. If you can see the dichotomy there, off-the-shelf, single-chip or monolithic chip computers without a lot of self-test, used in triplicate today, finally evolving into built-in test dual-string systems. That's where the objectivity ends.

That brings me to the third point that I want discuss today. Based on today's experience and working with Centaur, many of us are advocating an asynchronous three-string system, non-synchronous being non-cross-strapped and not operating in lock-step to the maximum extent practical, to reduce common mode system failures. There are three reasons, primarily, in view of today's environment and where we are today with Centaur and Titan Centaur: There's a lot of proven standard hardware and software technology out there, and we are going to a standard 1750A instruction set; we're going to use standard Ada software that are supported by many people, and have multiple users, where you have built-in validation through wide-spread use in the industry. So it's proven hardware and proven software, as opposed to the unique systems we all have developed. The second point is that it eases the validation process, because you can use existing tools. You have existing software tools which prove that software set works right. You have individual single-string models that can easily model multi-string systems. The third point is social, in a sense. In today's government views of the commercial launch vehicles, it's perceived to offer some lower risks, because it's a more understandable system.

That's where we are today in the state-of-the-art technology when it boils down to looking at a dual or a triple-string system.

There are just a couple other comments I want to add to amplify what Mike and Jim have said, concerning the new avionics technology. We're just finishing a competition for the new components on Centaur. Indeed, for a single string system the weight is about a third of today's system, maybe even less. The cost is one-quarter, or even less than a quarter, of today's systems. The power is one-third to one-quarter. The technology is here and, as far as performance is concerned, Jim is right. Without pushing the state-of-the-art, today's computer is about 100k ops, 150k ops, for those of you who understand instructions per second. Our new computer is probably going to have no less than two million instructions per second capability. There are some out there at four or five. The computer performance is getting better and the same thing with the guidance system accuracy, therefore cost effective avionic system redundancy is achieveable today .... Thank you.

Kenneth Cox: Thank you very much, John. The second panel issue in a technical sense is the best utilization of rapidly advancing technology with emphasis on life cycle and some of the thoughts on guidance and control designs for the future. John Hardtla.

AAS 87-630

GUIDANCE AND CONTROL REQUIREMENTS FOR FUTURE LAUNCH VEHICLES

John W. Hardtla*

The cost of space launches has become a critical issue. The time and cost to prepare a launch vehicle for flight has steadily increased since the start of the space program. If national goals are to be met, a significant reduction in these costs must be made. The design of the guidance and control subsystems are key elements in achieving this reduction. Guidance and control contributes directly to the effectiveness of the launch system and therefore impacts how well the launch system meets the needs of payload customers. Also, guidance and control has a direct impact on the cost of conducting flight operations.

The process of guidance and control design and operation needs a major overhaul. In the past, the design emphasis has been on making guidance and control meet the system performance requirements; in other words, make the system work. In the future, designers will have to consider not only system performance requirements but also requirements for meeting cost goals over the life of the launch system. This includes costs for algorithm design, software design, design changes, configuration control, maintenance, test, and the cost of preparing the software data load for each flight.

Designing specialized guidance and control flight software for each new launch system is costly, but this has generally been the approach. This is in sharp contrast to ground based information processing systems that make extensive use of algorithm libraries. User friendly guidance and control software would also be of great benefit. There is vast room for improvement in all these areas.

* Manager, Guidance Technology, Boeing Aerospace Company, MS 8C-47, Seattle, Washington 98124-2499.

Another factor in the cost of onboard algorithms is the lack of integrated design of the software and hardware. Typically, the guidance and control algorithm design has started after the launch vehicle hardware components have been selected, particularly the flight computer. This can result in unexpected problems and reduced system capability. Trading guidance and control design choices against hardware component options offers the opportunity for substantial cost savings.

A major contribution to guidance and control software costs has been the need for mission specific flight software design changes. Because of limitations in the onboard computer memory and throughput, early flight software operated only for a specific, narrowly defined mission. When the mission requirements changed, the software was modified. As payloads became more expensive, and thus the penalty for a launch failure more severe, extensive pre-flight software testing was conducted in order to insure the accuracy of the software modifications. This in turn increased the time and cost of making changes to the flight software to accommodate mission or vehicle changes.

Typically, the structure of onboard algorithms has not been designed to accommodate change. It is usually assumed that the mission and vehicle will remain invariant and that all future payload requirements can be met with the original design or with simple modifications. This has never been the case for a successful launch system. Vehicle hardware changes are often required to fix design problems, improve performance, or introduce safety features. Experience has taught us that payload needs cannot be fully anticipated and that algorithm and software changes must be made to accommodate specific payload requirements. The guidance and control algorithms must be designed with the requirement that they may undergo continuous change.

As launch system requirements have become more complex, so has the flight software data load. The effort needed to prepare the data load directly impacts the cost of launch operations and the launch on demand lead time. This time period is critical to national defense because of the need to place satellites in orbit on very short notice. The process of preparing a launch vehicle data load for a flight has traditionally been an expensive, complex, and a time consuming multistep process. It is an iterative design process that starts with payload requirements and ends with extensive hardware in the loop testing. The process is highly constrained by numerous elements, including launch vehicle limitations, payload and upper stage interfaces, and by payload delivery requirements. The development of this process has historically been an afterthought and certainly not designed to minimize costs or flow time. In order to realize cost savings, this entire process must be designed to avoid duplication, facilitate transfer of information, efficiently accommodate late mission and payload changes, anticipate constraints far in advance of launch, and rely heavily on automation. The key elements in making this happen are a realizable set of goals, a well thought out set of design requirements, a plan, and a commitment to allocate and then protect the budget necessary to make it happen.

Adaptive Guidance and Control

I would like to add a few comments about the need for new technology in the design of guidance and control systems for launch vehicles. There have been many advances in guidance and control technology over the past two decades. For various reasons very little of this technology has found its way into the design of guidance and control systems for launch vehicles. However, we are seeing it find a use in the design of other aerospace vehicles and even in consumer products. For example, automatic or adaptive control technology is appearing on new automobiles. Most recently with the introduction of anti-lock braking systems. Many automobile drivers may never have to activate their anti-lock braking system but its value is priceless in an emergency or unexpected situation. The same can be said for launch vehicles, adaptive guidance and control is most valuable when the unexpected happens.

Adaptive guidance and control can be defined in many different ways. Perhaps the most common definition is adaptability to in-flight perturbations due to the launch vehicle hardware or by environmental conditions caused by winds, atmospheric density uncertainties, or even cosmic rays upsetting the flight computer. Adaptability is also the ability to adjust to uncertainties in vehicle data such as mass properties or propulsion characteristics that were either not measured accurately or not measured because there was neither the time nor the resources. It may mean being adaptable to many different types of missions without having to make software changes. Adaptability can mean changes in payload that have not been reflected in the flight software data load or it can mean adjusting to contingency situations, either planned or unplanned. In summary, I think of adaptive guidance and control systems as being robust to large variations in the multidimensional space that characterizes the vehicle design and the environment in which it must operate.

By taking advantage of proven technology, guidance and control algorithms can be designed to provide automatic adaptation to changing mission demands and environment. Current technology, with its reliance on expensive ground support and preflight computation, must be significantly improved to meet the challenge. The algorithms may also need to be designed with the anticipation that the launch vehicle will undergo evolutionary growth. Guidance and control algorithms must be computationally efficient, robust, self-starting, and capable of functioning independent of ground control.

Shifting the burden of adaptability to the launch system is not without risk. Because of the complexity of adaptive systems they will require careful design. It is my experience that the onboard system needs a sufficient level of adaptability consistent with the technology available at the time of launch system design. At the minimum guidance and control systems must be able to adjust to unexpected perturbations and off nominal conditions.

In conclusion I am an advocate of:

1. Making significant changes to the process of designing and operating guidance and control systems, and

2. Implementing adaptive guidance and control technology.

AAS 87-629

EXPENDABLE LAUNCH VEHICLE AND UPPER STAGE MODERNIZATION - BETTER UTILIZATION OF CROSS PROGRAMS

Lloyd Thayne*

Kenneth Cox: Thank you very much .... For the third panel member discussion, one of the basic issues is better utilization of cross programs and Lloyd Thayne will address this, relating ELV programmatic priorities and some of his guidance and control experience.

Lloyd Thayne: Thank you. As I read through the presentation and later listened to presentations by Jim and Mike, I was impressed with the fact that they both had a very common theme in them, and that is that we need to do something, now in the avionics or ELVs.

I started my experience in the ELV world in the Titan I and Titan II days in the early 60s. I gravitated away from that for several years. My first real experience in guidance and control systems was in an area that had very little to do with expendable launch vehicles, with the exception of the fact that the vehicle on which I was working used the guidance system that was designed for an expendable launch vehicle.

Let me tell you briefly about that, so that you'll understand where I'm coming from. I'm not an expert on guidance and control systems. I never pretend to be. I'm a user of that technology and my first experience in using that was with the Autonomous Land Vehicle that we built for one of the DARPA projects. The Autonomous Land Vehicle was a very interesting challenge to me. I don't know how many of you are familiar with the challenges of that program. It was very challenging, I think, for everyone involved in it, because DARPA wanted to push us to the very limit. Basically, we were assigned for a new technology breakthrough on a quarterly basis. We started out with a Vision System which allowed a robot to view the scene in front of him, move about two or three feet, stop for 15 to 20 seconds, review the scene again and once again move two or three feet, and then make the decision on where it was to go the next time. That was the condition of the Vision System technology at the time when we were given the Autonomous Land Vehicle program.

We ordered a vehicle to put it on. The vehicle weighed, when it was loaded, about 16,000 to 18,000 pounds. It was a truck, if you will. It was built to haul several laboratory 19" rack computer and electronic systems. It was also built to drive cross-

* Manager of Electronics Engineering, Martin Marietta Aerospace, Space Launch Systems, P.O. Box 179, Denver, Colorado 80201.

country with no roads. So it was an eight-wheel, skid-steering vehicle. Our challenge was, after having been given the contract in August, by the following May we were to be able to load the programs into any computers we had on board, turn on the television on the front of the vehicle, bid it good-bye and wait for it a kilometer down a back country road, and hope that it got there. That's the only control we had over it, except for a panic button switch that hung from a little wire back about 20 feet, until somebody walking along behind -- if he could keep up with it -- shut it down in case of an emergency. We received the chassis for that vehicle on February 1 and we demonstrated the full-up vehicle traveling a kilometer down a back country road without human intervention on the May 30. Now to me that was a challenge in navigation. It was a challenge in guidance. It was a challenge in mixing various computer technologies to achieve an end product. It also presented to me many of the programmatic problems....

From a programmatic viewpoint, the problems are very challenging. We can step forward to the podium at any time and say we need a new avionic system for all of our ELVs. That doesn't solve the problem .... Let me just take the Titans, for example, because I'm more familiar with them. We're using the Delco computer, which was mentioned earlier, a 1970 vintage computer, very good, very useful which has been very successful for us. The performance of that computer has been phenomenal over the years in which we've used it. Our dilemma, as we try to build more of these, is that the parts that we have for them have corroded leads. They've been in storage for 15 years, and you can't buy anymore of them. It's heavy. It's single-string. And now the cost of the payload is such that you can't afford to lose a payload. So we're driven to the problem of, "Where do we go from here." The dilemma is that you have enough parts for 10, 15, or 20 of these guidance computers, when are you willing to junk all that you spent millions and millions of dollars for and move on to new guidance systems that we now have to pay for again.

There are conflicting priorities that have to be solved in this management dilemma. The first one that I see are the budget limitations for the projects. The Titan 4 program has serious budget limitations, as do all of the others that I know of in the ELV arena. So where do you spend your money? We can go through all kinds of analyses to prove that we can scrap the existing guidance system, replace it with a new one and over the life of the program save money. That doesn't solve the problem because there's no money this year or next year to do the development work to replace those items while we're still trying to make the launch dates that are scheduled. So you have not only an overall fiscal budget problem for the total program, you have an annual budget which is limited that you can't exceed. Those are some of the dilemmas that we have.

The second one is that the obsolete hardware still works. Do you really throw away things that still work? That's a terrible, emotional, as well as a scientific, trade-off to throw away something that you think might just do the job for you and get another 9 added onto the reliability number by spending several million dollars to get to the new system. So that's another dilemma that we're faced with.

The third dilemma is the desire to upgrade immediately to avoid future problems. The availability of future systems in the old design is very nebulous, very precarious, because they are obsolete in many ways. Well, the statement was made by Jim, that there is a tremendous cost avoidance to be found in going to newer, much more reliable systems, as the payloads get high into upwards of a billion dollars, that's not unreasonable at all. For some of the payloads and some of the costs of the launch vehicles you have to add the cost of the sites, the cost of any failures that occur and the research that goes into that. So you would say that cost avoidance is a very valid justification for spending the money now. However, it's been my experience, and perhaps the experience of some of you, that there seems to be a tremendous amount of motivation to fix problems that have already happened rather than to prevent them from happening. And that's the dilemma that we're in right now in this area. And I'm afraid that if we wait to fix the ELV avionics problems after they have occurred, we will once again end up in an overkill dilemma where we will eventually be buried in paper.

Some of the requirements as I see them from a programmatic standpoint for ELVs are, first of all, we must assume, as was stated earlier this morning that the ELV is a mandatory companion to the manned vehicle. Neither the manned vehicle nor the expendable launch vehicle is a total answer to our space needs. A second observation I would like to make at this time based on our discussions here is that an absolute knowledge of downstream hardware shortages does not automatically create the budgets to solve those shortages.

I would submit that our challenge in this arena is that everyone first of all claim ownership to his own problems, rather than passing them off to someone else. That includes us as contractors and others as government agencies and the scientific community as a whole, to accept ownership to our own problems and get on with them. We also must be allowed to find an acceptable cost solution to the problem as we see it, as Dr. Duncan suggested earlier this morning. One of the possible methods, as was alluded to earlier, is common usage of new advanced avionics hardware. This sounds very simple. It sounds very straightforward. Just develop one set of avionics hardware and use it everywhere. Go develop one and try to get anybody else to use it! There's an NIH factor there, Not Invented Here factor, that is so powerful it's very difficult to overcome. Nobody wants to use my hardware. And I don't want to use anyone else's. That's the dilemma that we get into constantly in that area. We have to step forward to that and grow up a little bit and get the job done. We must also eliminate all of the no-value-added processes. Any step that we go through that adds no value to the end product must be eliminated from our process and from our program. Only then can we implement, with a great deal of enthusiasm, the advanced avionics that we propose here today. Thank you.

Kenneth Cox: At this point I'd like to congratulate the speakers, both of them, as well as the panel members, for giving a lot of information in a very short time. We weren't sure we were going to meet the time restraints. This session has two parts, and the second part has a different group of panel members .... So we'd like to pause for about two minutes and stand up, but please don't leave! We'll try to start the second part of it immediately. At this time I think everyone that served on the first part deserves a hand.

FLIGHT MECHANICS/GUIDANCE AND CONTROL:

ASTRODYNAMICS AND PLANETARY MISSIONS

PANEL THREE -- SESSION II

Chairman: Joseph J.F. Liu
Chief of Space Station
Directorate of Operations Analysis
Headquarters
U.S. Air Force Space Command
Peterson AFB, Colorado

The following paper was not available for publication:

AAS 87-628 "Astrodynamics and Planetary Missions -- The J2000 Reference System," by Victor Bond, McDonnell Douglas Astronautics Co., Houston, Texas

AAS 87-620

ASTRODYNAMICS AND PLANETARY MISSIONS - INTRODUCTION

Joseph J. F. Liu*

There are two major subjects to be addressed at this second part of Panel 3: the planetary missions and astrodynamic standards. The planning of future planetary missions, the exploration of Solar System and their scientific significance are of great interest and therefore we are happy to include these topics in this Panel. The remaining portion of this panel will be focused on astrodynamic standards for space operations. To clarify the importance of such standards, I would like to briefly outline some background information for your reference.

As a consequence of the outgrowth of civilian and military space endeavours since the launch of Sputnik in 1957, numerous space operations systems have been developed and established. Due to this proliferation of different independent space systems, the interface and compatibility problems in space operations have continued to grow with the advancements in technology and astrodynamic sciences. At Air Force Space Command and U.S. Space Command, we have experienced increasing operations difficulties with a great number of product users while the interfaces have became unnecessarily sophisticated and complex in order to cope with these problems. The end results are compromises on accuracy, efficiency and the critical capability of reliable real-time responses. Recognizing these problems, the Deputy Chief of Staff for Operations, Air Force Space Command sponsored an Astrodynamic Standards Panel during the Space Operations Workshop III, September 1987. That panel recommended that an effort be initiated to examine the possibility of establishing a set of feasible and flexible astrodynamic standards for space operations. Specifically, the operations astrodynamic standards should include the basic definitions, coordinate systems, physical units, physical models, orbit theories, data formats, and others. My colleagues at Air Force Space Command are currently undertaking such an effort to formulate a flexible recommendation to serve as a strawman. It is also well noted that this is not only an extremely difficult task but also certainly a tremendous and necessary challenge. We would like to use this opportunity to introduce some theoretical and engineering discussions in this panel, to share our concern and hopefully to generate some common interest. Since the astronomical system is an absolute important issue in space operations, the new sys-

* Chief of Space Division, Directorate of Operations Analysis, Headquarters Air Force Space Command, Peterson AFB, Colorado 80914-5001.

tem published by International Astonomical Union, J2000, is also addressed as a subset of astrodynamic standards.

Finally, it is my great honor to introduce our three distinguished speakers.

Dr. William McLaughlin is our first speaker and he will present the exploration of the Solar System. Dr. McLaughlin received his Ph.D. in mathematics from the University of California at Berkeley in 1968. He immediately went to work for Apollo and was involved in numerous landing programs. He joined Jet Propulsion Laboratory (JPL) in 1971 and has worked on many planetary missions. He was the engineering manager for Voyager's Uranus missions and presently he manages the Mission Profile and Sequencing Section at JPL. He is a member of the International Academy of Aeronautics and has received NASA Exceptional Service and Outstanding Leadership medals. He also writes a monthly column on space at JPL for Space Flight Magazine.

Our second speaker, Mr. Victor Roger Bond, received his bachelor's degree in physics and engineering from the Washington and Lee University of Lexington, Virginia in 1956. He has a long standing career in scientific services for the U.S. Government. He served in the U.S. Army, NASA/Langley Research Center, NASA/Johnson Space Center. He is presently a senior staff engineer for McDonnell Douglas at Houston. Mr. Bond has been extremely active in professional societies. He is an associate in the Division of Dynamic Astronomy, the American Astronomical Society. He is also an consultant to the Commission on Celestial Mechanics of the International Astronomical Union. Mr. Bond will address the astronomical system J2000.

Professor Victor Szebehely is our third speaker. Dr. Szebehely received his undergraduate and graduate degrees from the Technical University of Budapest, Hungary, including his Ph.D. in 1945. He came to United States in 1947 and became a citizen in 1953. He has held many distinguished positions in research and teaching in many prestigious institutions. Currently, he holds the R. B. Curran Endowment Chair in aerospace engineering and theoretical mechanics at the Unversity of Texas at Austin. He has authored/edited 15 books including his famous book *Theory of Orbits* and authored more than 200 articles in professional journals. Dr. Szebehely is a renowned expert in advanced methods in celestial mechanics, dynamical stability, dynamics of artificial and natural bodies, undeterministic dynamics and has been the director of NATO Advanced Study Institute for many years. He received many distinguished awards from the AAS, AIAA, U.S. Air Force and he is a member of the U.S. and National Academy of Engineering. He was knighted by Queen Juliana of the Netherlands in 1956 and was the first recipient of the D. Brouwer Award of the Dynamical Astronomy Division of the American Astronomical Society. Dr. Szebehely will discuss the theoretical aspects of astrodynamic standards based on classical celestial mechanics.

AAS 87-624

PLANETARY MISSIONS

William I. McLaughlin*

Some of the missions of planetary exploration, planned for the near-term future, are presented and briefly discussed as to their scientific and engineering highlights.

INTRODUCTION

The NASA program of planetary exploration has suffered recently under funding restrictions and the lack of launch delivery systems. Solution of the launch-vehicle problem depends upon: 1) return of the Shuttle fleet to active status, 2) availability of expendable launch vehicles for certain missions, and 3) employment of planetary gravity assists to compensate for the loss of the Centaur G-prime as an upper stage for Shuttle. Solution of the funding problem depends upon a commitment from government to support this extraordinarily productive program: a plan for future exploration is in place.[1,2,3,]

The missions which are presented in this paper are listed in Table 1. Lack of space does not permit extensive discussion or inclusion of the many other missions which are under consideration. All of the planetary projects

Table 1

SELECTED MISSIONS

Mission	Launch Date	Prime Mission	Prime Data Return Begins
Voyager/Neptune	8/77	Neptune Flyby	1989
Magellan	4/89	radar survey of Venus	1990
Topex	12/91	oceanographic altimetry	1992
Mars Observer	9/92	multi-experiment survey of Mars	1993

* Manager, Mission Profile and Sequencing Section, Jet Propulsion Laboratory, California Institute of Technology, 4800 Oak Grove Drive, MS 301-250D, Pasadena, California 91109.

Mission	Launch Date	Prime Mission	Prime Data Return Begins
Galileo	10/89	Jupiter Orbiter & Probe	1995
CRAF*	2/93	Comet Tempel 2 rendezvous	1996
MRSR**	c.1998	Mars rover and sample return	c.2001
Cassini+	1995 1996	Saturn orbiters and Saturn and Titan probes	2002 2003
Daedalus++	21stC.	probe to Barnard's Star	L+50yr

* FY89 new start sought
** Joint JPL/JSC study
\+ FY91 new start sought
++ British Interplanetary Society study (1978)

approved for launch are included, plus some reasonably good possibilities for future missions. The Daedalus interstellar study is listed as a reminder that the future is unbounded.

THE MISSIONS

Voyager/Neptune

After a highly successful encounter with Uranus in January of 1986[4], the Voyager extended mission (beyond the Jupiter and Saturn encounters) continues with a scheduled encounter with Neptune in August 1989 (Fig. 1).

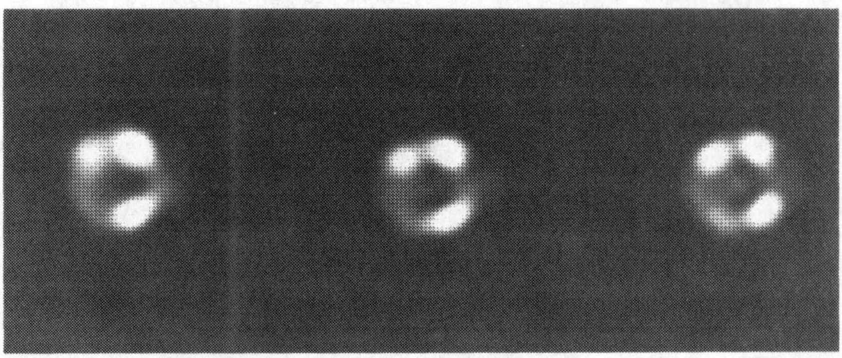

Fig. 1 Earth-Based Photographs of Neptune (taken with the 2.5m telescope at Las Campanas Observatory in Chile in 1983 and showing possible clouds of methane ice crystals high in the atmosphere)

Despite some problems experienced since launch in 1977, the Voyager 2 spacecraft is in good health. Continuing engineering work begun for the Uranus encounter[5], the capability of the spacecraft to function in the dark and distant outer solar system is being upgraded.[6]

The encounter will feature a very close overflight of Neptune's north pole followed by an approach to within 40,000 km of the large satellite, Triton. Considerable interest has been generated concerning Neptune's system of rings, which seems to possess irregular properties.

After the Neptune flyby, no further planetary encounters remain for Voyager 2, and it will join Voyager 1 in exploring the fields and particles environment of the outer solar system, possibly encountering the heliopause, and making stellar observations with (their) ultraviolet spectrometers. It is anticipated that the Voyager spacecraft will continue to function into the next century. Their missions may end with the decline of the electrical output of their radioisotope thermoelectric generators to unacceptably low levels or the inability of their Sun sensors to lock onto the ever fainter Sun.

The Voyager project manager is Norman R. Haynes (JPL) and the project scientist is Dr. Edward Stone (Caltech).

Magellan

Magellan will carry a synthetic aperture radar (SAR) to Venus and conduct a survey of the planet to a resolution better than one kilometer[7] (Fig. 2).

The orbit of the spacecraft will be highly elliptical with mapping taking place from altitudes of 250 km to 2113 km around periapsis: representing a span of time up to 37.2 minutes. The remainder of the 3.15 hour orbit will be spent orienting the spacecraft so as to point its 3.7m high-gain antenna to Earth and play back (at 268.8 kb/s) recorded SAR mapping data.[8]

During its prime mission of 243 days, about three quarters of the surface of Venus will be mapped. Assuming an extended mission, the opportunity arises to map the remainder of the surface and conduct investigations in detail of objects of special interest.

The Magellan project manager is John Gerpheide (JPL) and the project scientist is Dr. R. Stephen Saunders (JPL).

Topex

The Ocean Topography Experiment (Topex) mission to planet Earth is being conducted jointly with the French.[9] It will focus on achieving an understanding of the global circulation of the oceans through precision altimetry conducted from a 1334km, near-circular orbit (with an inclination of 63.1 degrees).

A desired accuracy of 13 or 14cm in sea-level measurements should be accomplished by use of a precision NASA altimeter. The altimeter's portion

Fig. 2 Magellan Radar-Survey Mission of Venus

of the error budget is 3 to 4cm, with uncertainties in orbit determination, largely due to uncertainties in the Earth's gravitational field, contributing the remainder. The project has a Gravity Improvement Program underway, and by launch it could yield a factor-of-two improvement in knowledge of the local gravity for Topex.

The Topex project manager (U.S.) is Charles A. Yamarone, Jr. (JPL) and the project scientist is Robert H. Stewart (JPL).

Mars Observer

The Mars Observer mission will conduct geoscience and climatological investigations for a full Martian year of 687 days.[10] The planetary survey will be accomplished from a near-polar orbit with an altitude in the range of 350 to 400km.

The geoscience objectives are:

1) Determine the global elemental and mineralogical character of the surface materials.

2) Define, globally, the gravitational field and topography.

3) Establish the nature of the magnetic field.

The climatology objectives are:

1) Explore the structure and aspects of the circulation of the atmosphere.

2) Determine the temporal and spatial distribution, abundance, sources, and sinks of volatile materials and dust over a seasonal cycle.

The extensive suite of instruments includes spectrometers, a magnetometer, and a camera. No scan platform is present, and the remote-sensing instruments are nadir pointed. However, their lines-of-sight can be adjusted internally by means of mirrors.

The Mars Observer is the first of the Observer Class missions, designed to low cost, utilizing relatively simple spacecraft and restricted flight-team support. Extensive participation in observation planning and noninteractive commanding will be done by principal investigators operating from work stations at their home institutions.

The instruments onboard Mars Observer contain extensive programmable capabilities resulting in a shift in the centroid of command traffic (compared to previous missions) from the spacecraft's central processor toward those contained in the instruments.

The Mars Observer project manager is William I. Purdy (JPL) and the project scientist is Dr. Arden L. Albee (Caltech).

Galileo

The Galileo mission, approved in 1978, will carry out an extensive exploration of the Jovian system with an orbiter and an atmospheric probe.[11]

Due to the cancellation of the high-energy Centaur upper stage for use with the Shuttle, Galileo will employ a gravity assist from Venus and two gravity assists from Earth in order to acquire sufficient energy for the traverse to Jupiter[12] (Fig. 3). The spacecraft has had to undergo modifications to be able to survive the more demanding thermal environment at Venusian distances from the Sun.[13]

At 150 days before encounter with Jupiter, which takes place in December 1995, the probe will be released from the spacecraft. The spacecraft will overfly the probe entry point into the Jovian atmosphere, relaying data to Earth for up to 75 minutes as the parachute-equipped probe descends toward the surface.

Shortly after this event, the spacecraft will ignite its engine to burn into orbit about Jupiter (with, also, a gravity assist from the inner Galilean satellite, Io).

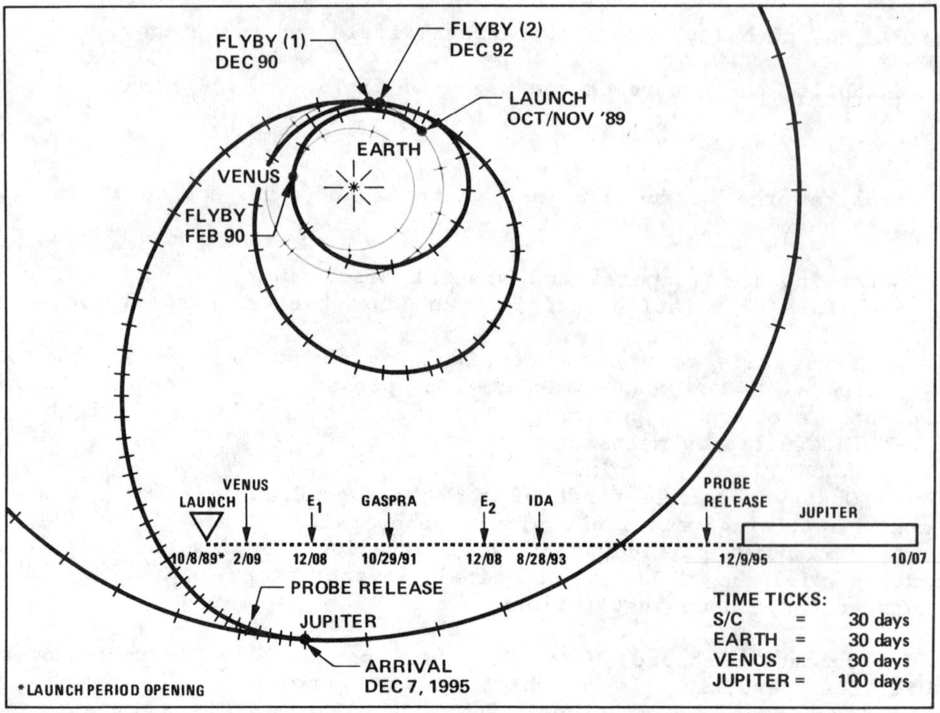

Fig. 3 Galileo's Gravity-Assisted Trajectory to Jupiter

For 22 months, the Galileo Orbiter will conduct an exploration of the Jovian system: planet, satellites, ring, and magnetosphere. Gravity-assist mechanisms will be utilized to implement the tour of the Galilean satellites; ten close encounters are planned during the prime mission, and a large orbital excursion will send the spacecraft to explore the tail of the Jovian magnetosphere.

The Galileo project manager is John R. Casani (JPL) and the project scientist is Dr. Clayne M. Yeates (JPL).

Comet Rendezvous Asteroid Flyby (CRAF)

The CRAF mission, if approved, would be the first of the Mariner Mark II series of missions, designed for more ambitious solar system explorations than those contained in the Observer Class.[14,15]

After launch in February 1993, the spacecraft would, like Galileo, proceed to Venus for a gravity assist and back to Earth for a single gravity assist before going on toward its rendezvous with Comet Tempel 2.[16] En route, the spacecraft would fly by asteroid 46 Hestia (136km diameter) at 11.4km/s for the purpose of gathering scientific data.

After rendezvous with the comet in late 1996, near aphelion of Tempel 2 (the comet has an orbital period of 5.3 years), the spacecraft will conduct a systematic examination of the nucleus and the coma and tail, as they evolve (Fig. 4). A penetrometer will be fired into the nucleus for in situ studies. After perihelion of Tempel 2 (September 8, 1999), CRAF will conduct an excursion into the tail of Tempel 2. End-of-mission is December 31, 1999.

Fig. 4 CRAF Observes Comet Tempel 2

The modularly designed CRAF spacecraft will carry on the high capability tradition of the Mariner class of vehicles, and, with minimal modifications, can be adapted to other solar system missions.

The CRAF development project manager is Ronald F. Draper (JPL) and the project scientist is Marcia M. Neugebauer (JPL).

Mars Rover Sample Return (MRSR)

The MRSR mission is currently in the study phase with JPL and the Lyndon B. Johnson Space Center (JSC) defining the trade-off space from which point designs can emerge.

The basic objective of the mission would be to return a five-kilogram sample from Mars to an Earth laboratory[17]. The sample would have been collected by a rover from numerous sites on Mars over a period of 500 days of surface operations[18].

The principal technical challenges facing MRSR are: (1) safe descent, (2) rover traverse, and (3) orbital rendezvous of the sample-return vehicle with the Mars orbiter. Although the two Viking landers descended successfully, significant possibilities for upset were revealed and will have to be guarded against. Rover traverse, to maximize scientific value, should allow coverage of several hundred kilometers of the Martian surface. This involves a judicious balance of remote and autonomous operation for hazard avoidance and sample selection. Finally, the autonomous rendezvous of the sample-return vehicle with the Mars orbiter, prior to return to Earth, presents an intricate docking problem.

If approved, an MRSR mission would further NASA objectives in the unmanned exploration of the solar system and as a precursor to the manned exploration of Mars. A target date for seeking a new start is fiscal year 1993.

The study manager is Dr. Donald G. Rea (JPL), and Jay H. Greene (JSC) is the deputy study manager.

Cassini

Cassini, if funded, would follow up the Voyager flybys of Saturn with two orbiters and two atmospheric probes (Saturn and Titan), much in the way that Galileo will further the Voyager looks at Jupiter[19]. Giovanni Cassini (1625-1712) was an Italian-French astronomer who discovered four satellites of Saturn and in 1675 called attention to the gap which bears his name in the system of rings. The mission would be the second in the Mariner Mark II series, after CRAF.

Following a Jupiter gravity assist, for the prime trajectory option, the spacecraft would deliver its probe and go into orbit around Saturn[20]. The tour of the Saturnian system would rely heavily upon gravity assists from Titan, the only Saturnian satellite large enough to be employed for this purpose.

Discussions have taken place with the European Space Agency (ESA) with regard to a joint NASA/ESA mission.

The U. S. study manager is John C. Beckman (JPL) and the U.S. study scientist is Dr. Wesley Huntress, Jr. (JPL).

Daedalus

One approach to mounting an interstellar mission was developed by a study team of members of the British Interplanetary Society, initiating their effort in 1973[21]. The objective was to produce a conceptual spacecraft and mission design for a first (unmanned) mission to a star. The target stellar system, Barnard's star, was selected in the hope that it might possess planets.

Daedalus was celebrated in Greek mythology as a talented craftsman who listed among his engineering credits the construction of the Labyrinth for Minos on Crete and wings for human flight (Daedalus and Icarus legend).

At six light years distance, Barnard's star could be achieved after about 50 years of flight, with speeds up to 12% of that of light, using a nuclear pulse rocket with external ignition. This technology, proposed by S. Ulam in the 1950's, is clearly beyond present capability but might be feasible for use in the next century, A payload of 500,000kg is envisaged.

On approach to the target system, onboard telescopes would search for planets and, if found, their orbits would be determined. Daughter probes would then be targeted for closer investigation.

The study manager was Alan Bond.

ACKNOWLEDGMENT

The work described in this paper was carried out by the Jet Propulsion Laboratory, California Institute of Technology, under contract with the National Aeronautics and Space Administration.

REFERENCES

1. Planetary Exploration through the Year 2000 - Part One: A Core Program, National Aeronautics and Space Administration, Washington, D.C., 1983.

2. Planetary Exploration through the Year 2000 - Part Two: An Augmented Program, National Aeronautics and Space Administration, Washington, D.C., 1986.

3. Planetary Exploration through the Year 2000 - A Core Program: Mission Operations, National Aeronautics and Space Administration, Washington, D.C., 1986.

4. "Voyager 2 Mission to Uranus," Science, Vol. 233, special Voyager issue, July 4, 1986.

5. R.P. Laeser, W. I. McLaughlin, and D.M. Wolff, "Engineering Voyager 2's Encounter with Uranus," Scientific American, Vol. 255, November 1986, pp.36-45.

6. W. I. McLaughlin, "Imaging Neptune", in "Space at JPL", Spaceflight, Vol. 29, November 1987, pp.390-391.

7. R. S. Saunders, "Questions for the Geologic Exploration of Venus", J. Brit. Interpl. Soc., Vol. 37, 1984, pp.435-441.

8. W. I. McLaughlin, "A Survey of Venus," in "Space at JPL", Spaceflight, Vol. 29, September 1987, pp.349-350.

9. W. I. McLaughlin, "Oceanography from Space," in "Space at JPL", Spaceflight, Vol. 29, June 1987, pp.258-260.

10. E. L. McKinley, "Mars Observer Mission", AAS-86-332, AAS 33rd Annual Meeting, Boulder, October 26-29, 1986.

11. G. L. Bennett, "Return to Jupiter", _Astronomy_, Vol. 15, January 1987, pp.6-15.

12. J. R. Casani, "Galileo's New Route to Jupiter", AAS-86-330, AAS 33rd Annual Meeting, Boulder, October 26-29, 1986.

13. W. I. McLaughlin, "Galileo Spacecraft Changes," in "Space at JPL", _Spaceflight_, Vol. 29, May 1987, pp.218-219.

14. R. F. Draper, "The Comet Rendezvous Asteroid Flyby Project," IAF-87-446, 38th IAF Congress, Brighton, October 10-17, 1987.

15. M. M. Neugebauer, "Comet Rendezvous - The Next Step", _Sky and Telescope_, Vol. 73, March 1987, pp. 266-270.

16. S. L. Miller, D. F. Bender, D. S. Stetson, and M. R. Myers, "Trajectory Options for the Comet Rendezvous Asteroid Flyby Mission", AIAA-87-0641, AIAA 25th Aerospace Sciences Meeting, Reno, January 12-15, 1987.

17. W. I. McLaughlin, "Mars Rover Sample Return", in "Space at JPL", _Spaceflight_, Vol. 29, November 1987, pp. 392-393.

18. B. H. Wilcox and D. B. Gennery, "A Mars Rover For the 1990's", _J. Brit. Interpl. Soc._, Vol. 40, October 1987, pp. 484-488.

19. D. F. Robertson, "Cassini", _Astronomy_, Vol. 15, September 1987, pp. 20-24.

20. A. B. Sergeyevsky, S. J. Kerridge, and D. S. Stetson, "Cassini - A Mission to the Saturnian System", AAS-87-423, AAS/AIAA Astrodynamics Specialist Conference, Kalispell, August 10-13, 1987.

21. A. Bond, _et al_., "Project Daedalus", _J. Brit. Interpl. Soc._, supplementary issue, 1978.

AAS 87-623

ASTRODYNAMIC STANDARDS

Victor Szebehely* and Joseph J. F. Liu†

This paper intends to establish the need for standardization in astrodynamics and discusses the basic underlying technical approaches to accomplish standardization. For many significant details see a recent report by J.J.F. Liu (1987). A preliminary astrodynamic standards for space operations has since been presented by H. B. Wackernagel (1988).

INTRODUCTION

The purpose of astrodynamics is to offer accurate orbit predictions for space probes and artificial satellites. The basic techniques used are the same as in the classical science of celestial mechanics which concentrates on accurate orbit predictions for natural bodies of the solar system, such as asteroids, comets, satellites and planets. While the basic techniques and scientific principles are the same in celestial mechanics and astrodynamics, the applications are often quite different for several reasons. First, in celestial mechanics emphasis is placed on long-time predictions, while in astrodynamics the usual periods of orbit propagations are much shorter. At the present time the predictability horizon in celestial mechanics is established by computations concerning the behavior of the solar system for 10^8 years. For artificial objects the requirements regarding the length of orbit prediction without propulsion is usually one month. This represents a ratio of 10^9 between the lengths of predictions in the two fields. Both fields, of course, might rely on observations during the duration of the orbits to rectify and update the ephemerides but this is not always feasible.

The concept of predictability horizon was introduced by Sir James Lighthill (1986) referring to the length of time for which meaningful predictions can be obtained.

The required precision of predictions in celestial mechanics and in astrodynamics are also different but these requirements might merge when interplanetary orbit calculations are considered.

---

* Dr. Szebehely is Professor, R.B. Curran Centennial Chair, Department of Aerospace Engineering, University of Texas, Austin, Texas 78712.

† Dr. Liu is Chief of Space Division Directorate of Operations Analysis, Headquarters, U.S. Air Force Space Command, Peterson Air Force Base, Colorado 80914-5001.

Another fundamental difference regarding the techniques used in celestial mechanics and in astrodynamics is related to the different types of orbits encountered in these two fields. Planetary orbits and the motion of natural satellites may be considered as slightly disturbed two-body orbits. Artificial objects, on the other hand, might be on orbits which vary from near-circular low earth orbits to earth-moon impact trajectories, to deep space probe orbits and to interplanetary trajectories with close approaches.

This paper is dedicated to astrodynamics and the above comments concerning celestial mechanics were brought forward to describe the similarities and the fundamental differences in the techniques involved.

The aim of astrodynamics is not only to offer orbit predictions with high accuracy but to perform the computations in a short time with often limited computational facilities. The last two requirements are of considerable importance in practical applications and are strongly related to helping us to establish standards in astrodynamics. For the purpose of this paper standards are those properties of an operation which have been selected as models to which performances may be compared.

In the following, the basic needs to satisfy the goals of astrodynamics are outlined and are related to standards.

FACTORS INFLUENCING PREDICTIONS

To predict orbital behavior we must perform three basic steps:

1. Obtain information, mostly by analyzing observations, to establish initial conditions. This is a considerable field itself having some fundamental difficulties and it offers us initial conditions with uncertainties.

2. As the second step we must formulate our equations of motion which are usually nonlinear sixth order differential equations. The formulation involves physical constants with values known only approximately. Some of the terms in the equations of motion become dominant for certain types of orbits and their effects might be neglected for other types. A detailed analysis of this problem of formulating the equations leads to another field of considerable difficulty and uncertainty.

3. Once the initial conditions are "established" (with errors) and the equations are "formulated" (with uncertainties of the constants and of the effects included) the propagation consists of solving these equations. The solutions may be established by series expansions with questionable convergence properties or by numerical integrations. The first approach is known as general perturbation method and the second as special perturbations. Both of these approaches involve approximations. The series in general are divergent and the numerical integrations involve error accumulations at every step. Consequently, error-free propagations are not realistic expectations. The time limit of meaningful predictions depends on the types of orbits to be propagated.

Therefore, because of the errors involved in the above three steps (initial conditions, formulation of equations, solution) no error-free predictions are possible in astrodynamics.

By using astrodynamic standards the effectiveness of the methods which are used to obtain predictions is increased. Since the predictions can be very sensitive to the accuracies of the above three steps, it is mandatory that the methods with which these steps are performed be standardized.

Unless standardization is accomplished, various orbit predictions made of the same probe are almost impossible to compare. Using different initial conditions, different formulations of the equations, and different solution techniques the results can vary widely and might have limited relation to the actual orbit.

Once again, the need to establish standards is clear in order to compare propagation techniques.

In the next section a method is proposed to evaluate the efficiencies of astrodynamic operations in order to establish standards.

EVALUATION OF ASTRODYNAMIC EFFICIENCIES

An approach to establish astrodynamic standards is to first investigate the input parameters affecting the results. The second step is to formulate a measure of efficiency. This measure is to be optimized and the corresponding standards should be introduced as the final and concluding step.

It is well known in astrodynamics that the selection of the "best" numerical integration technique strongly depends on the variables used and on the type of orbit to be propagated. In other words, the input parameters to our efficiency evaluation may be divided into three groups:

1. The type of orbits to be propagated:
   a. Near earth satellite orbits with low eccentricity;
   b. Near earth satellite orbits with eccentricities in the range of 0.1 to 0.3;
   c. High eccentricity, large semi-major axis, deep space probe orbits;
   d. Other orbits of practical significance, such as lunar trajectories, interplanetary missions, etc.;
   e. Long-time predictions or short arc propagations;
   f. Stable or unstable orbits.
2. The type of computational techniques to be used:
   a. Various numerical integration techniques;
   b. Semi-analytical methods;
   c. Analytical techniques.
3. The set of variables to be used:
   a. Cartesian rectangular coordinates;
   b. Orbital elements;
   c. Delaunay's elements;

d. Regularized time;

e. Stabilizing variables.

Regarding the measure of efficiency, the principal considerations are accuracy, time of computation, small computational facilities or simplicity of programs. These components may be combined many different ways to give a measure of efficiency. These possible combinations might be affected by the specific application being considered.

For any established measure of efficiency the above three parameters must be varied and in this way the "best" approach established. Th best approach will be the standard astrodynamic computation for a given problem. In this respect standardization is equivalent to optimization.

SUMMARY

The final conclusions and recommendation of this paper are to introduce standards in astrodynamics which optimize the performance. The standardization includes:

1. Formulation of equations of motion and election of values of physical constants;
2. Establishment of initial conditions;
3. Selection of dependent and independent variables with stabilizing properties;
4. Selection of methods of solutions properly suited for specific trajectories;
5. Uniform input and output formulations.

REFERENCES

1. Lighthill, J., "The Recently Recognized Failure of Predictability in Newtonian Dynamics", *Proc. Royal Soc. London*, A 407, 35-50, 1986.
2. Liu, J.J.F., *AFSPACECOM/DO Panel Report on Astrodynamic Standards,* Air Force Space Command, October 1987.
3. Wackernagel, H. B., Proposed Astrodynamic Standard, *DOA Technical Note 88-02,* 29 February 1988, Directorate of Operations Analysis, DCS/Operations, HQ Air Force Space Command, Peterson AFB, Colorado 80914.

ROCKET PROPULSION

PANEL FOUR

Chairman: Marc T. Constantine
Vice President and General Manager
Advanced Systems Division
Aerojet TechSystems Company

Host Chairman: A.B. Eisenberg
Regional Manager
McDonnell Douglas Astronautics Company

The following papers were not available for publication:

AAS 87-635 "Rocket Propulsion -- Rocket Propulsion In Space," by Dr. Clark W. Hawk, Chief, Liquid Rocket Division, U.S. Air Force Astronautics Laboratory

AAS 87-636 "Rocket Propulsion -- The Advanced Launch System," by Colonel John R. Wormington, U.S. Air Force Space Division, Los Angeles, California

AAS 87-631

ROCKET PROPULSION - INTRODUCTION

Marc T. Constantine*

Henry N. Bowes: Good Morning ladies and gentlemen. I'm Hank Bowes, the General Chairman for the 34th National Conference of the American Astronautical Society, and your chief host for this occasion .... Let me introduce Marc Constantine, who is taking charge of this session.

Marc T. Constantine: Thank you and welcome ladies and gentlemen .... I'm Marc Constantine, Vice President and General Manager of the Aerojet TechSystems Company. My co-host today is Al Eisenberg, who is Regional Manager for McDonnell Douglas here in Houston. We have three distinguished panelists to talk about propulsion: J.R. Thompson, who is Director of the Marshall Space Flight Center in Huntsville, Colonel Jack Wormington, who is Director for Advanced Studies for the Systems Division of the Air Force in Los Angeles, and William Piland who is the NASA National Aerospace Plane Program Manager and works out of the NASP Program Office in Washington. Our fourth panelist, Dr. Clark Hawk, of the Air Force Astronautics Lab came up with a health problem just yesterday and canceled his visit. But I think we can manage with the three panelists. There will be an opportunity for questions. Each of the panelists will make a presentation and then afterwards we will take questions from the audience and discuss any issues that you might have. To welcome you, we'd like to show you a video tape which will set the stage for our discussions today.

VIDEO TAPE BEING SHOWN HERE:

Only propulsion will take us there. Now a simple statement that explains why we are here today. I don't know how many of you were here for our Keynote Address yesterday by Mr. Willis Hawkins. But he talked about sharing a dream. While we in the propulsion community share this vision, with countless others throughout the nation, it is we in the propulsion community who can make it happen. I feel very fortunate to be a part of this community. After some thirty years in aerospace I continue to be excited about our work and the challenges that we face, particularly the challenges of today. I truly believe that we, together, have the talent, the determination and the spirit that we need to open a new era in space.

Our Keynote Speaker also talked about turning technology into reality, and mentioned several past challenges including the rise of our common ground

* Vice President and General Manager, Advanced Systems Division, Aerojet TechSystems Company, P.O. Box 13222, Sacramento, California 95813.

transportation systems of today, the automobile. Comparing this with the space transportation business, we're about at the same point in time as we were slightly before Henry Ford came on the scene. As you all know, there were maybe 30 to 50 various automobile companies that were each producing one or two of a kind, each one different, and very expensive. Henry Ford came along with a vision that said he wanted to turn this into transportation for the masses so that all could afford to travel about in automobiles. I think right now that's where we are in the space transportation business. We've done a lot of good technology, and we've come up with a lot of systems that have met a lot of different applications. But it seems that we have a different system for each application, and we are, in a sense, "onesies" and "twosies". What we now need is transportation for the masses, to make transportation to space inexpensive.

Sitting on this podium with me, I think we have three Henry Fords, and out there, a lot more Henry Fords. We need them to achieve the space transportation systems of the future. And the only way we can get into space in mass is to reduce the cost of transportation. All these gentlemen are going to talk about various ways of doing this. I am pleased to introduce these three gentlemen, whose vision and dedication are well documented.

As I mentioned before, following their individual presentations they will convene as a panel to answer any questions you may have. We'll begin with J.R. Thompson, Director of the Marshall Space Flight Center, and J.R. will talk about the evolution of the Shuttle propulsion. He's well qualified to talk about it. Several years earlier he was NASA's Project Manager for the Space Shuttle main engine, which is probably the most complex rocket engine that has ever been developed. So he has the background and experience to talk about the evolution of that system into a newer, low-cost system. Prior to returning to NASA/MSFC as Director, he served as Technical Director for Princeton's Plasma Physics Laboratory.

AAS 87-634

ROCKET PROPULSION - THE SHUTTLE-C

J.R. Thompson, Jr.*

Thank you very much, Marc. It's indeed a pleasure to be here and to address this group on propulsion. Specifically, where I'd like to address my remarks this morning is on our space Shuttle. I will give you a quick update on that, because I think that is a key to our getting back into space. Then I'll discuss the evolution of some of those systems.

First of all, let me state right at the outset, that I believe that the Space Shuttle, as we know it today, with some of those critical elements, is going to be the backbone, certainly of NASA's Space Transportation System and is going to serve us for another two decades. But frankly, we have to go to work on reliability, maintainability, and cost. I say, reliability, and to be very specific, I really believe that the unreliability that we've got in today's Space Shuttle focuses in two areas. Both are in propulsion.

The first is the Shuttle main engine and the second is the solid rocket motor, and I believe that they are in that order of importance. I've spent a lot of time over the last year thinking about how we can get the reliability in these systems that we need. I've gone back and reviewed a lot of data and I think, if we're fair and honest and if you look at the smaller, simpler systems, put them all together, without disregarding failures, you're talking about reliabilities on the high-side of the mid-90%, about 95%. Certainly our Space Shuttle got 96% before we had the first failure. But that's not nearly good enough. We can't use that as an excuse at all. As a matter of fact, I think it has been stated that, once we get back to flying again on a steady-state basis, we hope to launch about 12 times a year. Ninety-six percent reliability at that launch rate is going to put us back on the beach every other year. If we increase it to 98%, it's every four years, or to 99%, it's every eight years. So it's an area that we almost can't spend enough time, attention, and, yes, even money on.

We've certainly got to work on the cost of the Space Shuttle. Over this last year I've seen, and been a part of, some cost creeping back into the Space Shuttle. I think, once we start flying again, we're going to have to take some of the actions that we've implemented now and put them in two piles, those that are paying off for us and those that aren't. And we've got to be smart enough to separate the two.

Just a few comments about what we're doing to try to improve the reliability. On our Space Shuttle main engine, the real focus this last year has been in testing -- test-

---

* Director, NASA Marshall Space Flight Center, Huntsville, Alabama 35812.

ings for margins, making sure that we know where our margins are and where our margins are not. Just this year we've completed over a hundred tests. Most of these have been full-duration, which is about 500 seconds. Some have been abort missions which didn't get up in about 800 seconds. We've completed the equivalent of about 23 missions successfully, with no criticality #1 failures during this calendar year. We're now running the engine on some occasions, approaching twice the normal burn time of a mission to make sure that we've got adequate margins there. And about 75% of all of our ground tests now are dedicated and allocated to running malfunction tests, where we build in anomalies, so that we make sure that we know, upfront, how the engine is going to respond. I think we still have a lot of work to do but we are progressing quite well.

I've seen a few old friends here today who will remember some of the early problems in the Shuttle engine. I think if they came back and visited today, they would recognize what has taken place regarding the problems of the turbine bearings and blades. The margins are a lot higher than we used to have.

We've got to make sure that we build into our maintenance plan safeguards so that we don't try to cut any corners and fly with too fine a line between trying to push the life and having adequate margins. But I think the test program we've laid out is going to be in the direction to do that.

On the Solid Rocket Motor, the focus is in the same area. I'm quite satisfied with, our redesign. It's coming along very well. If you ever have the time and you're in Huntsville and want to hear more about the results of our last firing, we'll be glad to provide them. It was highly successful. We've got a lot of margin in that design. We're doing a lot of testing, sub-scale, full-scale, full-duration firings, where we pressurize those joints, under load. And we're starting to build some anomalies in there, so that we know how the motor is going to react to anything that slips through our NDE in processing. Then we have a number of qualification firings. There is a minimum of three left before we start flying again. I think those tests are going to go well. The thing that worries me most in the solid rocket motor is how you build them. They are fairly simple in design, what about the processing? And that's where we're really going to be looking at in the future.

We're starting to look at what we call an advanced solid rocket motor. First of all, let me state that we're going to be using our existing motor design, probably another five years or so. And we can make it work. We are going to have to pay close attention to all the areas that we need to be sensitive about in the processing of this solid rocket motor. But when you talk about using the Space Shuttle for two decades plus, you're talking about building four to five hundred of these motors, for a launch rate of about 12 a year. And so we need to stand back and make sure that we're geared up to do it over the long haul, and get the reliability in these solid rocket motors.

Our advanced solid rocket motor we hope to have operational in about 1993, even in a tight-budgeted environment within NASA. This is NASA's number one new initiative for FY89. It has the full support of everybody in the agency behind it. It's not only a matter of redesigning this motor that's going to be important, including, for example, bolted joints, and making sure we get away from pin joints but making sure that we have the facilities, the processing, the automation in place where it makes sense to build it right by the numbers everytime. I think if you look at where you have most of the problems in solid rocket motors, it's in something in processing that slipped through the crack.

At the same time we're going to try to recover about eight to twelve thousand pounds of payload capability for our Space Shuttle that we've lost primarily in the reaction to the *Challenger* accident, where we put more beef and meat into some of the other elements on the vehicle. This will get us back close to the 65,000, perhaps 70,000 pounds of payload capability on the Shuttle.

We are starting to conduct some very serious studies on a liquid rocket booster in parallel. We're not on the same fast track as we are with the advanced solid rocket motor. But at a later date, if we prove ourselves in following our present course, for example, then we want to be able to exploit some old technology, the big liquid boosters. We're starting to trade studies, including design studies. The key is going to be the simplicity; we're not going to try to be fancy in terms of recovery; they will probably be expendable. We're looking at both pump-fed and pressure-fed systems. I personally would lean in the direction of pressure-fed systems. The key to make it work is, since we've currently designed the Shuttle, how big we can let the diameter and length of this grow before we create a whole host of new problems. So this is coming along in parallel. We hope to evolve it, perhaps in some breadboard tests, either at Marshall or at one of the contractor sites over the next several years. Regarding propellant combinations, I'm certain we're talking about lox hydrocarbon for which NASA has a lot of interest.

The obvious advantage of a liquid system like this is that if you have proper diagnostics on board and you detect something going amiss, you can shut it down. So we would probably have a cluster of engines in the base, ranging somewhere between four and six, probably a half-million pounds of thrust or higher.

That's what we're doing over the short term in our Space Shuttle. The focus is on reliability. We've got to rebuild our highways to space. I think we've all seen, over the last two years, that there's nothing more unreliable than launch vehicle systems that don't work.

The illustration shows the evolving the elements of the Space Shuttle. We call it the Shuttle-C, C for cargo. To me it would be the very earliest real heavy-lift launch vehicle that we could get in this country. What I like about it is that we can build on what we have. I want to state at the outset here that this is the complement our current Space Shuttle, not in competition with it. I think that manned systems are always going to be the centerpiece of NASA. The Shuttle-C, right now, we envision as an unmanned system, a cargo system. The Shuttle would be used to take the people and some of the payloads to orbit, but Shuttle-C would transport the cargo. The key, of course, is to build on what we have; that is, use the existing Shuttle propulsion elements. The reliability that I talked about earlier is a benefit; otherwise, we have bigger problems, because they would be the same elements common to the Space Shuttle. The common elements, of course, are the Shuttle engines. Then there are the solids, and the external tank. Hopefully, this will be in the timeframe for a Shuttle-C of about 1993....

The payload capability will be more than double the capability we've got today in our Space Shuttle. Out of the Eastern Range over a hundred thousand pounds, from the Western Test Range, I think the number is somewhere in the range of eighty thousand pounds of capability.

The Shuttle-C will have a lot of uses. Although we at NASA believe we have a way to to get the Space Station in orbit with our existing systems, with a Shuttle-C or an unmanned cargo carrier it would be much easier. Shuttle-C, with a capability in excess of a hundred thousand pounds, would make significant reductions in the launches required to get this Space Station in orbit. There are a number of planetary missions that we could operate if we had a Shuttle-C.

I think the launch facilities are a key element. We've already got them in place with the side-mounted Shuttle-C, where we replace the Orbiter with a cargo module. Then there would be essentially no modifications at all required to any of the pads. At least in theory, the major elements in processing at the Cape would be common, both for the manned and the unmanned Shuttle, manned being our Orbiters that we have today and the unmanned, the Shuttle-C. The synergism with Shuttle, if we can keep it there and if we move in this direction, I believe, will even go a long way toward helping us with the launch cost of our Space Shuttle.

The next illustration shows the schedule. I don't believe it's overly ambitious. We're in the definition stage now. We have some very modest study money, and contracts are in place. We hope over the next year or two to crystallize this in terms of how it stands with other launch systems that the country's looking at. We'll have a better feel for where we stand with our Space Shuttle. I just think it's a natural and a great complement. And then with the go-ahead, at not too ambitious a pace, I believe we could have it in the air by 1993.

In summary, I think the real work we've got in NASA is on our existing Shuttle. We've got our job cut out for the next several years in getting the reliability into the systems. And again, the unreliability in our Space Shuttle is focused on propulsion. Certainly that falls under the responsibility of us at the Marshall Space Flight Center. I believe, we've got the team and place to do it. We have the support, I think, throughout the agency; it's recognized that right now it's our weak link in holding up our total agency. We can move on to that with elements like the advanced solid rocket motor. We can build on what we've already got with our Shuttle-C. I'm sure you don't have to change that picture much to recognize that what you're really looking at is Energiya. We'd be doing in reverse order what is very obvious that the Russians are now doing. That is, starting with the unmanned Shuttle-C and then evolving to their manned Shuttle.

But this does not take the place of the things we need to do in the future. I'm sure there will be some questions along those lines. How does this play into ALS? You're going to be hearing that in a moment. I, frankly, view them as complementary. ALS has much broader objectives, over the longer haul to try to drive the cost down substantially. But at the same time we've got to keep the reliability every bit as high as for our Space Shuttle in the improvements that we build into it. To me it's not an either/or. I see an early heavy-lift launch vehicle being a Shuttle-C, which came from the building block of the Space Shuttle. We will then start to put some of the technology into the design and in place for the future at about the turn of the century, which would be the Advanced Launch System, which I believe that Jack is going to cover next. Thank you very much.

AAS 87-633

ROCKET PROPULSION - THE NATIONAL AEROSPACE PLANE

William M. Piland*

Good morning. It's a pleasure to be here this morning to talk to you about the most challenging technological program to come along in quite sometime. That's the National Aerospace Plane Program. I want to give you a little bit of history. You may recall that in the late '70s, the Supersonic Transport activity in this country declined to almost nothing, and with that went most of the hypersonics technology activity in this country. That included the use of facilities and people. There was a small part of NASA that pursued hypersonic propulsion technology. I mention that because this is a propulsion session. That activity in the late '70s, early '80s, along with some advances in computational fluids, and other reasons which I'll discuss in a minute, led to a decision in the mid-80s, that we were ready to embark on a technology program to focus on aeronautics and space disciplines, a program that could provide the technology, and technological basis for future transportation systems using hypersonic propulsion.

President Reagan announced this program in the State of the Union Address in 1986 and called it the Orient Express. Many people misunderstood what he said, as they usually do. At least, we didn't hear what we thought he should say, that this is a technological program. He said it was a program to provide an aircraft that would fly from Dullus Airport to Tokyo in two hours, and have the ability to fly to orbit. I want to stress what this program is not. The National Aerospace Plane is not an operational system. You will recall the X-series of aircraft which many of us were involved in, especially the X-15 program. This is the same kind of vehicle called the X-30. It's a flying test bed, to test or demonstrate technology. The goals of this program are to provide the technological basis for the future, not only for hypersonic aircraft but for vehicles which can fly our payloads to orbit at a significant increase in efficiency and lower cost. Therefore, the specific objective of our program is to develop this flying test bed with which to demonstrate the requisite technology for future vehicles.

I'm going to say just one thing about these derivative vehicles. First of all, in the vein of President Reagan's announcement, the Orient Express could have the capability to close distances, in terms of time, on the globe. For example, a flight at Mach 6 from London to Tokyo, could take one-fourth the time that the 747 takes to fly the same distance; that is, substantially shorten flight times on the globe, and open

---

* Program Director, Program Management Office, National Aerospace Plane, NASA Headquarters, 600 Independence Avenue, S.W., Washington, D.C. 20546.

up new markets, say, in the Pacific area. As an orbital payload delivery system, we quote the potential of an order of magnitude reduction in cost. I don't know what the reduction will finally be, but that's our goal. It's going to be difficult for any of us to achieve that goal. However, in view of our advancing technologies, and reasons I'll mention later, we are optimistic that something close to this goal is achievable. Future systems will provide routine, on-demand access to space, probably at first applied to small, priority payloads delivered to orbit, getting there in a more routine aircraft-type operation, at lower cost, and afford operation on multiple aircraft runways.

Specifically, the goals of the X-30 program are to demonstrate those technologies associated with single-stage-to-orbit-vehicles. They include air-breathing propulsion across the speed range from start of roll on the runway to orbit velocities of Mach 25, and demonstrating this capability using horizontal takeoff and landing using integral takeoff and landing gear. We don't use a sled like some other foreign concepts, and we take our landing gear to orbit. The X-30 will have a two-man crew and be fully reusable for 150 flights to demonstrate the technology associated with full reusability which reduces cost. The payload of the X-30 will include extensive instrumentation of the systems. Since this is a research vehicle, we need to understand what and how these systems are operating and try to get research-quality data from the flight test. For those technologies associated with sustained hypersonic cruise, we're going to demonstrate air-breathing propulsion to accelerate and cruise. The X-30 will have a continent-to-continent range. We will demonstrate substantial cross-range at reentry, using air-breathing propulsion for landing, not a glide-back. It will have a powered go-around capability and demonstrate the capability of the vehicle to ferry itself from airport-to-airport, rather than flying piggyback on some other system.

Just a word about our schedule in 1985, we surveyed all past research results and concluded that we were ready to focus our activities on this program. The present phase, which ends in 1990, includes design and ground test of engine and airframe components and conceptual design and, by the end of this phase, preliminary design of the X-30.

Parallel to this activity is a technology maturation or technology development program, focusing primarily on government labs with appropriate facilities and expertise. We've had people come out of retirement, reactivated old facilities and applied national assets, where they exist, focusing them on the technical issues associated with the program. In 1990, we will make a decision whether we will have advanced those technologies sufficiently to build the X-30 and conduct a flight research program. The first flight is intended to be sometime in 1994. The revolutionary technology advances required to achieve the goals of this program include unrestricted revolutionary air-breathing propulsion technology and high-strength, lightweight materials that are able to operate in very high temperature environments. Because we're not capable of simulating the entire flight regime on the ground, we are relying very heavily on advances in computational fluids and new computer machine capability to allow us to predict the performance of the systems and design those systems in the flight regimes we're concerned with. High-efficiency energy management is needed to effectively integrate controls as well as other energy systems in the aircraft for a lightweight, efficient system. And probably the key to low-cost transportation, in my opinion, is the ability to reduce the manpower associated with operation of the launch vehicle. One of the ways of doing that is to have more intelligent, automated checkout systems, on these vehicles. I won't say anymore about the subject of control systems today because NASP -- and I will use that acronym for the National Aerospace Space Plane Program -- is not spending a lot of money in this area. We're relying on advances in other programs in the industry and the government, and we will borrow from those efforts and apply those technologies to meet our goals.

I want to say that although this audience is primarily interested in propulsion, please understand, that in terms of the National Aerospace Plane concept, propulsion is the whole vehicle. The propulsion system is an integral part of the airframe. And the structures, and materials, and everything else associated with this aircraft are a function of the propulsion system. The hypersonic propulsion activities associated with this program include both analytical and experimental research and component development. Specific impulse is a measure of propulsion efficiency, the thrust of the engine divided by propellant flow rate in "seconds". The efficiency of LOX/hydrogen rockets, is on the order of four to five hundred seconds. Recall what Colonel Wormington mentioned about the sensitivity to cost associated with small changes in I_{sp}. They were on the order of 15 seconds. We're talking about major increases in specific impulse performance of these air-breather engines ramjets and scramjets with average specific impulse across the speed range of 800 to 1200 seconds.

Let me pause a moment for those who may not be familiar with ramjets and scramjets to explain what they are. They are basically, engines with very few moving parts, no turbo machinery. A ramjet was demonstrated in 1937 or 1939 by Dr. Sänger in Germany. He put a sewer pipe fueled by a mixture of kerosene on a truck, and drove as fast as he could and sustained combustion in this pipe. He later demonstrated the concept on an airplane. For the X-30 you're basically forcing air into a tube at, say, Mach 3 to 6. The air is forced into an internal configuration where a normal shock occurs, and on the other side of this shock is subsonic flow. In the combustion chamber where subsonic flow exists, it's relatively easy to mix fuel, sustain combustion, and get thrust. But as the Mach number increases, the pressures and temperatures within the ramjet increase substantially, and that implies increases in weight of the engine. Somebody had the right idea by redesigning the internal configuration of the engine. In a scramjet, or supersonic combustion ramjet, you swallow the shock and you achieve supersonic flow throughout the engine. At Mach 20, the flow through the scramjet approaches free stream conditions, and trying to mix the fuel efficiently, sustain combustion is a real technical challenge. Somebody has equated this to trying to light a paper match in a 90-mile-an-hour hurricane. And that's probably an easy challenge, as compared to what we're trying to do here.

You can get substantial improvements in specific impulse using ramjets and scramjets. The result is that payloads can be placed in orbit for very small takeoff weights and the implication is smaller vehicles. If you have a smaller vehicle, it's easier to operate and probably will be much more efficient in terms of the cost of operation. Substantial advancements in scramjets and ramjets have been made. A scale model of a ramjet was tested in the Langley Propulsion Facility at Mach 4. The purpose of the test was to demonstrate that we could predict flow fields within the module, as well as the sustained combustion. We were able to predict the flow characteristics; we sustained combustion and we achieved an average specific impulse level of 4,000 seconds. A scramjet model was tested at Mach 10 in the Calspan shock tunnel. The purpose of this test was to study mixing of fuel in the combustion chamber by injection at different angles, different pressures, as well as trying to reduce the skin friction within the scramjet. That's important to overall engine performance. They used a film-cooling technique which did reduce skin friction. We were able to sustain combustion very well in that test. There have been numerous other tests that give us encouragement that ramjets and scramjets are possible for operation at higher Mach numbers.

Returning to the test that the engine is an integral part of the airframe, the engine is no longer a bolt-on component. The entire forebody of the airframe provides the compressive surface to force the air efficiently, into the inlet. The afterbody of the aircraft is the expansion surface or nozzle. In this simplistic way of looking at the design engineer's dilemma, the efficiency of a propulsion system is a function of the aerodynamics of the flow. It's a function of the structural design of the airframe as well as the engine and the tremendous thermal load. All these disciplines must play together in an efficient design of a vehicle such as the X-30.

A word about materials: High specific strength, yield strength, divided by the density of the material at high temperature range, 1400°F or more, is required for a vehicle, like the X-30. Some alloys and super-alloys, and carbon, maintain their strength over a large temperature range. There are significant challenges in achieving materials that will work on this vehicle, since the characteristics of the above materials are inefficient.

Advances are now being made. The rapid solidification rate technology is a case in point. From a vessel full of molten alloy, the alloy is dropped onto a spinning disc. The disc throws the droplets into a very cold helium environment, and the droplets are cooled at a very high rate on the order of one million degrees per second, and the dust falls to the floor. By cooling the droplets at such a fast rate, you lock in properties that would otherwise be lost in a slower cooling rate. Through a powder metallurgical process that dust is vacuumed up and put together and we're able to achieve materials with substantial increases in strength for low density which are able to operate at very high temperatures.

Another example is the use of a plasma deposition process used to place tialuminide on a silicon carbide matrix for advanced metal matrix/composite materials that are further improving the characteristics for operation in this high temperature range.

Regarding the computational fluids that I mentioned earlier, I can't stress enough that we are very much dependent on the ability to predict the performance of the systems, and the phenomena associated with these vehicles. There are eight Cray machines now working, at least part time, almost full time in some cases, on this program. The numerical aerodynamic simulator at Ames, for example, has 50% of its capabilities devoted to the National Aerospace Plane Program. One of the things we need to do in the code development is to verify how well the code predicts performance or predicts a phenomenon, and then extrapolate with that code to some other condition we can't simulate on the ground. We use data from many wind tunnel tests to verify the prediction flows simulated by the computer.

Organizationally, the program is in place, and it has been for almost two years. The National Aerospace Plane Program is overseen by a NASP steering group, which is chaired by the Under Secretary of Defense for Acquisitions and the Associate Administrator for the Office of Aeronautics and Space Technology in NASA. It is a joint NASA-DoD program. On the DoD side there's the Air Force, the Navy, the Strategic Defense Initiative Organization, and the Defense Advanced Research Projects Agency (DARPA). DARPA has the responsibility for the present Technology Development Phase. At the Wright-Patterson Air Force Base, there's an implementation organization called the Joint Program Office composed of Navy, NASA, and the Air Force Personnel. The NASP Team is spread all the way across the country. We have roughly 3,000 people working on this program in the United States today.

Now, back to the schedule. We've just completed our Engine Concept Review and Aircraft Concept Review. We down-selected, from three contractors working on engines to two. The two who remain working on this phase are Pratt and Whitney and Rocketdyne. For the aircraft design for the rest of this phase, we have selected General Dynamics, McDonnell Douglas and Rockwell. The engine contractors, during this period, will refine their concepts, build components, test them to verify the design soundness. Then they will build a close-to-full-scale module of their concepts, which will be tested at free-stream conditions of Mach 8 in two facilities that are being put together by Aerojet and Marquardt. The aircraft airframe designs will be refined during this period, using data from the engine contractors and, at the aircraft design review, we will have preliminary designs of at least three concepts to consider for the X-30. One of the things that we've designed in this program are milestones including the assessment milestone in early 1990. At that point before we enter Phase 3, we will make a decision as to whether the technology has advanced sufficiently to give us confidence, before we build the airplane, that we are on a reasonably low-risk path to success in this program. In addition, as you have heard, earlier design margins are very important. They are measures of the performance of the various systems. We will decide if we have sufficient margins in performance to assure success. And of course, based on cost and schedule, we'll make a decision at that point whether to build the X-30. If we're not ready, we won't. We'll continue the technology development program until we are.

In summary, then, I hope that you understand why I'm enthusiastic about this program. It is a technological challenge. We have put together a national effort, which I think is oriented toward achieving success. It is a technology development program. We're not developing a prototype or an operational system. It is unconstrained by specific operational requirements and we have research flexibility. We're going to build a vehicle to demonstrate technology for future planners who will decide what the replacement shuttle will look like, for example. These technologies should lead to much lower operational cost. And lastly, we're back in the hypersonics business. I thank you very much.

AAS 87-632

ROCKET PROPULSION - PANEL DISCUSSION

Panel Moderator: Marc T. Constantine
Panelists: William M. Piland
J. R. Thompson, Jr.
John R. Wormington

Marc T. Constantine: Thank you, Bill. The plan was to use these blueprints to get you into low Earth orbit in the next couple of years in an inexpensive manner and my last speaker was going to tell what we will do after we get up there. Unfortunately, he wasn't available, so I've got you into low Earth orbit fairly inexpensively. I think, you will have to wait until next year to find out what you're going to do while you're up there. We have about a half an hour in which these gentlemen will answer questions, and discuss issues. Here's a question that our missing panelist was going to answer. To any or all of our panel members and the issue or question is:

Question: What is the current status for research and development of ion propulsion? It seems to me that this would be an excellent power source for interplanetary and deep space vehicles.

Marc T. Constantine: This is presented by John Taylor of Boeing. Any of you want to take a crack at that? You'll have to wait until next year. Sorry about that. I know Dr. Hawk could have answered that question. The Rocket Lab, or now, the Astronautics Lab, is continuing to work on this type of propulsion, as is NASA/Lewis. There are some programs in progress that are active but I can't summarize the status at this time.

Here's a question or comment that's addressed to Colonel Wormington.

Comment: You mentioned instrumentation as a method of failure detection. The ever-decreasing size, cost and power consumption of electronic processing suggest the feasibility of real time monitoring of engine functions, together with dynamic reconfiguration. Possible benefits could be incipient fault detection, in-flight correction to avoid failure, and data to put into the design process to improve subsequent versions of the system.

John R. Wormington: I agree with the comment. It turns out that the high reliability that we're after in the ALS and the demands we put on the avionics system are probably one of the more significant drivers in the requirements for Health and Status Information, and the need to be able to use that Health and Status Information in a dynamic sense, as you say, to reconfigure the vehicle to account for failures. Two things we need to concern ourselves with. One is, a high-reliability system must have some inherent -- I'll use the term, "robustness" -- defined as the ability to withstand unplanned actions, either failures or environmental impacts, and be able to cope with them and then continue the mission.

Regarding the engine-out capability, if you lose an engine, the avionics systems must be able to detect the fact that we are losing an engine, not that we have lost an engine, shut it down, reconfigure the propulsion system, maybe reconfigure the trajectory, in a way to account for that failure and allow us to press on with the mission. That kind of dynamic capability has to be in a high-reliability system in order to be able to continue the process of flying the mission.

There's another side to it with respect to failure rate and down time after failures. One of the things that we worry about after a failure is what was the real cause of failure? If any of you have been associated with a Failure Review Team, the process concerns itself with what we see, what could we see as an actual failure, and then, what is the string of events that led to the visible effect of that failure? And where did that string of events start? What was the trigger that caused that string to occur until you finally saw something? In this case, knowledge is power. When you have a failure, the more information you have, the quicker that process can become, the more likely it is that you will be able to categorize a failure immediately, or very quickly, in terms of the systemic or design problem of a random failure.

As an example, somebody stuffed a rag in a turbo pump -- we had one of those in an Atlas. We can't necessarily detect rags but we can certainly find out what happened in the turbo pump. We have to determine whether it's a random failure or the process of a failure that we can identify and inspect for, or whether it's a design failure, where we have to redesign a part, in which case, we have a management decision. If it is a random failure, we can inspect and fly. If it's a design failure, now the question is do you fly in an area where you have determined that it is still successful? Or do you stand down and fix it? But in all cases, knowledge is power. And so the instrumentation system, the health and status information, have to work to provide that information to you, in such a way as to prevent a failure or, if you do have a failure -- and launch vehicles are not immortal, there will be failures -- you have to determine what happened, why it happened, then make the management decision as to whether you should continue flying.

Marc T. Constantine: Thank you. I have a question here for Bill.

Question: What is the anticipated flight envelope for the X-30 research craft? What Mach number in the atmosphere? Will this craft actually be going to orbit? Will there be only one craft? And will there be a payload?

William M. Piland: There will be two research aircraft built. As I mentioned, the payload is a part of the instrumentation system. The weight of that instrumentation system is the payload in the research airplane. The flight envelope will be a gradual expansion. Opening up the envelope we will first taxi down the runway, then we'll take off and go around the pattern. And we'll expand that envelope to achieve orbit. The plan is to go all the way to orbit to demonstrate the technologies associated with single-stage- to-orbit systems. There are some interesting problems associated with the flight of such an aircraft. It does have to fly in the atmosphere, if you're using air. For the scramjets, of course, it flies up past Mach 1 at about 125,000 feet, so you don't get a sonic boom, you get a sonic pop. That's not a concern to us right now. One typical envelope is to take off from Edwards, fly toward Maine at a cruise Mach number of say, 5, 6 or so, slow down a little and make a relatively low G-turn, and then accelerate back across the United States, probably not going to orbit, but maybe getting close to it. We'll gradually increase that envelope up through Mach 10, 15, and 20, and when we finish that particular flight, we'll have a glide-back to Edwards and then start the engines again for power to touchdown. Eventually, we hope to extend that acceleration curve to go directly to orbit.

Marc T. Constantine: Thank you. We have a number of questions on the Shuttle-C for J.R., and J.R. will address those questions and give you to answers.

Question: You said that the most important aspect of our launch systems is reliability, yet the Shuttle-C incorporates the (SSME), Space Shuttle Main Engine. Can you comment on this and the possibility of a more simple SSME?

J.R. Thompson, Jr: First of all, let me state that the design goal on the Shuttle Main Engine was to operate it at what is called 109% of the rated thrust. It started out as being an emergency power level but its performance became more critical than that early on, so that it became the operational power level of the engine. The design goal, in terms of life, was 7-1/2 hours. We're actually operating the Shuttle Main Engine today, and flying it derated. All of our missions that are planned in the near future are at 104% of rated thrust. Basically, we're looking at getting half life -- about three hours out of the engine parts before we go into extensive refurbishment. The reliability of the Shuttle Main Engine, if you're fair, and you include all failures, at the 109% level, is really unacceptable. It is all right in an abort condition, when you're faced with an emergency and your only alternative is to ditch. So we do use it in the abort mode in very limited cases. Regarding the reliability of a three-engine cluster of a Shuttle Main Engine, operated, derated at 104%, and with very extensive refurbishment -- and there's cost associated with that -- essentially there's half life, for the three-engine cluster of about .985. We're working to get it higher, but those are the facts.

The way we plan to use this configuration in Shuttle-C, one configuration that I happen to like is two Shuttle Main Engines operated at 100% thrust. The reliability of that creeps up to a little over .99. And they would be expendable, at least early in the program. So I think it is compatible. With the Shuttle Main Engine, for the flights we've had to date, we've had no safety critical in-flight failures. As a matter of fact, we did shut one engine down early, but that was actually because of the malfunction of a safety sensor that told us we had an unhealthy engine, which was in error. We shut it down, and went ahead, achieved orbit and completed a mission successfully. So I really don't see an incompatibility in that. It is a very high performing engine. We're pushing it to the limit in the design. Now we have backed off and are currently operating it in its derated mode for our manned Shuttle flights. I think that's very workable if we implement the maintenance provisions that we have today.

For example, we have planned to reblade the turbo pumps after every flight, at least for the first two flight years, replace the bearings after every flight for the first two flight years. That gives us some time to further incorporate some improvements in those areas, and while operating the same engine. Later, the derated mode drives your reliability a little higher for Shuttle-C. What's the possibility of making a more simple SSME? We're studying that. Obviously, the balance we're trying to achieve is that nine times out of ten, when you make changes, they're in the right direction. You have to worry about the tenth time. And so we're embarked on an improvement program to try to simplify the engine. But we've got a way to go on that. We think we have a safe engine now. And we're going to continue the ground test program while we're flying, to make sure that we maintain adequate margins for safety.

Let me just follow up with another question; then I'll pass it on to one of the speakers.

Question: What specific steps will be taken to improve the reliability of the Shuttle and it derivatives?

J.R. Thompson, Jr.: I personally believe that the real key is to test, and then to test and then to test again. Earlier, Jack mentioned the reliability of the RL10. I happen to have started working on that engine at Pratt-Whitney, right after I got out of the Navy. That was in the early 1960s. So the message is, that that engine has been around some 25 years. It has accumulated probably eight to ten times more ground test experience than the Shuttle Main Engine. Now it's not as expensive to test, but there is, I think, a very good message there, and that is that for any of these propulsion systems, critical as they are, we've got to force our failures on the ground and have to really understand where our margins are. That applies not only to our Shuttle Main Engine but, I believe, to our Solid Rocket Motors as well. We at NASA, I think, got caught up in becoming operational. Was the Shuttle launched too early? We tended to back off from the supporting ground test program. And believe me, we've reinstituted that, and when we come under budget pressures, I think the trick is going to be, to make sure that we don't fall back into that trap again. But that's our present course.

Marc T. Constantine: Jack, there are several questions on the ALS. Could you respond?

Question: Do you really believe the United States will be successful in building a simple, highly reliable booster capability? Have 51-L and the Titan-down events provided sufficient incentive?

Why Important and of Interest: The United States has a long history of building in more and more complexity and ending up with a single "do-everything" unit versus multiple simple units at mass production costs.

Recommendation: Determine some way to insulate against budget/schedule pressures that drive designs to "put everything in one basket."

John R. Wormington: In essence the question is, what's changed? If it was good then, why? If it's not good now, why can't we get there? Let me talk a bit about that.

The key, I think, to what we're trying to do is making a commitment to a consistent approach toward the launch vehicles. Our history in launch vehicles, in the past, has been that we end up being performance-driven, or performance-limited. We end up asking the launch vehicle to do more than the launch vehicle was originally designed to do. Payload designers, and I have been one, end up designing for whatever the launch weight is plus 10%. And the solution to that has been, traditionally in the past, to scrape a little bit more out of the launch vehicle, to the point of doing some things that essentially increase the risk. We have an opportunity in a new vehicle to get out in front of that performance-limited, performance-driven activity. If we can get away from the idea that we're building a launch vehicle to equal the largest size payload that we want to launch -- and I don't care particularly what number you pick, 60-80,000 pounds is typical -- and build a vehicle on the basis of providing what I call a pipeline to space, instead of sizing it to meet exactly those requirements that we can identify. If we can get beyond that concept of an oversized vehicle, we then have an opportunity to put back into the system the kinds of capabilities that we need, the kinds of margins, the kind of excess design capacity, to allow us to operate efficiently and economically.

I'm not preaching revolution just in launch vehicles; I'm preaching revolution in payload operations as well. We haven't done that in the past. Typically in this industry the payload people get what they ask for.

Try that with Federal Express! Federal Express does a very good job if you agree to live within their limits. You give them a package that's wrapped so, and you don't tell them about the environmental condition that the package has to survive. If you put any extra conditions on it, they tell you to go to somebody else. If what you're really after is reliable, low-cost, robust transportation, then you have to revolutionize, not only the way we're approaching launch vehicles, but the way we approach spacecraft. The revolution comes on both sides. If the revolution occurs only on one side, then the old method of operation comes back and impacts the launch vehicle activity. Instead of having a very clean, simple launch operation, there are gantries, there are platforms, there are people all up and down the vehicle, there are interconnections between the payload and the launch vehicle that make the launch processing complex, requiring a very large team of people, and there are unique operations that go with each launch vehicle. The launch vehicle does, in fact, again take on the characteristics of being an engineering extension of the payload itself. And each launch vehicle becomes a unique launch vehicle paired with the unique payload. That is costly. If you want simple, low-cost operations, you have to standardize. And that means standardize on the payload side as well as on the launch vehicle side. That message is just now getting across to the payload people.

I'll be running a seminar in Space Division for all comers, NASA and Air Force, regarding the payload side, to preach that gospel. You'll hear the yell of anguish all the way in Houston. It is not a comfortable thing for the payload people to think about that kind of standardization. It puts requirements back on them. It forces them to add, on their side of the interface things that they want the launch vehicle to provide. But that is necessary, if you want to routinize, standardize space operations and thereby get low cost.

Marc T. Constantine: Bill can we hand over the next question to you?

Question: Can we really expect to reach orbital velocity with a fully air breathing system? Should we really expect a composite air-breather/rocket system to do the job? Will the NASP program address this if required?

William M. Piland: I thank the author of this question, because it makes a point that I didn't. As in any technological adventure, in my opinion, our reach should exceed our grasp and certainly that's the case here. Although theoretically there's no reason why an air-breathing system won't work up to Mach 25 or higher, it's obvious that the efficiencies of the system are unknown, because we can't simulate it on the ground and we're not, at this point, sure of our code. That's the reason for the X-30 -- to test scramjets at higher Mach numbers. Yes, we do expect to go to orbit, and you will find a rocket on the X-30, for a number of reasons. If we increase that envelope and get up to, say, Mach 16 and the thrust gets very close to the drag and the air-breathing system's going down, we want to know why. We've got to fly at those Mach numbers to be able to get the data. So we will have a rocket there to assure that we can reach those test Mach numbers. In addition, there's a safety factor involved with putting a rocket on. If something happens to the air-breathing system, there's a margin having a rocket on there to allow the pilot to get home safely. So yes, I believe there will be a rocket on the X-30. In a derivative operational system, I don't know. It depends on the success and how well scramjets perform on the X-30. I think we'll always have to have a rocket on there to deorbit. So we'll need to put that rocket to other good use if we possibly can. Another question.

Question: Could you briefly discuss handling qualities and control systems? Is a remotely piloted mode being considered?

William M. Piland: Handling qualities and controls in avionics is, in general, an area to which we have not paid enough attention in the NASP program. From this point on we're going to have to pay more attention, because of the very complex integration nature of the system. Not only that, which I showed in my briefing to you, but also in controls. The control of the propulsion system and the aerodynamic controls have to be integrated very efficiently. Handling qualities are of concern, for safety as well as other things. But in the next type of vehicle we're interested in safety and we're interested in efficiency of the test program. But we still need to put more emphasis on handling qualities of the vehicle for obvious reasons. We have had a number of advisory groups question whether a remotely piloted vehicle is the way to go. You have to fly the X-30 at full scale. We don't know how to scale these engines with the flow field in front of the compression surface. If we flew one, a remotely piloted vehicle would have to be something on the order of the size of the X-30. Whether to put man in there or not, the jury is out. The decision has been made to fly a two-man crew. Whether that decision changes in 1990, I'm not prepared to say. But the present plans are not to fly a remotely piloted vehicle.

Marc T. Constantine: The next question was actually directed to Jack. But I think probably requires a combined answer between Jack and J.R.

Question: Is there a role for the United States Air Force in the development of Shuttle-C? Might it not delay ALS development? Has Congress reacted favorably to Shuttle-C and ALS concepts?

J.R. Thompson, Jr: Just speaking from the NASA side, I think there is a role for the Air Force. I know Jack and others are trying to get somebody at Marshall to participate in those studies. I personally don't see competition in this. One is much more near term. The other is going to take more time. We should certainly start on the development. There's some work that needs to be up front on the ALS. So from that standpoint, I don't see them being competitive, at least not over the long haul ....

John R. Wormington: We believe there's a role for the Air Force in Shuttle-C, and we've already started exercising it. We had people there for the selection. And we'll have people there for the program when it does start. So the answer is yes. Congress has been reacting favorably to the ALS and Shuttle-C. I think the real point there is not Congress. I think the community, Congress, you all, the various agencies of the government, industry, the concerned citizenry that are interested in the space program have all come to the realization, or are coming to the realization, that what this nation lacks is the real, honest to god, bulk carrier, a truck, a vehicle to get large packages, mass, to orbit. I think that we collectively are working on the question of, "What is a the way to do that for the Nation?" The vehicles that we have right now, aren't oriented toward that. The Shuttle is a very fine manned vehicle and it ought to be devoted to those things that need man's activities involved in it. The Titan is a good vehicle. But it's small. It is performance-limited now. We used to talk about the Titan as a very large launch vehicle. No more. We need a large launch vehicle, something that can support what we as a nation want to do for the next 20 or 30 years. That launch vehicle, in its various forms, call it Shuttle-C, which is an early version available with existing hardware, or the ALS which may or may not use existing hardware depending on the cost ramifications associated with it, but certainly will be available some time in the mid to late '90s and then on. The need, I think, is the common thread through all this. It is the need to get large bulk cargo to space efficiently and cheaply. If I had to characterize Congress now, I'd say that, "I think that the need is recognized or is being recognized." And they're looking to us to say, "What is the proper solution?"

Marc T. Constantine: I think we will address one more question. J.R. would you take the next question:

Question: Is Shuttle-C really proposed to be a Shuttle or expendable launch vehicle. That is, will it be returned for reuse? Is it meant to be part of the ELV fleet or will it be the ELV fleet?

Why Important and of Interest: The role of this vehicle in NASA's budgets for ELVs in 1990-1995 time frame is of major concern.

J.R. Thompson, Jr.: It's going to be expendable. The solid rocket motors will be recoverable, like they are today. The external tank will be expendable like it is today. Certainly the cargo carrier itself will expendable. The Shuttle main engine, at least the scheme that I would propose early on, is to fly used SSMEs, good ones that have perhaps flown eight or ten times, then expend them. I think that is a good way, with a low flight rate of a Shuttle-C, to help amortize the cost. I think it's just proposed to be part of our ELV fleet. Both the DOD and NASA are looking again at ELVs and I think they've got a large role to play. The cost of the whole system in NASA's budget, obviously, is something undergoing a lot of discussion now between us, the Department of Defense, and the Congress, and certainly the Administration. Just one passing comment: The reason I'm really bullish on Shuttle-C, is that it's an opportunity to build on what we've already got. I think we, in aerospace, and certainly in NASA, have not done that to the extent I think we should have. Cost is very important and we need to continue to work to get it down. Today's Shuttle Main Engine, a brand new one, is around $40, 44, 45 million. I think with some work that we have underway we can drive that cost per unit down to somewhere in the low to mid $30 millions. That's still expensive. Transportation costs are high. We need to work to get them down. But the cost, depending on how you do the accounting, is somewhere between $3000 and $6000 a pound payload to use the Shuttle. We've got to work on a transportation system. But the payloads, as Jack mentioned earlier, are the other end of it. Those costs per pound are $50,000, and climbing. And really, the transportation costs are only about 10% of the total. We need to work on both ends.

Marc T. Constantine: Thank you. I'm sure we have other questions that the panelists would be happy to answer afterwards. I want to thank the panelists. I think you have received some good comments from them. I want to thank you, the audience, for participating. In appreciation for their participation the AAS has asked me present these plaques to J.R., Colonel Wormington and Bill Piland. And with that I close this session. Thank you very much.

ASTRONOMY, ASTROPHYSICS AND SOLAR SYSTEM EXPLORATION

PANEL FIVE

Chairman:	George Newton Chief, Advanced Programs and Technology, Office of Space Science and Applications NASA Headquarters
Host Chairman:	William Phinney Chief, Experimental Planetology Branch Astrophysics Division NASA Johnson Space Center

AAS 87-637

ASTRONOMY, ASTROPHYSICS & SOLAR SYSTEM EXPLORATION - INTRODUCTION

George Newton* and William Phinney†

Dr. William Phinney: I am the host chairman for this panel, Bill Phinney from the Johnson Space Center. It's my task to introduce now the chairman of this session, Dr. George Newton. He's an old NASA hand who joined NASA Goddard in 1961, where he has conducted experiments on atmospheric density and composition. Among his accomplishments was developing state-of-the-art ultra-high vacuum measurements and calibration standards. Then in 1978 he moved to NASA Headquarters, where he served as Manager for Advanced Programs in the Solar Terrestrial Division. In 1981 he was assigned to the Astrophysics Division in the Office of Space, Science and Applications, and is Chief of the Advanced Programs and Technology Branch, where he is responsible for the management of all advanced mission studies for the astrophysics program. He got his Ph.D. in physics at Catholic University after a master's degree at the University of Kansas. Dr. George Newton.

Dr. George Newton: Thank you, Bill. I'd like to welcome you to Panel 5. I think that you may find the content of this panel somewhat different from that of the previous panels, in the sense that much of what you hear today, in this panel, will be the hopes and aspirations and planned activities of the space scientific community of the nation.

We have one change in the panel membership. Dr. Michael Duke was unexpectedly called to Washington for a meeting, and Dr. Douglas Blanchard has kindly stepped in on short notice and will deliver the talk on Planetary Exploration in the 1990's.

I will introduce our two other speakers, and then after that they will speak sequentially, and then I will speak. The question and answer session at the end will be similar in the format to previous panels.

Dr. Douglas Blanchard is Chief of the Planetary Materials Branch and Lunar Sample Curator at the Johnson Space Center. He obviously is very familiar to many of you in the audience, having spent 14 years at the NASA Johnson Space Center. He has a Ph.D. in chemistry from the University of Wisconsin in Madison.

* Dr. Newton is Chief, Advanced Programs and Technology, NASA Headquarters, Office of Space Science and Applications/EB, 600 Independence Avenue, S.W., Washington, D.C. 20546.

† Dr. Phinney is Chief, Experimental Planetology Branch, Astrophysics Division, NASA Johnson Space Center, Houston, Texas 77058.

Another person who needs no introduction to you, I'm sure, is John Aaron who has, since his graduation in 1964, spent his career at Johnson and currently at NASA Headquarters. He has held increasingly responsible positions at the Johnson Center involving many of the key programs there including Gemini and Apollo. He served as Chief of the Spacecraft Software Division, which had responsibility for the software for the Space Shuttle Orbiter, and served as Acting Manager for the Space Station Program Office here at Johnson. In August of 1987 he was appointed Acting Assistant Administrator for the Office of Exploration at NASA Headquarters. In that position he is responsible for coordinating agency activities that expand the human presence in space beyond the Earth, particularly in missions to the Moon and Mars. The order of our speakers today will be Dr. Blanchard, who will speak on Planetary Exploration in the 1990's, followed by John Aaron who will speak about Future Exploration Activities. Following Mr. Aaron, I will talk about the Great Observatories in Space. Doug.

AAS 87-639

PLANETARY EXPLORATION IN THE 1990'S

Douglas Blanchard*

Thank you. I'd like to talk today about the planetary program. It's sometimes hard to know the players without the scorecard. There are lots of names of missions, and it's hard to remember who's going where. So I'll try to put that in some order. I also want to convey the underlying philosophy that guides the planning and the execution of the planetary program. Primarily, I'm talking about unmanned, unpiloted planetary science missions.

There's a general philosophy about the planetary exploration program in the U.S., and that is that it goes in stages. It goes from a reconnaissance to an exploration stage, to a detailed study stage. And this is done with some sort of balance across all the planetary opportunities. Reconnaissance or flybys and hard lander explorations are in the realm of orbiters and perhaps simple soft landers. Detailed studies get more complex. Viking was a detailed study. Sample returns have had detailed study.

Reviewing the history of all the missions that have been flown to the nearest bodies to us, Moon, Venus and Mars, clearly the lunar days were days of very intense activities. We did the Moon in a hurry. The Russians also were very intent upon exploration of the Moon. They spread their program out a little bit and ended up bringing back three samples in an automated sense. After we were finished with lunar exploration, the philosophy changed somewhat into our current philosophy of trying to obtain information about all the planets before we commit to a detailed study of any of them. You can see the effect of that on the U.S. program which explored Venus with a certain amount of space between the missions. Mars exploration hit a temporary peak -- an impressive one was Viking.

On the other hand, the Soviet Union seems to take on themes. In the mid-1970s they were intent on getting to Mars, but met with a lot of failure. And then their attention focused on Venus after a fairly substantial exploration start. Now, by all that we know from talking to them, their focus has gone to Mars, and their intent upon exploring Mars. What we have now lined up in the future is a Phoebus mission, an unmanned lander, and finally, a sample return mission.

We are intent on Mars too. So we are roughly on a parallel course there. There are are two mission concepts about the plan that evolved in the early 80s for planetary science, and that is, it was hoped that we could maintain a certain level of activity in what were recalled the Core Missions. And once we had that level of ac-

* Planetary Material Branch, NASA Johnson Space Center, Houston, Texas 77058.

tivity in planetary science established and on-going, then we would think about Augmented Missions. The Core Missions were intended to be low cost. In order to make them low cost we accepted a higher degree of risk. Also we limited the objectives. The idea was that if you had regular access to planetary exploration, you could do one or a few objectives at a time. You would not have to put all things on every spacecraft. There was a strategy developed for the inner planets and one for the outer planets and missions were lined up according to science priorities. The augmented missions are high-visibility theme missions, such as Mars sample return, comet nucleus sample return. These are very challenging missions, but they pay for the challenge, in terms of the very high scientific return. These are not regular missions, but would be new initiatives and major programs.

The first set of missions, Pioneer and Voyager, are missions that are already flying, and we're familiar with them from their results. The next set of missions are those that are waiting. They would be flying except for the *Challenger* accident. They are now awaiting the next opportunity and they are on the Shuttle manifest. The third category of missions are those that are in various degrees of planning. The Mars Observer is a pretty solid craft. We're working on a new start. I'll go through each of these in a little detail to explain the objectives for those missions, and their status. So far we've gained acceptance to the idea of the Core Mission concept. But it has not been committed to completely, in a sense that we've been told to go ahead and manage our own missions. Every mission is still a new start. The Mars Observer is approved and the Core Mission concept will also take on some credibility when the next new start in the Core Mission is approved. The Mariner Mack II is a spacecraft that is intended to do the Core Missions to the outer planets. Our first new start in that program is the CRAF Mission comet asteroid flyby; it is the highest priority for a new start for the planetary program. We're hopeful that that will get underway very soon.

The Augmented Missions are in the early study phases. They're very complex missions and need a lot of thinking through to make sure we do them the right way. They offer singular opportunities, so it's well worth the additional planning. The accident caused us to put three missions that were ready to go on the sidelines. In addition, the vehicles that were going to take them on the inner planetary trajectories, Shuttle Centaur, were canceled. So that was a serious blow. In the meantime, other strategies have evolved, other upper stages, different trajectories. Making use of gravity assists in the trajectories extends the lifetime of the missions, and lowers the energy that you need.

On the upside, the planetary missions are again on the manifest. And there seems to be a commitment in NASA of renewed emphasis on the science goals of NASA and keeping those up front. What's happened with the blueprint that was put together at the beginning of the 80's is that it has moved more slowly than we had hoped. But this has increased the need of planetary scientists' points of view to load up the missions. The opportunities are not coming as regularly as the plan has projected. Also it has raised the stake in terms of risk and reliability of the spacecraft. So what has happened is we're seeing an evolution of the risk philosophy in the planetary program. A single launch, single-spacecraft kind of scenario is good if you have frequent opportunities and if you're dealing with relatively low-risk missions. If you are in a high-risk mission situation, then there's justification to going to a dual launch in a two-spacecraft philosophy. Somewhere in between is a single launch with a backup spacecraft. We also take note of the mixed launch fleet. As those opportunities come available, I'm sure that the planetary program will diversify.

Following are the major missions planned:

Magellan is the first of the inner planet missions, it's waiting. It's a radar mapping mission for Venus. It's going to map the globe of Venus at about one kilometer resolution. That's about a factor of 10 better than the resolution obtained from Venera 15 and 16 mission. The interesting thing about Venus is that the Russian images have given tantalizing evidence that, in fact, there may be plate tectonics active on Venus. This is quite a surprise. We know that they're not active on the Moon. We are pretty sure that they're not active on Mars. But the radar images give every indication that there are structures that look a lot like ocean ridges, spreading ridges, and that's very exciting. So the factor-of-10 increase in resolution is an excellent science mission. The 15-month trajectory is again probably a result of the somewhat lower energy stage.

Mars Observer is the first of a class of spacecraft called the Observer Class. The Observer Class is intended to service the planetary needs for the inner planet missions. Its job is to map Mars with information about the composition, mineralogy, and atmosphere, to understand the atmosphere dynamics and the effects of the dust on the climate of the planet, look for the volatiles and try and solve the volatile problem which is a high-priority issue for Mars. This is scheduled for launch in September of 1992. Again, the backup to this, and the new philosophy that is gaining some acceptance, is that we will buy a second spacecraft to back up Mars Observer. This spacecraft will be configured to be a Mars Observer, should we fail on the first Mars Observer attempt.

If we succeed in the Mars Observer attempt, the spacecraft then becomes a Lunar Geochemical Orbiter and the second exploration mission. The Lunar Geochemical Orbiter has been an apple in planetary scientists' eyes for a very long time. It turns out that by virtue of the fact that we have Viking Orbiter imagery of Mars and much more limited imagery of the Moon, that we probably know more about the globe of Mars than we do about the globe of the Moon. An interesting time factor here is that we obtain digital imagery of the Moon. So we think there's a real need to get to the Moon, and do a very thorough job of mapping the Moon by looking again at the composition and the mineralogy, and also the resources that are available. This has an eye toward making John Aaron's job a little easier. This is a second in the Observer Class and presuming that the Mars Observer is a success on the first try, then we're looking at a mid-1990 launch for Lunar Observer.

Now let us look at the outer planet missions. Pioneers are missions that apply. Some of them have done most of their work already, but Pioneer 6, 7, and 8 remain operational. Pioneer 7 was pressed into service to look at Halley's Comet and got a fairly close look, and I recall that the results of that pass were quite interesting and perhaps controversial. Pioneer 10 and 11 are exiting the Solar System in opposite directions. Part of their role was to find the edge of the Heliosphere; one is going right and one is going left. One of them is out already and the other one is nearly out of the Solar System. Pioneer, the Venus Orbiter, in Phase I of its mission took radar images, low-resolution radar images of Venus and did other observations of Venus. It's now in Phase II where it's basically transitioning its orbit and taking particle and field measurements. Phase III begins and ends in 1992 when the orbit is phased into an orientation that we can get imagery of parts of Venus that we've never had before. Its demise will be a collision with the atmosphere of Venus.

The Voyagers have been extraordinary missions, by any measure. Their intent was to reconnoiter the outer planets. We have enjoyed their imaging, but in addition to that they've carried quite a complex contention of space physics instrumentation. Voyager I has completed it's encounter with Jupiter and Saturn, and it's exiting the Solar System. Voyager II encountered Uranus) in January of 1986 and it has one last stop in the Solar System and that's Neptune, at the very end of this decade. It appears to be flying in very good health, so the anticipation is keen that we will get excellent data.

Galileo is one of the Core Missions for the outer planets, one that is waiting to be launched. Its task is to orbit Jupiter and study the rings and the satellites of Jupiter. It will probe the Jovian atmosphere. It, too, carries a number of space physics experiments. Its launch is set for November of 1989. It does a VEEGA, Venus-Earth Earth Gravity Assist. It takes a while to do that, but it allows you to steal just a little bit of momentum from the planets and accomplish the mission.

There's an interesting backup philosophy with this mission, Cassini, of which one of its configurations is a Saturn Probe Spacecraft, Saturn Orbiter and Saturn Probe. Should Galileo have a problem, its backup is Cassini, which will follow.

Ulysses is the International Solar Polar Mission -- a joint European-U.S. mission. It's intent is to look at the Sun, the Heliosphere outside the ecliptic. It is scheduled now for launch in October of 1990. It will fly to Jupiter, go over the top of Jupiter and, with the gravity assist, be swung into a plane that's more or less normal to the ecliptic and we'll be able to look at the Sun at a solar latitude from which we've never been able to look before.

One of the high-priority science needs, one of the real burning interests in the science community is in primitive bodies. CRAF is a mission that is intended to get us close to some primitive bodies. CRAF stands for Current Rendezvous Asteroid Flyby; another name for it these days is Mariner Temple II, until we change the target perhaps, and then it will be a Mariner. Its objective is to fly by a main-belt asteroid. Incidentally, all of the outer planet missions that are planned, will now fly by an asteroid on their way out. We'll try to piggy-back that science and all the missions that we can. It will rendezvous with a short-period comet, short period being in the six-year range, and it will land a penetrator on the comet nucleus. We've never seen a comet nucleus. The closest we have come is the Giotto mission imaging showing a rapidly degasing ball, and you've all seen the pictures of that. This may be our chance to obtain a clear view of a comet nucleus. For the Mariner Mark II missions we're planning to use Titan-Centaur launch and use gravity assist, Venus to Earth Gravity Assist, and we hope to have this launched in 1993.

Cassini is a mission of high interest to the Europeans and we're interested also. It's a Saturn flyby Titan probe. Titan is one of the more interesting satellites in the Solar System. It's in many ways amazingly analogous to Earth except that it has a methane atmosphere and it operates at an entirely different temperature range. This is an interesting mission both to understand Saturn and to probe the mysteries of Titan. What we may end up with is a European probe and U.S. spacecraft to do the mission. This is a relatively important high-risk mission. We'll probably do a dual launch with dual spacecraft. But in this case, we'll probably have two different spacecraft. One will be a Titan probe and the other will be a Saturn probe. You'll recall the discussion of reliability on Galileo; it's that Saturn probe that is the backup for Galileo, so that if we have a failure in Galileo, we can retarget the Cassini spacecraft which has the Saturn probe and accomplish much of the mission that Galileo would have accomplished.

Now we'll talk a little about the Augmented Missions. The Augmented Missions are highly ambitious missions, but, at the same time, they're extraordinarily interesting.

Mars Sample Return Mission. The plan here is to bring samples back from Mars that have been carefully selected. It turns out that it takes you a year to get there, and a year to get back, but that trajectory allows you almost a year, or in other opportunities, more than a year, on the surface -- a lot of time to do very careful exploration on the surface. We want to bring back samples, so that we can analyze them intensively here on Earth. Also we want to put a Rover on the surface of Mars that will help us select those samples. But after the samples have been selected and sent back, that Rover will then become an exploration tool in itself, and it goes on what we call an Extended Mission.

We want to understand what is happened to Mars. We have received some tantalizing evidence from Viking. We think we have some meteorites that have relationships to Mars. We want to understand what has happened to its climate. One of the major points of interest on Mars is its volatiles. Where have they gone? What's happened to its atmosphere? Was there ever a time in Mars history when perhaps there was potential for life. It looks unlikely now that there are viable life forms on Mars. But there's enough evidence for water in a milder climate. That's one of the high priorities that we're going to study with the sample return and with the on-surface science. There are a perplexing number of variations for this mission, and they're being studied and weighted one against the other and traded one against the other. We're hoping that we can pull this mission into a form that can qualify for a new start in the 1992-1993 timeframe. That would get us to a launch in 1998 and the sample would come back in 2001.

The last mission that we want to talk about is the Comet Nucleus Sample Return. This is also an ambitious mission. The idea is to go a comet and visit the nucleus, but bring some of that nucleus back. A nucleus is estimated to be 100-150 degrees Kelvin. We want to bring it back more or less intact. So it's a challenging mission from that respect. This is a mission that is of high interest in Europe. ESA has called it one of its cornerstone missions. Chances are that we will investigate thoroughly the possibility for a joint mission on this opportunity. This is a mission that doesn't launch for a while. We're projecting something like the year 2000 for launch. It's a long mission as well. It will chase a comet that probably has a period in the realm of five to six years, so to make the round trip, we're talking about six years. This really does end up moving us out of the planetary exploration in the 1990s. But, nevertheless, certainly all the preparation will be done during that time.

In conclusion, these are the best of times and the worse of times, it depends upon your time scale. We've been in the planetary exploration business for 25 years. It has been a phenomenal 25 years. The last ten have been less phenomenal than the first 15. So you have to keep things in some sort of perspective, and I would like to end this in something of an upbeat note. I think we have the planetary exploration strategy in an orderly plan. The *Challenger* accident came at a critical time for the planetary exploration program. But the exploration program is again rescheduled and put back on track. There are tremendously exciting science missions, both in the Core and the Augmented Program. We're eager to go.

Dr. William Phinney: Thank you, Douglas. I would ask that if you have questions at the end of each talk, you forward them over to the collector and that they be assembled so that we can utilize them for our question-and-answer time at the end of all the talks. The next speaker will be John Aaron.

AAS 87-640

FUTURE EXPLORATION ACTIVITIES - ASTRONOMY, ASTROPHYSICS & SOLAR SYSTEM EXPLORATION

John Aaron*

Good morning and thanks to my host and to you all and for the opportunity to speak to you today. I'd like to spend a few minutes talking about the plans and activities associated with the Office of Exploration and, in a particular, the planning associated with the Office of Exploration in context with the Agency's total planning effort. In the Agency there is planning for the near term and the long term; it covers the range all the way from the very near term aspects of returning the STS safely to flight to trying to accomplish and encompass the very visionary goals of the recently released report of the National Commission on Space.

Within the near term there are significant activities within the Agency involving the planning and implementation of various new starts, some of which Doug has just mentioned, as well as the resumption of our launch manifest and the release of related information as well as the aggressive development of the Space Station. In setting up the Office of Exploration, we have in context the fact that there's planning at all levels within the NASA, covering the total range and time span. This planning is aimed at both returning and maintaining an overall healthy approach and posture for the various themes of science and exploration that NASA has in place today. If you look at our themes of science and exploration you will note that we have great programs which, in addition to astronauts, expand the programs associated with great observatories, Earth sciences and observations, the exploration of the outer Solar System, as well as the focus on the exploration of the inner planets that Doug mentioned, and the scientific utilization of the Space Station Freedom.

To return and keep those themes of science and exploration healthy within the Agency, we're also doing major planning in the near term to provide and assure that we have the appropriate infrastructure, in terms of a viable transportation system, and that we return to healthy technology and scientific research programs as prerequisites.

In addition to planning for the near term, the Agency has stepped up its thrust to understand potential bold new initiatives or major new goals so that NASA can focus on, both for the long range as well as for use to steer and guide our near-term decisions and capabilities. That's what's involved in the process of setting up the Of-

* Assistant Administrator for Office of Exploration, NASA Headquarters, 600 Independence Avenue, S.W., Washington, D.C. 20546.

fice of Exploration. Our particular objective is both to plan the possibilities as well as to understand the engineering feasibilities of setting up major new initiatives of long range that would involve extending the human presence in the exploration of the Solar System. These we need to do from a standpoint of both focusing on our near-term program but also understanding the feasibility -- given that the Nation, in the interest of maintaining permanent leadership in space exploration, provides additional resources and has the national will -- of proceeding to accomplish the goals of exploration.

Today I'll discuss in some detail, the activities relative to planning the initiatives as well as doing the research of the feasibility of those initiatives of primary interest to the Office of Exploration. The planning for these major kinds of initiatives for the overall aspects of extending human exploration has been going on for two or three decades within this Agency and around the nation, particularly the extended exploration of Mars, as well as setting up a permanent base on the Moon. The most recent studies associated with that, I think, were envisioned in the report of the National Commission on Space (NCOS). They covered a very wide breadth of planning which, as it expands in the next 50 years, offers an aggressive projection involving an incremental approach encompassing all the programs from a near-Earth concentration to expanding into the Solar System and to establishing an outpost on Mars.

Now with that as a backdrop the Agency, by way of Dr. Sally Ride, set up a near-term planning activity that would concentrate on the next 15 years, with particular emphasis on understanding the aspects of space exploration which still stands for U.S. pride and leadership as well as concentrating on some of the near-term bridging capabilities required. The results of this activity were documented in her report "Leadership and America's Future in Space."

By way of background, I'll go over briefly the content of the report. That set of syntheses of Dr. Ride was established with an *ad hoc* team involving the entire Agency. They actually picked four areas to evaluate: (1) A concentrated Mission to Planet Earth to understand global change, to understand Earth as a dynamic system as well as collect data and knowledge that would support effective decision making for the future. (2) Robotic exploration of Solar System with the objective to maintain U.S. leadership in the planetary exploration that we have enjoyed for the last 25 years. (3) A human expansion initiative with a serious look at establishing a lunar base, both as a significant science research as well as looking at ways to utilize the lunar resources and the space infrastructure to support a lunar base. I think it is very important to the overall human exploration of space to establish an understanding in the long term of operations research experience of living and working on a body remote from Earth, so very necessary for the operations experience that we need. (4) The most challenging of all was the initiative involving human exploration of Mars, establishing human presence on another planet and increasing the science and understanding of the Solar System.

With the four initiatives that were selected to evaluate, I must point out a confusion factor that I ran across in my interchange with the space community. These four candidates that were selected were not meant to be all inclusive of all the initiatives that the Agency could and should focus on. However, they were four initiatives that were representative of a set that could be evaluated against a certain particular set of criteria and merit. One aspect of that was determining to what degree they would help us maintain the U.S. leadership in space exploration, as well as to what degree they were technically challenging in general and technically challenging to NASA. Another aspect was to understand to what degree the benefits of science and knowledge, the technology, the spin-off, could keep our overall support base wide, in terms of maintaining our competitiveness in the world market place.

Another dimension that we need to understand is the bridging of capabilities and investments we need in research and technology within the Agency and the country that would make these initiatives happen. That's what I refer to as the appropriative infrastructure of capacity and prerequisites.

The conclusions of the Ride report, I think, are worthy of mention. First in the leadership area, you have to look at it in depth. Of course, I think it's obvious that our leadership in the exploration of space cannot simply be any more proclaimed. It must be earned and be earned on a continuous basis and demonstrated. I think important to the Agency is that leadership is also a process. It involves not only selecting the key elements of leadership but very clearly and distinctly enunciating those priorities and then building, maintaining, and defending the resources of capabilities as well as the associated technology. As to the technical and programmatic conclusions, I think what we always find, when we look at these kinds of bold new initiatives, is that we must put together a program that begins with today's capabilities. With these significant bold new initiatives there are major investments involved in long leads to attain our prerequisites for missions with our technologies. And until we get on with those kinds of investments and our prerequisites and our capabilities within the infrastructure of exploring space, they will always be ten or fifteen years in front of us.

One particular thing that was reaffirmed as part of this study is that Earth-to-orbit transportation remains as one of our most pressing problems. These major initiatives, particularly the lunar base and the human exploration of Mars, involve delivering, on a routine basis, millions of pounds to low Earth orbit each year, certainly indicative of a robust Earth-to-orbit transportation system.

Also reaffirmed was the need to rebuild our technology base, in particular investing in technologies that are required as long leads to enabling these next levels of exploration missions. By looking at all the common ingredients of the technology investments that are required across the total range of initiatives, it has been formed into a program called Pathfinder and has been submitted as part of an 1989 new start, a very critical program to get started on if we're ever going to accomplish these kinds of initiatives.

The third item mentioned in the report is a pacing item in that what science research must be pursued in addition to understanding the human adaptation problem? That involves the whole question about whether we need to develop a capability, from an engineering standpoint, of providing artificial gravity. The impact on the Space Station on these kinds of initiatives, particularly the human exploration initiatives, needs to be understood. To the level we understand them now, they have a significant driver and characterization on Phase II of the Space Station. These vehicles that are required to be utilized for space transport must be refueled in space and have to be a major transportation node to provide the turn-around and maintenance for the space base.

Another conclusion of this report, which will require further detailed planning, is the need for additional science and engineering data precursor missions and as prerequisite to extending the exploration of humans either to Mars or to a lunar base. We probably need those in the form of additional Mars Observers for better mapping and remote sensing of Mars as well as the surface exploration to understand the characteristics of Mars. Also needed is a return of samples to understand the real biological and physical characteristics of the surface conditions. For the Moon we will need additional lunar orbiters to understand where additional resources of interest may be, as well as to possibly return samples from the Moon to seek out the potential resources utilization capability of the Moon.

In terms of what is happening next with this report, I think it is significant that the results have been presented to the NASA management. The Agency has now made a commitment to continue with more detailed planning and study of the initiatives and to make the commitment to formalize the human exploration initiatives and form line organizations spearheaded by the Office of Exploration. That will transition these studies from an *ad hoc* study approach to an actual firm planning association with a line organization working toward a formal organization. The Office of Exploration will act as the central catalyst to leading those activities, but use the major line organizations of NASA because these major initiatives have implications in a significant way on the near-term NASA programs. It also gives us provisions for additional resources and sets up the management mechanisms to be the forerunners of these initiatives. I think it does something else within the Agency itself for all of our near-term and evolving programs. It provides a mechanism that the line organization and program managers can use as short-term steering on their day-to-day decisions about the near-term program associated with Space Station, transportation, or technology or science and our operations systems capability development.

The Ride Report and various other reports give us an excellent data base for our objectives in the Office of Exploration and are a point of departure for the next phase of our work. We're well into the implementation of the forward plan. The office is established. We have the Headquarters Program Offices supporting our activities. We are defining the next level of the work and we have made the assignments to the research centers within NASA. In FY88 we will refine initiative techniques and continue reviewing existing studies to evolve a better plan for the prerequisites of science missions, precursor technology, and life sciences research that we need to put in place. In FY89 and 90 we hope to be able, in addition to continuing aggressive work on planning, to actually make sure we get the Pathfinder Program under way, including life sciences research and establishing the evolutionary requirements for the Space Station. Our overall objective during the next three years is to plan and support the decision-making process so the Agency will be in a position to make a mature decision on a bold new initiative by 1992.

In summary, the Office of Exploration and the related activities associated with major new initiative studies is well under way. Those within NASA who are working on these plans are very enthused and excited. I think the interaction and cooperation across the NASA organizations is at an all time high. We have planned and submitted our FY89 budget request, which was enthusiastically supported by the Agency and we're willing to defend it in a very aggressive posture with the OMB and the Administration. We're on a concentrated path of defining our long-range initiatives, as well as placing particular emphasis on our near-term program. This is a very significant effort which the Agency is seriously undertaking and is the key to a viable long-range civil space program.

AAS 87-641

GREAT OBSERVATORIES IN SPACE

George Newton*

Today, the astronomical world is poised at the start of a revolution. Ground-based and space-based equipment can provide unprecedented power for observing the Universe. These instruments will increase by orders of magnitude the volume of space that is observable. In fact, the power now available for observation can extend back to the opaque curtains created shortly after the big bang origin of the Universe. These instruments can also provide spectro-coverage across much of the electromagnetic spectrum and provide contemporaneous observations at many wavelengths. Furthermore, the confluence of ideas and concepts from high-energy physics on the fundamental structure and nature of matter, when combined with astronomical observations, will provide fundamentally a new understanding about the origin of the Universe. Indeed, perhaps the only way that some of these physical theories can be tested is through astronomical observations, simply because the densities of matter and the temperatures are so great that the only place known where they may exist, or conceived where they may exist, was in the creation of the Universe itself. The next ten to 15 years promise to determine profoundly how we view and understand our Universe and how we view our world, and perhaps how we view our place within it.

This morning, I would like to describe to you the NASA Great Observatories role in this revolution of observation and understanding, and show a video tape which describes the next observatory after the Hubble Space Telescope; that is, the Advanced X-ray Astrophysics Facility which has high priority in Fiscal Year 1989, a new-start candidate. I will conclude by describing and addressing the role that we would like to see the Space Station play in implementing the Great Observatories Program. The Great Observatories Program is designed to address very fundamental questions, such as, our concern and interest in how the Universe began. We're interested in knowing whether the laws of physics, as we understand them, are applicable throughout the history of the Universe. Current understanding suggests that they are. We're interested in how matter aggregated into the large clusters of galaxies that exist, for each galaxy contains billions of stars in itself, and each cluster can contain hundreds of galaxies. Why did Nature choose this particular structure? What does it tell us about how Nature operates? We're interested in the life cycle of stars. How are they born? And in their lifetime, how are planets formed and how does life itself begin on the planets? And of course, the stars die, some going into Supernovas, some going into White Dwarfs, and some perhaps, going into Black Holes. We would like

* Dr. Newton is Chief, Advanced Programs and Technology, Astrophysics Division, Office of Space Science and Applications, NASA Headquarters, Code EB, 600 Independence Avenue, S.W., Washington, D.C. 20546.

to understand these types of questions about why Nature operates the way it does. The pace of discovery in Astronomy is nearly exponential....

The Great Observatories consist of the Gamma Ray Observatory, which is currently in development and scheduled for launch in 1989, the Hubble Space Telescope, which is also in development and scheduled for launch in 1988, the Advanced X-Ray Astrophysics Facility seeking development start and following the AXAF, the Space Infrared Telescope Facility, which is a telescope cryogenically cooled to liquid helium temperatures. Then there is the very large baseline array, which deserves status; it is the National Science Foundation funded array of telescopes, which are currently in development for radio astronomy.

Matter in the Universe can be characterized by temperature which determines where the photonic radiation from that matter will occur in the spectrum. When matter is very, very hot or highly energetic, it occurs in the gamma ray region; typical activities there are nuclear processes. Still very hot matter can emit X-rays and these typically go down to wavelengths of a few angstroms. The Hubble Space Telescope observes in the visible region where the eye is sensitive in near infrared and the ultraviolet. The Space Infrared Telescope Facility observes in the infrared region of the spectrum and the far infrared out close to about a thousand microns, 700 microns or so. Then, of course, there is the meter class of wavelength ranges covered by radio astronomy.

There are some important points in the concept for the Great Observatories Program. One is that, because we're interested in understanding the physics that's operative, we need to look across the electromagnetic spectrum, because different physical phenomena and different physics, operative in many of the objects like a quasar, cover a broad spectrum of physics which then is characterized by different temperatures. To obtain the total picture of these objects, one needs to look across this electromagnetic spectrum. The other powerful concept in the Great Observatories Program is the contemporaneous observations that they provide, not necessarily simultaneous observations, although they can be, in part, and for rapidly changing physics with time in order to follow that. But the contemporaneous nature is a revolution in the culture of astronomy. Earlier, a discovery was made, was understood, and then followed up tens of years later or perhaps by the next generation of astronomers. When we have this array in space, we can turn those discoveries around, disseminate them across the community and bring to bear the other diagnostic tools necessary to understand those phenomena in a matter of days, months or years, not generations....

I'd like to close by addressing the relationship with the Space Station. The long life of the observatories we're talking about -- ten to 15 years -- is accomplished primarily through on-orbit servicing of these facilities. We feel that the servicing is a proven capability, and one that we're planning to exploit to help us achieve the objectives discussed. We feel that it's been demonstrated as a viable concept through the Skylab, Solar Maximum, Westar and Paulapa missions and through the Skylab Apollo Telescope Mount activities. We have been able to fix instruments and exchange modules. In the future, what we would like to do is use these for maintenance and upgrading of the Great Observatories so that the current plans of science can be accomplished. We also see this as an opportunity in our Explorer Program where we have serviceable spacecraft currently in development, and in the future, for Spartan class opportunities for small payloads to be sent out from the Space Station and for the scientific instrumentation to be replaced at the Station.

All of this is pointed toward the next generation, beyond the Great Observatories, also by establishing the capabilities and the technology and developing the experience needed for us to move on to the next stage, illustrated by such concepts as interferometers in space in both the optical and radio region of the spectrum, larger aperture devices which cannot be made on the ground and launched into orbit, but are probably going to be assembled on orbit. Included would be large deployable reflectors.

Our concept is to develop an evolutionary path, which would start by using the STS and our Great Observatory concept for Hubble and Gamma Ray Observatory and/or AXAF as a backup, and to use the STS servicing capability. But in our opinion, this should be evolved to the early Space Station and that capability should also be present, even in Phase I, from our viewpoint. We would then like to see that capability built up to provide greater thermal and contamination control, to be supplemented by having a refueling and cryogenic resupply capability added to the Station, and further supplemented by using the semi-automated and the robotic activities that the Agency will be developing. We would like to develop this full line of capabilities in the late 90s and early 2000s to take the next step, so that the Great Observatories, by providing new knowledge, will enable us to accomplish the science that the astronomers will want to do. The benefit to the Great Observatories is that while we have a viable concept for them in the STS system, the STS situation in the future during the Space Station development, and the loading and manifesting and establishing the accessibility, must be, at this point, an unknown. We would like to have a low premium term insurance policy, if you will, generated by having the real rudimentary capabilities for STS servicing in the early Space Station. As I indicated earlier, this will provide the capability for an evolutionary path to even greater utilization by astronomers in the future. We see the Great Observatories also as a permanent presence in space, one that can be used by the world's scientists, one in which this nation can lead. Thank you.

ASTRONOMY, ASTROPHYSICS, AND SOLAR SYSTEM EXPLORATION - PANEL DISCUSSION

Panel Moderator: George Newton
Panelists: John Aaron
Douglas Blanchard

Dr. George Newton: Are there questions for any of the talks? I'll take them in the order in which I received them. I have one here directed to both Dr. Blanchard and Mr. Aaron.

Question: Do you see any chance of joint U.S.-Soviet planetary missions, particularly to Mars? What are the advantages and drawbacks to joint efforts?

Douglas Blanchard: Joint space missions, planetary missions, with the Soviets have lots of advantages. One is you don't have to pay all the price. The other is, they have capabilities in planetary science that we could use. The problem is, you please half of the people and you don't satisfy the other half. Technology transfer is a difficult situation to deal with. The other aspect that's difficult to predict is whether you can sustain a scientific collaboration in a politically turbulent world. I think there's great potential to collaborate and it would have scientific return. The collaboration would have return in a lot of other theaters as well. Whether it's going to happen? I really can't venture a guess at this point. I think there are possibilities. The Soviet scientists, that we talk to, look for those opportunities as well as we're looking for those opportunities.

John Aaron: I might add something to that regarding the characteristics of the major initiatives involved in both the precursor aspects of the science and data acquisitions, as well as the actual human exploration initiatives themselves. They are massive undertakings involving significant numbers of systems and capabilities, which, when you lay out the scenarios and look at the capabilities and infrastructure and systems that one needs to accomplish them, there are many techniques and many avenues that would lend themselves to cooperation in many ways. So it's one of the things we're keeping open as we lay out the scenarios and do our studies. In the near term we will look for places and options where the capabilities of both nations, as well as other countries, could be leveraged into such a program, because those options will remain open for some time.

Douglas Blanchard: I might add to that. There are actually discussions in progress and work being done regarding collaboration. Brown University and Burnasky Institute have had a long-standing agreement to communicate. The Russians have been very generous with their Venera data. Likewise, Cal Tech and a corresponding Soviet organization have recently initiated discussions at a scientific level. So there is actually quite a lot of scientific collaboration going on. And quite a lot of generous sharing of space information.

<u>Dr. George Newton:</u> John, I have one for you.

<u>Question:</u> The U.S. lacks a national commitment to a long-range plan for space. Why has the NCS plan not been put before Congress and the nation for approval? Whom can we count on to fight for a long-range commitment?

<u>John Aaron:</u> The recent activity on National Commission on Space happened just last week, in fact, as a result of the signing of the Appropriations Bill in Congress. A formal request, as far as that bill was concerned, was that NASA should respond to the National Commission on Space. So we're in the process of putting together a response that asks for an implementation plan to that report. Our report is due in 60 days. So there is an active process to respond to the National Commission on Space.

<u>Question:</u> How often do you expect the Hubble Space Telescope to be refurbished once it's in Space?

<u>Dr. George Newton:</u> That depends on several situations. Of course, it depends on the function of the observatory and from a technical and engineering viewpoint. It also depends upon the solar activity and the atmospheric drag that the spacecraft would experience, and therefore the decay of the orbit, so that it periodically would have to be reboosted. It also depends on the modules that can be replaced and what their engineering lifetime is, what the experience failure rate is, and the nature of the specific faults that do develop and can be expected to develop over its operating lifetime. As a generality, one plans on from two to five years, depending upon these variables. And, as I recall, Space Telescope has base-lined into its planning for refurbishing to be about every two to two-and-a-half years.

Douglas I have a question for you.

<u>Question:</u> Explain what you said about what we've learned about Venus from Russia's mapping activity. What are the implications?

<u>Douglas Blanchard:</u> What we've learned recently is that there are land forms, ridges, and transform faults that look a lot like the kinds of structures that we see at the spreading centers in the middle of the Atlantic Ocean, for instance. We don't have complete global coverage, so we don't see the other half of the plate. We think that perhaps there's some indication that they're spreading about an equatorial band. The intriguing question is, Where is the plate going at the other end? We don't have that kind of data. So, as a result of sharing the Venera data, this has come to our attention that it's a relatively significant scientific discovery. Would you repeat the second half of the question?

<u>Dr. George Newton:</u> It said to explain what you meant and then what are the implications?

Douglas Blanchard: I guess the implications are that the planet has a thermal engine, much like that of the Earth, and that's significant, in terms of how we structure our theories for how planets formed and how they evolved. The fact that you actually have enough energy, enough motion among the plates, tells you a lot about the internal thermal budget of the planet.

Dr. George Newton: John, I have two questions for you.

Question: Does NASA have a mechanism to synthesize proposed long-term national objectives for the civilian space program, integrating the objectives of manned exploration, space commercialization, science, etc.?

John Aaron: That's one of the mechanisms that we're putting in place with the formation of my office. One of the things that we're defining is not only to look at the near-term missions and themes for science and exploration that I mentioned, but also to determine to what degree we take the human initiatives and then develop criteria for long-range objectives, and actually balance and blend those into an appropriate mix of the many themes of NASA, including human extension and exploration in space. There's also an active process within the organization of the Office of Space, Science and Application. The Associate Deputy Administrator of the agency is working very closely with both of our initiatives to fill options for the national objectives and goals and aspirations. We're very much looking for a mechanism that will propose a balanced approach of both manned and unmanned programs.

Dr. George Newton: John, I think you've answered the next question. But perhaps you'd care to comment on it.

Question: What is the mechanism, and is it adequate? Do you consider it to be an adequate mechanism?

John Aaron: Where are we in the mechanisms? These mechanisms that we're talking about here require a planning capability across the NASA Headquarters organizations and Field Centers that we have not had in place before. We have a new management structure and new planning structure. I sense there's intense spirit of cooperation to enable us to proceed.

Question: Does and should the mechanism include input from the segment of the general public that closely follows the space program, and whose support does it require?

John Aaron: The mechanism is intended to do that. One of the things that was impressive about the National Commission on Space Report, in the way they accomplished that, was that they went around to all the places in nation, with all kinds of different forms, indexing in various ways. And we have the advantage of all that data. I think it's a very effective mechanism, so we're planning those kinds of mechanisms, both for the general public, as well as for the scientific and government communities whose inputs we require. These are major initiatives and will therefore require national will, a national commitment from all levels to accomplish.

Dr. George Newton: This question is addressed to myself or Douglas Blanchard. Perhaps I can make the "or" an "and".

Question: What are the prospects, other than "tea leaves", for sufficient space science funding to achieve a truly balanced space science program?

Dr. George Newton: It's very hard to assess the prospects for significantly increased funding in the present climate, in my opinion. I happen to be a believer that an increased NASA program will result in an increase in the space science program. The historical balance of between about fifteen and twenty percent of the program has been devoted to space science. Therefore, the over-used but very apt analogy is that, "a rising tide raises all ships". I see that as perhaps the most viable approach to generally increase the NASA funding level. And the space science funding level will then also be increased. I think to approach it the other way is certainly an alternative, but involves much more of a change sociologically and in a number of other ways. It would change the fundamental balance that traditionally has been within the Agency. I think that is probably a more difficult task. So I would augur for the increase in the NASA funding approach. Would you like to add anything to that, Douglas?

John Aaron: If you look back across the 25 years or so, history will bear out exactly what you're saying, that we had a very aggressive, overall space program involving human exploration. We also had a very aggressive program in the 60s and 70s in the form of unmanned science, as your charts and Doug's charts indicate. So I think history bears out the fact that those two are very much connected phenomena and therefore require balance. The key is balance and is part of a very important objective. The last question is directed to Dr. Blanchard.

Question: Have you looked at using multiple IUS engines for Galileo, Ulysses, etc.? Could you launch the additional stages on expendables and assemble them using the Shuttle?

Douglas Blanchard: You're asking the wrong person this question. I think there have been a variety of ways that have been examined for assembling and launching spacecraft. In the near term, on-orbit assembly is going to be an issue for quite some time. We have to learn a lot more about how to work in space before we can take on some of these tasks of putting together stages with payloads. I really can't offer informed opinion upon multiple IUSs. I just learned what IUS meant a couple of years ago.

Dr. George Newton: I have a question here that I'm going to have to interpret more than I have the others.

Question: What interest do the astronomers who use the Great Observatories have in using research labs converted, or created from, used external tanks from the Shuttle.

Dr. George Newton: There is interest in that approach, particularly in the very high energy areas, in gamma ray astronomy. In fact, there have been some studies funded by the Office of Space Flight or Code M, and to some degree by Code E, which are examining the feasibility of some of these questions. Using these external tanks, taking them into orbit and then doing some rework on them, putting counters inside the tanks so you have an enormous aperture by normal standards, putting a segmented telescope inside to catch the synchrotron radiation, focus it, then using it for our Gamma Ray Telescope that would have significantly high resolution is very hard to deal with in energy, gamma rays. There are some very preliminary studies that would characterize them as being conceptual pre-Phase A. We're not looking at such concepts on a systematic basis at this point. But I think there is interest and, in fact, it might be a very good utilization of some of that material that could be used in space. We see this as something after the Great Observatories, perhaps in the 90s or early 2000s.

That's all the questions I have received. If there are no others, or if John or Douglas have no other comments they would choose to make, I'd like to thank the panel.

Henry N. Bowes: I may add my thanks to the panel because it's you who have made this conference what it is. I appreciate the support.

LIFE SCIENCES

PANEL SIX

Chairman: Arnauld E. Nicogossian
Director, Life Sciences
Office of Space Science & Applications
NASA Headquarters

Host Chairwoman: Carolyn L. Huntoon
Director
Space and Life Sciences
NASA Johnson Space Center

The following paper number was not assigned:

AAS 87-643

AAS 87-642

LIFE SCIENCES - INTRODUCTION

Arnauld E. Nicogossian*

Ladies and gentlemen, good afternoon. I would like to welcome you to the first Space Life Sciences Session of the AAS Annual Conference. We are very pleased that AAS has decided to include us in this session. I hope it won't be the last. Life Sciences within NASA is a broad science program and has basically two major components. The first and best known deals with enabling research and technology to assure human health, well-being, and productivity in space exploration. The second component, less known to engineers, but well familiar to basic scientists and astronomers, deals with research, exploration and development aimed at studying the origin, evolution, and distribution of life on Planet Earth and in the Universe. This also includes the search for intelligent life in the Universe.

I wish we had more time today to acquaint you with all the different facets of the Space Life Sciences Program. But we are limited to only one hour and 50 minutes, which gives the speakers only 20 minutes each, so we could have some time for discussions and some concluding remarks from Dr. Huntoon, the Host Chairwoman. We hope that at the next sessions we will be able to tackle all the issues associated with the first goal of the life sciences research and development program.

After considerable discussion, we decided to focus on three major areas of life sciences which deal with future strategies. In planning for the 21st Century, and missions beyond low-Earth orbit, to the outreaches of the Solar System those issues deal with physiological adaptation of humans to weightlessness, including optimal performance and work in that environment; the delivery of medical care in remote areas in space; and will limit the major enabling technology of the closed ecological life support system, designed to provide for the sustenance and support for space travelers. These are obviously topics of great importance and constitute the technological base needed between now and the end of the century to enable any reasonable excursions by humans into the Solar System.

I would like, finally, to thank the Program Committee for helping us with this session, especially, Dr. Huntoon, on this panel, who put in the long hours to make

* Dr. Nicogossian, M.D., is Director-at-Large, AAS and Director of Life Sciences, NASA Office of Space Science and Applications, NASA Headquarters, 600 Independence Avenue, S.W., Washington, D.C. 20546.

sure everything goes right. She is the Director of Space and Life Sciences at NASA Johnson Space Center. I'm not going to dwell any further on her background, because you will know her scientific and managerial accomplishments and her contributions to the field of life sciences. She will be our first speaker and will talk about the Physiology of the Human System in Space.

Our second speaker is Dr. Donald F. Stewart, one of our physicians and researchers at NASA Johnson Space Center, who has been devoting some of his efforts in the past few years to helping design the Health Maintenance Facility for the Space Station. Don is going to share some of that work with us.

Our third speaker is Dr. Robert L. Helmreich, Professor of Psychology at the University of Texas in Austin. He has had long association with NASA. His expertise is not only in space psychology, but also in aviation. He's a member of the Space Science Board of the National Academy of Sciences. His topic will be Human Performance in the Aerospace Environment.

Our last speaker is Dr. Robert D. MacElroy from NASA Ames Research Center. Dr. MacElroy is a scientist and a manager. He has been working for many years on life support systems and also on what we came to know as the closed ecological life support system sometimes called the bioregenerative life support system or closed environmental life support system.

AAS 87-644

PHYSIOLOGICAL EFFECTS OF SPACE FLIGHT

Carolyn L. Huntoon*

We have a unique job in life sciences, one that we have had in the space program since its inception. In this audience are those who preceded me in the space life sciences program. You would think that, by now, we would understand what happens to the human body in space. But indeed, as in a lot of areas, we have found that the more we subject people to the space environment, the more questions we have. Today I want to review our plans and some of the issues we are facing in the coming years. Can we fly longer Shuttle flights than we are now flying, and what are the issues? How soon can we fly longer Space Station missions? Life sciences medical research and operations must answer these questions.

NASA has always planned to have a Space Station for people to go to and stay for long periods of time to do meaningful work. We are still planning to do that, but we have to keep them there longer than we had initially planned and we must do it sooner than originally planned. Programmatic requirements for supporting the Station with rotation of crews has forced this requirement. We are facing that challenge right now in our planning.

We have talked about lunar bases and manned Mars missions and of course, that is what people in life sciences dream about because those are the challenges. All of these programs cause us to start with a plan that will give us the technology and the studies, here on Earth and then in the Shuttle and on the Station, that will support us to do these advanced missions.

We are going to have the capabilities to do life sciences work on the Station. We are going to study people in a space environment and provide for their care. We will have much of the information we need by the time we need it, because we are going to acquire it from our Station work on maintaining healthy crews. Other members of this panel will speak about some of the problems we are facing in long duration space flight.

Our future plans include journeys to Mars where a different set of problems must be addressed. We are facing all of this and we are planning to assure the foun-

* Dr. Huntoon is Director, Space and Life Sciences, NASA Johnson Space Center, Houston, Texas 77058.

dation for our future. One of our problems has been the lack of the needed technology to support these studies. We are basing our information to date on Skylab, mainly for the long duration studies. Our Skylab work has given us data and experience for space exposure up to 84 days. Our Soviet colleagues used the Skylab legacy as the foundation for most of their extended duration studies.

From Skylab we learned many things, and of course one of the important ones was that man does very well in space. We did note some problems. There were some other findings in the endocrine, hematology, immunology areas, none of which made us too worried. Most of the problems had no lasting effects medically but were intriguing physiology.

We have been able to collect quite a bit of data on the 24 successful Shuttle flights. We answered some of the questions that we had not been able to answer from our previous space flight experience, particularly those addressing early responses to the weightless environment.

Some of these results are worthy of mention. Astronauts stay in a supine position with the legs elevated for several hours before launch. Investigations by Drs. John Charles and Michael Bungo have indicated that some of the fluid shifts that we had been talking about occurring in space flight were probably initiated on the launch pad. So, some of the physiological changes that we have always attributed to space flight are probably induced in an operational sense on the ground, and we should take that into account. If you find this hard to believe, put your knees up and lie on your back for two or three hours and see how well you feel when you stand up. In our Spacelab missions particularly, we were able to measure some of the acute changes that relate to adaptive responses. For the first time, Drs. Nitza Cintron and Lakshmi Putcha were able to examine kinetics of therapeutic drugs and they found that drugs may not be absorbed in space at the same rate as on Earth. Drs. Bungo and Charles and I found that postflight cardiovascular deconditioning and important endocrine/metabolic changes can be reduced by using a salt/H_2O countermeasure. And Dr. Millard Reschke and his colleagues have been investigating countermeasures against space motion sickness, which continues to be a problem in over 1/2 of the crewpersons.

I wanted to briefly mention the Soviet space program because the Soviets certainly are doing what we are going to be doing again and that is long duration space flight. We have obtained some good information from them in several areas. Most important, their reports show that there were no lasting effects from a 237-day flight. After several months the crew apparently is back to their pre-flight condition. The Soviets have reported that the length of time cosmonauts spend in space is not reflected in their post-flight condition. The Soviets are very concerned with cardiovascular and muscle deconditioning, and they use a series of countermeasures to try to prevent as much muscle loss as possible. They also have had some limitations in their flights and they worry about the same issues that our program is addressing. Unpredictable medical emergencies, bone mineral loss, and the psychological aspects of long-duration flight are areas which we and the Soviets agree need much more work.

We are addressing the issues for the Station and the extended duration Orbiter work in several areas. The major issues are motion sickness, fluid changes, cardiovascular changes, red cell changes, muscle mass, bone mass, immune system and, of course, neurosensory disturbances.

We know space motion sickness is an acute problem; it occurs early and only lasts for a few days in flight. It is not a long-term problem for us with the Station, but we continue to study it in the Shuttle program. We have done many ground investigations. One line of research involves prevention and treatment with drugs we have been using for years. Now we understand better that the drugs are not being handled by the body in space flight the same as they are on Earth. Some of our scientists, including Drs. Reschke and Harm, are busy studying ways to preadapt our crewmembers to weightlessness through a pre-flight training facility.

Body fluid loss and redistribution are probably normal physiology which present no problems of any magnitude, except that when one returns from space there is orthostatic intolerance. We have helped ameliorate this problem in the Shuttle program by using oral normal saline replenishment prior to deorbit. The Soviets have used lower body negative pressure and perhaps anti-gravity suits as well as saline. We are particularly interested in learning more about this, if we do extend the duration of the Orbiter flights beyond the seven to nine days that we now fly.

Currently, the crew who land the Shuttle are given salt tablets and water equivalent to a liter of physiological saline, a few hours before they land. Without this countermeasure blood sodium decreases about 1.5 percent. Astronauts who have taken the countermeasures have much less of a change. Some of the hormonal changes are more dramatic without the countermeasure. Use of the countermeasure reduced the flight-induced increase in aldosterone from 126% to 20%. We have shown that the body is responding to the salt water; it is helping to replenish the loss of fluids that occurred in flight. We have good evidence that it works. The problem is that some of the evidence suggests that it would not work as well beyond 8-10 days. This must be studied and perhaps the saline will be augmented for a more effective countermeasure after longer space flights.

Cardiovascular problems have been an issue since our very first space flight. The crew stand up post-flight and some of them feel faint. Arrhythmias have been observed during flight. Dr. John Rummel and colleagues reported some reduction in aerobic exercise or work capacity after the Skylab flights. Returning to 1g, cardiovascular deconditioning is clearly demonstrated and it takes a while for our crews to get back to their pre-flight condition.

Drs. Michael Bungo and Philip Johnson developed a cardiac index of deconditioning involving heart rate and blood pressure. The index is lower for crewmembers who used the saline countermeasure, meaning they did quite well post-flight. But Dr. John Charles found that on the longer missions, beneficial effects of the countermeasure are reduced. We certainly need some more data in this important area.

We have talked about the issue of altered blood flow. Drs. Cintron and Putcha have done some studies comparing the pharmacokinetics of acetaminophen ("Tylenol") before and after flight. The drug was measured in saliva. The peak of drug absorption was lower on mission day 4 than it was pre-flight. There are many factors involved in what happens to drugs appearing in the saliva, from absorption to blood flow, but apparently in this case less drug is presented to the receptor. We know that if we are going to use medications in flight, whether for medical treatment or to try to prevent adverse physiological changes, we must understand these mechanisms. We have had four successful tests of collecting saliva samples after drug administration during flight.

Drs. Fischer and Johnson first found that red blood cells were lost during space flight in the early Gemini program, and we are still working on the exact mechanism. Dr. Steve Kimzey continued to study this issue in his Skylab experiments and then with our Spacelab experiments we actually focused on a mechanism. But we are still in a quandary about why the body is not making the same amount of red blood cells in flight as it does on the ground. During a short-term mission it does not seem to bother anyone, but on a long-term mission, if a crewmember had an injury it could be a problem. Upon return to 1g, there is a significant loss of red cell mass, but the cells are replenished after several weeks. We have to deal with that as just being a normal response. Some countermeasures may be needed on very long flights.

In one experiment, on Spacelab 1, crewmembers drew and processed the blood and did some of the analysis on board the Spacelab and returned some samples to Houston for analysis. We measured erythropoietin for the first time on that flight; that is the hormone that controls red cell production, and we did find it to be decreased by the space flight exposure.

Regarding muscle mass loss, we have encouraged astronauts to exercise the weight-bearing muscles. A treadmill is part of the gear the crew request on the Space Shuttle because they like to exercise and it makes them feel better. On the short Shuttle missions, the crew decides upon their own protocols and exercise to the extent they wanted or had time for during the mission. We know that orthostatic intolerance is probably decreased by exercise. What we do know for sure is that, if we extend the Shuttle or when we go to Station and beyond, we are going to have to do a lot more serious planning with not only our exercise protocols but the devices we use and make sure we are correctly exercising all the right parts of the body.

Understanding bone mineral loss is still a major concern for long duration space flight exposure. It was hypothesized that people would lose mineral mass from their bones after exposure to weightless flight. Drs. Rambaut, Whedon and I found that the loss is about 2.6 grams over the 84 days of the longest Skylab mission. But the important finding was that calcium continued to be lost every time we measured it. We do not have data beyond that. The Soviets have not done metabolic balance studies, but judging from some of their bone density measurements, bone calcium loss does not occur to the extent that one would project in the bone sites measured.

We and the Soviets are using bed rest studies to investigate countermeasures to lessen bone mineral loss. Countermeasures studied by Dr. Victor Schneider and his colleagues are fluoride, various drugs, diet adjustment and of course, there is a component of exercise that we believe affects bone mineral content.

We have just started to study the immune system in space flight crews. Because medical science in this area has been advancing so fast, we at NASA have been trying to apply the new methodologies. Dr. Kimzey first noticed a change in lymphocyte function after space flight following Skylab. We have not been able to prove that the steroids secreted in response to stress are responsible for that finding. We do know that lymphocytes are not quite so active after exposure to the space environment, and people who have a suppressed immune system may be more susceptible to infections. We know it is true on the ground, so we are compelled to learn more about this process as related to space flight, and, of course, we will have some experiments in this area for the Station.

We have tried to study neurosensory disturbances in Shuttle flights. Dr. Reschke and his colleagues will continue to study this complex problem. Because of the reports we receive from some astronauts after space flight on disorientation and illusions, we know that we need to understand better the nervous system in space flight.

Artificial gravity is being considered for very long weightless exposures. The truth is, we do not know about the effect of artificial gravity on the human system. We do not presently know whether 1/2g is better than no g or 1g. Those studies need to be done; some of them can be done on the ground; most of them will have to be done in the Station environment.

Fortunately, radiation effects on our STS crews have been minimal. We have never approached, with any of the Shuttle flights, total crew dosages that are limiting. We do not anticipate approaching those limits with the Station either. We will always be very cognizant of radiation doses and we always measure and record them. But our concern, of course, comes with abnormal events taking place on the Sun. Also radiation is a greater concern when we leave Earth's magnetosphere as on journeys to the lunar surface or to Mars. Protecting our crews against radiation, timely alarms so that safe havens can be sought and more precise measurements of the environment must all be accomplished.

The areas of psychological and habitability factors have not been a major concern to date in our program, but they will become more important as we extend mission duration and crew size.

During our previous space missions we have found that the crews become very close and are obviously good friends. They live and work together in small quarters and that has been acceptable for the short Shuttle flights. Special attention will be directed toward privacy and other personal needs in the Space Station and on the other longer missions.

The Closed Ecological Life Support System is an extremely important understudied concern in dealing with the Mars mission or lunar base, because it will be difficult to undertake those kinds of missions on the same terms as our plans for the Space Station. Fortunately, the Station will afford us the opportunity to assemble these system components and to test them as well.

The environment inside the spacecraft has to be perfectly maintained 24 hours a day for the duration of the mission. We have met a real challenge to fly animals and humans on the same space mission. Because their health habits are not the same, that task is no small one. The cages had to be properly maintained and the crew had to tend to the animals without being exposed to them and their products unnecessarily. We tested this and it worked; it had some problems but the system worked. And we are going to fly it again on our dedicated life sciences mission in Spacelab. This is one area in which we have been somewhat more adventurous than the Soviets. We have flown animal experiments with humans. We think that superior experiments result from this combination, since they can be tended and humans can come to the aid of the animals, modify procedures in real time, and act in the usual role of observer and investigator. On Space Station we intend to do animal and plant work while it is inhabited by humans.

The challenge of taking care of humans in space is our most important responsibility at the Johnson Space Center. It is incumbent upon these humans to build that Space Station, get it working and occupy it. Our job is to make sure they stay healthy. This is certainly a paramount goal for us and we are looking forward to achieving it.

AAS 87-645

MEDICAL CARE DELIVERY IN SPACE

Don F. Stewart*

The origins of Space Medicine as a unique branch of Occupational Medicine are briefly discussed. The delivery of medical care in Space is reviewed, both from a historical perspective and in regard to future planning for Space Station.

INTRODUCTION

Space Medicine is actually an area of concentration within the broader field of Aviation Medicine, which in turn is one of the sub-specialities of Occupational Medicine. Now, like all practitioners of Occupational Medicine, there are certain unique characteristics of the aerospace environment which shape the nature of our medical practice. Aviation Medicine is unique in the sense that this speciality has a very synergistic relationship with the Space Sciences, although historically the Medical Sciences have not kept pace with the evolution of operational spacecraft. This was true in the time of the Mercury Program and is true to a certain degree today, as we look forward to extended duration Space Station missions.

Aerospace Medicine

Before addressing the topic of medical care in space, it is important to briefly review Aviation Medicine from a historical perspective. Back in the era of World War I, it as observed that greater than 50% of the aircraft that were lost in the early stages of the war were lost as a result of physical defects of the pilots. As a result of this, a specific medical branch, dedicated to the care of the pilot was established. By the end of the second year, as a result of better pilot selection, the fatalities attributable to physical defects of pilots had dropped to twenty percent. At that time, all military physicians were known as surgeons. Practitioners of this new speciality came to be known as Flight Surgeons.

World War II resulted in tremendous advancement in the operational capability of combat aircraft. This in turn, created in a whole new host of aeromedical problems that the physicians of the time had to address, including issues such as acceleration effects, hypoxia, and high altitude bail-out. Another product of World War II was a tremendous technology advancement in the form of the V-2 rocket, which, in essence, marked the practiced beginnings of the Space Age, and therefore, the true beginning of Space Medicine.

* Don F. Stewart, M.D., HMF Project Physician, NASA Johnson Space Center, Houston, Texas 77058.

The rapid advances in rocketry which occurred following World War II, were accompanied by a range of new operational medical problems. The primary biomedical issues, which defined the nature of Space Medicine, included the effects of sustained accelerations, protection of Man from the hostile environment of Space, the effects of microgravity on the human organism, which has had no evolutionary reason to evolve adaptive mechanisms to weightlessness. Therefore adaptation in Space is accomplished by using adaptive processes that were intended for the terrestrial environment. The last condition that makes Space Medicine unique is the fact that it is practical in a very remote environment, with limited support from ground based medical facilities.

Early Space Programs

Looking now at the early Space Programs in terms of the medical care, the Mercury Program was more notable for what it disproved than what it proved. Prior to the initial flights of the Mercury Program, there had been a very vocal group within the medical community which had predicted a number of catastrophic effects from exposure to micro-gravity. These predictions included such problems as cardiac arrythmias, collapsed lungs, and space motion sickness. As we now know, a certain number of these predicted effects were in fact accurate, but, none of them were of a catastrophic nature.

The primary biomedical accomplishment of Mercury then, was to demonstrate that man could survive and function effectively in a space environment. With the Gemini Program the medical questions were expanded to ask "Can Man survive and function in Space up to fourteen days?", which was the amount of time required to accomplish a Lunar mission. This question was answered by the Gemini 7 mission which demonstrated that man could tolerate weightlessness for up to two weeks. Another of the Gemini missions of biomedical significance was the Gemini 4 mission. During this mission, the first American EVA was accomplished by Ed White, which was most notable for the unanticipated high metabolic rates which were observed.

In terms of inflight support, the medical capabilities on both Mercury and Gemini were very basic. In the case of Mercury, this consisted of only three medications; in the Gemini Program, the medical kit contained about ten medications. Neither of the missions had even rudimentary medical equipment such as a blood pressure cuff.

Medical support for the Apollo Program posed a new set of problems for the medical community. The most important of these was the lack of a rapid return to Earth in the case of a serious medical contingency. While weight and volume limitations prevented the inclusion of any sort of comprehensive medical system, the kits which were flown were substantially larger than those included on the Gemini missions, and included approximately 30 items. A slightly smaller kit of similar nature was included in the stowage for the Lunar Module. The actual inflight medical assessment was similar to earlier missions, based on biotelemetry, television and voice communications. In terms of actual inflight illnesses, the Apollo Program was characterized only by a number of minor medical problems. A number of these, however, including one case of mild decompression sickness, or the bends, and several instances of mild cardiac arrythmias would have had potential for mission impact if the symptoms had been more pronounced.

The Skylab Program represented the transition to a longer duration Space Program. As a result of the longer duration, greater potential existed for mission impact resulting from a medical problem. Skylab was characterized by a fairly comprehensive medical capability which included diagnostic, therapeutic and preventive capability. This system, known as the Inflight Medical Support System (IMSS) was designed to provide treatment for ill or injured crewmembers, as well as, for support of certain inflight medical experiments. Not inclusive of the exercise hardware, the weight dedicated to the IMSS was approximately 100 lbs., which was a significant increase from any of the previous missions.

Skylab was also notable for being the first program in which a physician was included as a crewmember (Dr. Joe Kerwin). In addition to having a physician onboard, each of the Skylab crewmembers received about 100 hours of medical and dental training. In terms of medical events, once again, Skylab was similar to the Apollo Program in that no major injuries or illnesses occurred inflight. The medical events which did occur were minor in nature, and appropriately treated by the inflight medical capability.

STS Medical Care

Inflight medical care for the STS Program did not involve any new support capability relative to the earlier missions. In that the planned missions were relatively short, and considering the capability of the orbiter to land at a number of prime and contingency landing sites, a relatively modest medical kit was designed to support the mission. To administer medical care inflight, two crewmembers are designated as Crew Medical Officers (CMOs) and receive a limited number of medical briefings to familiarize the CMOs with the SOMS Kit and basic medical procedures. Inflight management of a medical contingency would be done with the support of the crew surgeons. The crews must initiate such support by requesting a Private Medical Conference (PMC).

The actual medical kit has two sub-kits; the Emergency Medical Kit (EMK) and the Medications and Bandages Kit (MBK). The SOMS Kit, while capable of providing care for a wide range of medical illnesses or injuries, would not be able to provide adequate care for most serious life threatening medical problems.

In terms of actual medical problems during the Shuttle Program, we have only relatively mild inflight illnesses. The most serious problem has been space motion sickness. This occurs in about 50% of the first time fliers, although space motion sickness classified as severe, occurs in about one out of every ten first time fliers. The symptoms of SMS are quite varied from individual to individual, however, serious cases are almost always characterized by a profound malaise and lethargy, and repetitive episodes of nausea and vomiting. Treatment for SMS onboard the Shuttle has been based on the prophylactic use of motion sickness medications such as Scop/Dex and symptomatic treatment once the illness has begun.

The typical clinical course of SMS is relatively brief, usually lasting no longer than 48-72 hours. In most cases, the crewmember has been capable of performing their assigned duties.

Another capability that we do have inflight is the limited capability to treat decompression sickness. This is accomplished by utilizing a modified helmet for the spacesuit, which allows an over-pressure of approximately 4 pounds per square inch.

Space Station Challenges

Space Station, again, presents the Life Sciences community with new biomedical challenges. The Space Station operational scenario is unique, relative to anything NASA has done before. This will be a relatively permanent facility, with mission duration of approximately 90-180 days.

Crew composition will differ from the previous missions, in the sense that as the program matures, more of a "mixed population" will fly, i.e., some of them older and less rigorously screened than in previous programs. Crew size will initially be four to eight and could expand to a crew of sixteen.

Quantifying the medical risk of extended Space Station habitation is difficult, however, the cumulative medical risk results from three contributing factors: 1) Medical/surgical conditions which would be more likely as a result of a micro-gravity environment (i.e., kidney stones), 2) Medical/surgical conditions which would be more likely as a result of occupational activities onboard Space Station (i.e., EVAs), and 3) Baseline medical risk, which could be expected in any adult population with similar demographics. In terms of actually predicting the risk, this is difficult to do, in that the surrogate populations which have been studied all have significant differences from the astronauts corps, and it is difficult to extrapolate from these questionable analogue populations to determine a true medical risk. To a certain extent, medical capability onboard Space Station should be justifiable on humanitarian reasons. If one examines the Soviet long-term space flight experience, it would suggest a serious medical problem requiring evacuation could occur as frequently as every 2,000 man-days.

Plans for inflight medical care onboard Space Station center around the Health Maintenance Facility, or HMF. The HMF is a modular medical care facility which consists of 12 subsystems mounted in two Space Station functional racks. The goal of the HMF is to provide medical care which is comparable to terrestrial health care, within the context of programmatic limitations. Other objectives of the HMF include the following: 1) Avoidance of an unnecessary rescue, by providing adequate inflight care; and 2) In a scenario where a medical rescue is required, maximize the possibility that the rescue will be successful.

A new class of down-sized medical instruments designed for use in the physician's office has been advantageous in selecting hardware for the HMF. The design philosophy for HMF subsystem hardware is based on modifications of state-of-the-art, off-the-shelf hardware. A piece of hardware which is most typical of this class of medical instruments in the Kodak Ectachem Chemical Laboratory Analyzer. This device is compact, simple to use and service, and is based on a dry chemistry technology which is gravity independent. The wet chemistry analyzer which it replaces occupies a volume 20 times as great.

The subsystem hardware being developed for the HMF, in most instances, is selected on the basis of three considerations: 1) Weight/Volume, 2) Adaptability to micro-gravity, and 3) Capability.

Another key piece of hardware in the HMF is the patient restraint. This device will be used to restrain a patient during medical procedures and will also function as a bed should it be necessary to have an "in-patient". The third major capability this allows is that it functions as a litter in case of an emergency medical transport.

The preventive component of the HMF, consists of exercise countermeasures. There are a number of operational justifications for providing an adequate inflight exercise capability. Some of these are related to performance of the inflight crew duties which include frequent, repetitive EVA's. In addition, it is important that a crewmember be able to deal with any sort of any IVA contingencies which might occur. It would be preferable that each crewmember be able to conduct an emergency egress without assistance from other crewmembers. Another justification for providing exercise onboard Space Station is that exercise is a normal part of the astronauts routine, and it benefits both crew moral and physical well-being. Looking at this from a different perspective, the physiological perspective, there are physiological reasons for providing exercise. One of the key reasons among these is maintenance of musculoskeletal integrity. We know that crewmembers lose approximately one percent of total skeletal mass per month while inflight. This loss of bone is more concentrated in the weight bearing skeleton, and it is possible to see on the order of a 10% loss per month in the weight bearing bone such as the calcaneus.

Another area of concern, which has not been adequately addressed is the fact that the ligaments and tendons, which are the interface between bones and muscles, also tend to atrophy (lose strength) during spaceflight. These are the weakest point in the musculoskeletal unit, and might be more subject to injury if they have lost a significant degree of structural integrity.

In terms of exercise on the Shuttle, a treadmill has been included for most Shuttle missions. The reason for providing an exercise capability onboard Shuttle are not as convincing as they are for the Space Station. The changes of short duration flight are less significant, however, they are of some operational use in keeping the lower extremities fit in case of the need to do an emergency egress. The primary egress route from the Shuttle also involves a fair degree of upper body strength. This system is being redeveloped, enabling a new egress route from the Shuttle, which will require less bodily strength.

One other issue of interest in is exercise compliance. Our best estimate now for the amount of time required to maintain fitness onboard the Space Station is around 90 minutes per day. We have developed a number of displays to help promote exercise compliance, and to provide "diversion displays," either video or music that the crewmember can listen to while he/she is exercising. A prototype display has been fabricated, where workload on the bike is ergometer automatically varied by computer to keep the exercising crewmember's heart rate in the target heart rate zone.

AAS 87-646

PERSONALITY AND ORGANIZATIONAL INFLUENCES ON AEROSPACE HUMAN PERFORMANCE

Robert L. Helmreich*

Individual and organizational influences on performance in aerospace environments are discussed. A model of personality with demonstrated validity is described along with reasons why personality's effects on performance have been underestimated. Organizational forces including intergroup conflict and coercive pressures are also described. It is suggested that basic and applied research in analog situations is needed to provide necessary guidance for planning future space missions.

INTRODUCTION

The loss of Shuttle Challenger and the crash of Northwest Airlines Flight 255 share a common causal factor - failures of individuals in work groups to make appropriate decisions and perform effectively. In both instances the proximal determinant of the outcome can be considered to be psychological. The importance of psychological factors in crew performance is being recognized increasingly (Ref. 1; Ref. 2). In this presentation I will focus on only two factors, personality influences and organizational effects. Discussions of other psychological factors can be found in Hackman (Ref. 3), Helmreich (Ref. 4) and Helmreich & Wilhelm (Ref. 5).

PERSONALITY CONSTELLATIONS

Two core dimensions of personality have been isolated: _instrumental traits_ relating to achievement and goal seeking, including aspects of achievement motivation, and _expressive_ traits relating to interpersonal behaviors and orientation. Measurement of these attributes is achieved through psychometrically reliable self-report instruments that assess both positive and negative aspects of these dimensions (Ref. 6; Ref. 7). Results in the prediction of performance have been robust in a variety of situations ranging from academics to scientific research to jet aircraft. In the case of pilots, positive performance in command of jet transport aircraft with multi-person crews is related to high scores on positive instrumental traits including a need for Mastery of new and challenging tasks, and low scores on negative instrumental attributes including such traits as arrogance and hostility. Also

* Professor of Psychology, The University of Texas at Austin, Austin, Texas 78712.

positively related to performance is possession of high scores on expressive traits including sensitivity to others. The latter finding reflects the fact that operation of a complex aircraft is a group endeavor requiring the close coordination of a crew more than the skills of the lone pilot wearing a white scarf.

Recently, the personality battery has been expanded to include aspects of what has come to be known as the "Type A Personality" (Ref. 8). The Type A is usually described as a driven individual with high levels of ambition, time urgency, impatience, and aggression. Earlier research has suggested that Type A individuals may be both prone to coronary heart disease and more successful vocationally (Ref. 9; Ref. 10). Our new formulation of the construct yields two moderately correlated factors that have been labeled Achievement Striving (AS) and Impatience/Irritability (I/I) (Ref. 11). Looking at the two factors in relation to behavioral criteria has shown a consistent pattern of outcomes: Achievement striving is related to positive performance including scientific and academic attainment but is unrelated to negative health outcomes (Ref. 12) while Impatience/Irritability is associated with a variety of health complaints including poor sleep quality, headaches, and digestive and respiratory upsets but is not correlated with performance (Ref. 13). These two factors are correlated with the instrumental and expressive traits described above but still increase the predictive power of the battery.

Important findings regarding personality and performance were obtained in a recent dissertation by Thomas Chidester (Ref. 14). Chidester replicated the finding that instrumental and expressive attributes were related to both technical and managerial aspects of flightcrew performance. He also found that Achievement Striving was a positive predictor of performance and that the Impatience/Irritability dimension was related to a variety of health complaints among flightcrew members.

An intriguing question arising from these data is why significant and replicable relationships between personality and performance are being found when the consensus in the literature is that personality and performance are weakly related if at all. One possible resolution of this seeming paradox may be in what we have christened the "honeymoon effect" of motivation on performance. In a recent study (Ref. 15), correlations between components of achievement motivation and performance over time were examined. At the end of training there were no significant correlations between the predictors and performance in a sample of airline employees. With the passage of time, however, the correlations increased in magnitude and stabilized. We have interpreted this as reflecting that most individuals, when selected for a desired position, will exert maximum effort to perform well during training and probationary periods and this level of effort may mask the influence of personality on performance. It is not until after the individual has settled into the routine of the position and the "honeymoon has ended" that personality influences on behavior begin to emerge strongly. Looking at the literature on pilot personality, it is notable that the criterion variable almost universally employed is performance in or completion of training while in the present research the criteria involved the performance of experienced crews in line operations. The changing magnitude of obtained correlations is shown graphically in Figure 1.

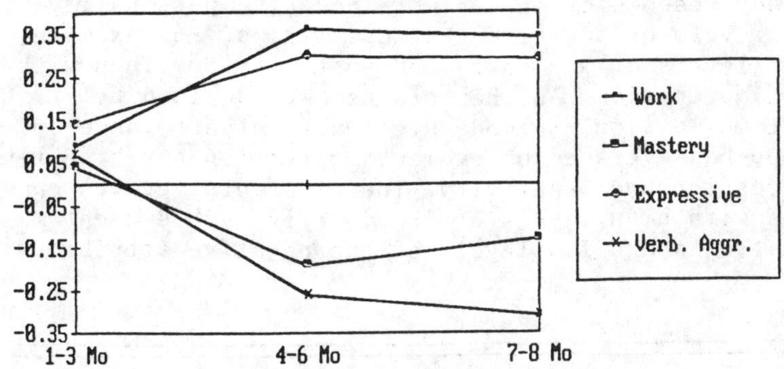

Fig. 1 Correlations of Personality with Performance Across Time

As the figure indicates, two attributes, Work motivation and Expressivity become significant, positive correlates of performance and two, Mastery and Verbal Aggressiveness become significantly negative after time on the job. The results for Mastery are particularly informative. This variable from the Work and Family Orientation measure of achievement motivation (Ref. 6) reflects a need for new and challenging tasks. The job in question, operating a simplified reservations computer terminal, is a repetitive and mundane activity. Clearly, those high on this characteristic do not find this need met after considerable exposure to the work.

Another characteristic of research on personality/behavior relationships may have served to hide meaningful relationships. This is a tendency of investigators to concentrate on limited aspects of the total personality and to look at them in relative isolation. This restricted approach fails to consider the distribution of combinations of different trait intensities - in other words, it fails to look at the constellations of personality combinations that exist in "real people" in the "real world". For example, with what frequency are individuals with both highly instrumental and highly expressive personality traits found in the population or research sample. Thomas Chidester, Steven Gregorich, and the author have been applying the technique of cluster analysis to determine the distributions of differing combinations of positive and negative personal attributes using the personality characteristics described above (Ref. 16; Ref. 17). Depending on the research population, four or five relatively common clusters of individuals with particular levels of instrumental and expressive attributes emerge from the analyses. These clusters reflect meaningful constellations of traits as they are distributed across individuals.

A highly innovative dissertation study by Gibson (Ref. 18) demonstrates the utility of this approach. Gibson's study was an examination of relationships between personality factors and ratings of managerial performance. Cluster analyses based on the instrumental and expressive trait dimensions gave five readily classifiable groups. One of these clusters nicely defined the "average" manager. Individuals in this group scored as average on both positive and negative instrumental and expressive dimensions. Three other clusters were marked primarily by the elevated presence of one or more <u>negative</u> personality dimensions and/or low levels of positive characteristics. For example, one group showed high levels of arrogance and hostility combined with low achievement motivation. Another cluster was defined by slightly higher achievement motivation and moderate levels of arrogance and hostility. A fourth group had average achievement motivation but high levels of verbal aggression and negativity. The final cluster was composed of individuals with high levels of the positive achievement motives and expressive traits and low levels of the negative attributes.

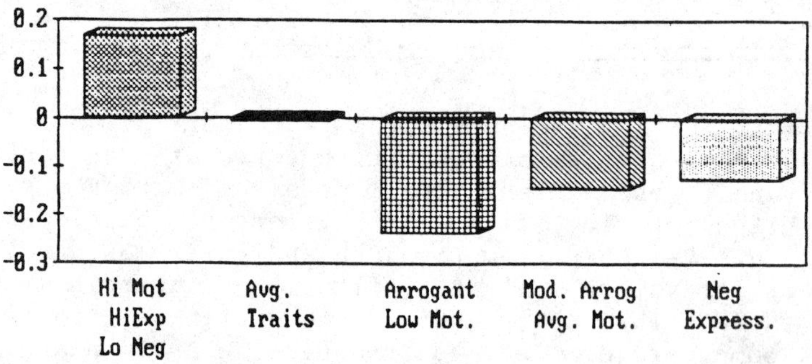

Fig. 2. Mean Performance Z-scores for Each Personality Cluster

Figure 2 shows the average <u>performance ratings</u> of the five groups where the ratings are expressed in terms of Z-scores with a sample average of zero. Ratings were obtained from each manager's superiors, peers, and subordinates. It is striking that the group with the "average" personality received <u>average</u> ratings with a mean of almost zero. Each of the groups characterized by one or more negative attribute dimensions received below average ratings. On the other hand, the group defined by high positive and low negative attributes received positive mean ratings for performance. In summary, the results obtained using these conceptual variables and the cluster analytic approach to determining the joint occurrence of these attributes in research populations would appear to have considerable theoretical and practical utility. Another application of cluster analysis will be discussed in the following consideration of attitudes and performance.

What about relationships between personality and behavior in isolated environments such as long duration space missions? I would argue that the relationships should be even stronger as the situation is more intense and the "honeymoon" likely to be quite brief. The question warrants extremely high research priority.

ORGANIZATIONAL EFFECTS

Much of our data on long missions have come from research in undersea habitats during projects such as SEALAB and TEKTITE (Ref. 19; Ref. 20). Investigators have tended to look at isolated groups such as space crews as microsocieties operating autonomously apart from the larger society. In reality, of course, these groups tend to be highly dependent psychologically and often physically on their parent organization or society. Lunar missions and undersea habitats were both highly dependent on extensive, earth based support activities for their survival. The organizational structure of undersea missions such as SEALAB and TEKTITE was military and hierarchical just as space missions in this country are. Ground Control had, and exercised, the right to dictate schedules and other activities of crews. This real and perceived control by an outside force caused considerable discontent in the undersea missions (and also, coincidentally, served to increase cohesiveness by creating an "us against them" atmosphere which varied as a function of differing exercise of authority by successive managers in ground control. In some instances, open conflict emerged between the two groups.

Similar intergroup disputations have occurred in both the U.S. and Soviet space programs. For example, during a Skylab mission the crew broke off radio contact and refused to work for a full day. The extreme form of this type of relationship between isolated groups and control has many of the earmarks of classic paranoia with perceptions of powerful outside forces controlling the destiny of small groups. An observed fallout has been an increase in group cohesiveness, but at the expense of building a confrontational psychological environment. This type of conflict has roots both within the isolated group and in the larger social system that encompasses both. It raises major researchable questions about optimum levels of autonomy for isolated groups and about issues of authority and command both between and within groups.

Most groups in restricted environments have, almost by default, a vertical, military style, hierarchy with a powerful, appointed leader. It is interesting to note that the Russians, at least by their reports, operate with a much more democratic organization within their spacecraft and with more shared authority between the isolated group and mission control. What is distinctly absent is valid, empirical data exploring intergroup relations in this kind of environment, not to mention data on mission control as a distinct social and operational setting.

One of the most critical organizational problems centers on subtle pressures to take risks. One need only consider the case of mission personnel induced to launch and aircraft crews subtly pressured to fly in marginal weather or with marginal equipment. The Rogers commission pinpointed this phenomenon in NASA (Ref. 20) while Nance (Ref. 21) has noted its operation at a major airline. Clearly social psychologists can and should play a role in developing recognition and countermeasures for organizations and individuals facing such pressures.

There is a critical need for basic and applied research focusing on individual and organizational issues, especially when faced with the problems associated with long term occupancy of space. I am cheered by the fact that groups within NASA are showing increased commitment using the results of important analog research such as that in aviation and are supporting a returning to undersea habitats as optimal research sites.

ACKNOWLEDGMENT

The research reported here was supported by Cooperative Agreement NCC2-286 between NASA-Ames Research Center and The University of Texas at Austin, Robert L. Helmreich, Principal Investigator. The Technical Monitor of the project is Dr. H. Clayton Foushee, NASA-Ames Research Center.

REFERENCES

1. Helmreich, R.L. Applying Social in Outer Space: Unfulfilled promises revisited. American Psychologist, 38, 445-450, 1983.

2. Orlady, H.W. & Foushee, H.C. (Eds.), Cockpit resource management training: Proceedings of the NASA/MAC Workshop. NASA-Ames Research Center: CP-2455, 1987.

3. Hackman, J.R. Group level issues in the design and training of cockpit crews. In H. W. Orlady & H.C. Foushee (Eds.), Cockpit resource management training: Proceedings of the NASA/MAC Workshop. NASA-Ames Research Center: CP-2455, 1987.

4. Helmreich, R.L., & Wilhelm, J.A. Evaluating cockpit resource management training. Proceedings of the Fourth Aviation Psychology Symposium, Columbus, OH.: Ohio State University, 240-246, 1987.

5. Helmreich, R.L. & Wilhelm, J.A. Human performance in aerospace environments: The search for psychological determinants, submitted for publication.

6. Helmreich, R.L., & Spence, J.T. The Work and Family Orientation Questionnaire: An objective instrument to assess components of achievement motivation and attitudes toward family and career. JSAS Catalog of Selected Documents in Psychology, 8, 35, MS 1677, 1978.

7. Spence, J.T., Helmreich, R.L., & Holahan, C.K. Negative and positive components of psychological masculinity and femininity

and their relationships to self-reports of neurotic and acting out behaviors. Journal of Personality and Social Psychology, 37, 1673-1644, 1979.

8. Jenkins, C.D., Zyzanski, S.J., & Rosenman, R.H. Progress toward validation of a computer scored test for the Type A coronary prone behavior pattern. Psychometric Medicine, 33, 193-202, 1971.

9. Friedman, M., & Rosenman, R.H. Type A behavior and your heart. Greenwich, CT: Fawcett, 1974.

10. Matthews, K.A., Helmreich, R.L., Beane, W.E., & Lucker, G.W. Pattern A, Achievement-striving, and Scientific Merit: Does Pattern A help or hinder? Journal of Personality and Social Psychology, 39, 962-967, 1980.

11. Pred, R.S., Spence, J.T., & Helmreich, R.L. The development of new scales for the Jenkins Activity Survey Measure of the Type A construct. Social and Behavioral Science Documents, in press.

12. Helmreich, R.L., Spence, J.T., & Pred, R.S. Making it without losing it: Type A, achievement motivation, and scientific attainment revisited. Personality and Psychology Bulletin, in press.

13. Spence, J.T., Helmreich, R.L., & Pred, R.S. Impatience versus achievement strivings in the Type A pattern: Differential effects on students' health and academic achievement. Journal of Applied Psychology, in press.

14. Chidester, T.R. Mood, sleep, and fatigue effects in flight operations. Doctoral dissertation, The University of Texas at Austin, 1986.

15. Helmreich, R.L., Sawin, L.L., & Carsrud, A.L. The honeymoon effect in job performance: Delayed predictive power of achievement motivation. Journal of Applied Psychology, 71, 1085-1088, 1986.

16. Chidester, T.R., Helmreich, R.L., Gregorich, S.E., & Geis, C. Isolating subgroups of pilots on performance-related personality dimensions: Implications of cluster analysis for training and selection, in preparation.

17. Romesburg, H.C. Cluster analysis for researchers. Belmont, CA: Lifetime Learning Publications, 1984.

18. Gibson, R.H. Managerial performance appraisal: Sex differences and correlates of ratings among leaders, peers, and subordinates. Doctoral dissertation, The University of Texas at Austin, 1987.

19. Radloff, R., & Helmreich, R. Groups under stress: Psychological research in SEALAB II. New York: Appleton-Century-Crofts, 1968.

20. Helmreich, R. The TEKTITE II human behavior program. JSAS *Catalog of Selected Documents in Psychology*, *2*, 13, MS 70, 1972.

21. Rogers, W.P. *Report of the presidential commission on the Space Shuttle Challenger accident*. Washington, D.C.: Government Printing Office, 1986.

22. Nance, J. *A splash of colors*. New York: William Morrow, 1984.

AAS 87-647

BIO-REGENERATIVE LIFE SUPPORT

Robert D. MacElroy[*] and Theodore Wydeven, Jr.[*]

The basis for and the potential uses of bio-regenerative life support are examined. Bio-regenerative life support systems are an alternative to physical-chemical regeneration techniques for use when resupply of a crew in space is expensive, or when the logistics of resupply are difficult. Many of the scientific studies required for bio-regenerative life support systems have been completed and preliminary development of some components will begin within the next 12 to 18 months. The focus of the work that lies ahead will be efficient power and mass use, long term system stability, component function, systems integration, and extensive testing in the space environment. Because of the advantages of bio-regeneration it is anticipated that human life support for long term space missions will evolve to include increasingly large amounts of biologically-based regeneration.

INTRODUCTION

Humans require food, air and water to sustain life. A typical 70 kg human, working moderately in space, each day requires about 2800 Calories contained in about 650 g of dry food, and also uses 836 g of oxygen and drinks about 1850 g of water. An additional 1560 grams of water are ingested in food and produced metabolically, and about 20.2 kg is used for sanitation and domestic use.[1]

Enough of these materials for an entire space mission, plus the packaging necessary to contain them, must be launched into space with the flight crew, or they must be resupplied by separate launches during the course of a mission.

Regeneration is the alternative to including all materials in the initial launch, or to resupplying them during the course of a mission. Various devices for regeneration

* Ecosystem Science and Technology Branch, 239-4, NASA Ames Research Center, Moffett Field, California 94035.

have been under study and development by NASA for the past two decades, and two major classes have been considered; one utilizes physical-chemical processes to regenerate oxygen[2] and water,[3] the other employs a combination of biological and physical-chemical processes to recycle materials and to regenerate food, oxygen and water.[4] The primary objectives of this paper are to describe NASA's approach to bio-regenerative life support (BLS), and briefly to discuss the application of the approach to future missions.

LIFE SUPPORT REQUIREMENTS IN SPACE

Human food fulfills two requirements: it supplies energy to maintain the body, and it supplies the organic materials that the body must have to grow or to keep in good repair. Much of what the body needs can be assembled or synthesized from the small molecules, or monomers, that arise from the digestion of larger, more complex molecules in food material. However, but some nutrients, such as vitamins, some amino acids and some lipids, cannot be synthesized in the body. These small molecules must be obtained from food or from supplements to the diet.

The air we breathe contains about 21% oxygen (160 mm Hg, partial pressure) and 0.034% (0.25 mm Hg) carbon dioxide, with about 79% nitrogen. Partial pressures of oxygen less about 80 mm Hg, or CO_2 concentrations greater than about 12 mm Hg cause deleterious metabolic responses in humans. Water is ingested by drinking, in food, and as a product of the metabolic oxidation of food, and must be microbiologically and chemically safe. Water is used in human metabolism for several purposes, including replacement of that lost through evaporation from the skin and lungs, and of that used to carry wastes from the body.

A significant fraction of the mass of life support materials is packaging. The tanks and containers needed for water, oxygen, food and waste materials must be considered, as must the methods for maintaining safe atmospheric levels of carbon dioxide (e.g. lithium hydroxide and its containers). An estimate of the input/output balance of life support materials is illustrated in Figure 1.[1] Table 1 summarizes the estimated per-person daily use of all materials needed for life support, including water required for sanitation and housekeeping uses, and the packaging required for transport.[1] Without regeneration, the total mass needed (43 kg/person/day), would total 46,440 kg over a 180 day period (a possible crew rotation period for a Space Station mission) for a crew of 6. Over a 6 month period this translates to 1 to 2 Shuttle payloads dedicated only to life support materials.

BIO-REGENERATION AND RECYCLING

Growing plants, whether higher plants such as wheat or corn, or algae (e.g. *Spirulina*[5]), remove carbon dioxide (CO_2) from the atmosphere and incorporate it into organic materials. This process of photosynthesis occurs in the green parts of the plant and

Table 1

Use of Consumables and Packaging for Life Support per person, per day

Material	Consumables Mass (kg)	Packaging Mass (kg)
Water	22.08	6.87
Food	1.18	0.45
Oxygen	0.83	0.36
Waste		
CO_2	—	2.90
Liquid/solid	—	8.24
TOTAL	24.09	18.82

is driven by the capture of light. The first step in the process is to split water (H_2O) into 2 H atoms and O atom. The O atom then combines with another O, and O_2 is released into the atmosphere. The 2 H atoms are used by the plant to reduce (CO_2) to form a sugar. The sugar is used by the plant to synthesize protein and lipids, and is thus a source of food materials for other organisms, including humans.

The process of photosynthesis in higher plants is accompanied by the transport of water from the roots to the leaves, and then into the atmosphere through the process of transpiration. The purity of transpired water is very high, and the rate of transpiration is fast because water brings essential ions to the leaves from the roots and also regulates leaf temperature. A crop of plants sufficient to supply 97% of the food for a crew will transpire 3 times as much pure water as the crew needs for all purposes.[1] Table 2 summarizes the relative proportions of food, water and oxygen produced by growing plants in BLS system.

Thus, growing plants can produce food, remove CO_2 from the atmosphere, produce oxygen, and purify water through processes that are driven by light. Plants, which form the basis of the Earth's ecosystem, can be used in space to reduce the need for resupply. However, regeneration of materials in a closed system, such as a spacecraft or a base in space, is not efficient unless it is accompanied by methods of recycling the major life support elements, such as carbon, hydrogen, nitrogen and oxygen. On Earth the recycling processes are complex, but for the most part they depend on the metabolic activities of heterotrophic organisms (e.g. people, fish, bacteria) to oxidize organic material. Neither human metabolism, nor the metabolism of the human gut

Table 2

Life Support Consumables Produced by BLS System
per person, per day, assuming 97% regeneration of food

Material	Consumables Mass (kg) (per person)	Percent of Crew Requirements
Food	1.2	97
Water	300.0	322
Oxygen	3.6	97
CO_2 removal	3.4	97

bacteria, are completely efficient in oxidizing organics. As a consequence an unoxidized residue plus bacteria is released as waste material (feces). In most situations on Earth, wastes are further degraded by bacteria. Metabolism in bacteria is similar to that of humans in that they oxidize organic materials to CO_2 and water.

The recycling of H, O and C can be summarized as follows:

$$\text{Light (energy)} + CO_2 + H_2O \longrightarrow \text{sugar} (\rightarrow\text{organics}) + O_2$$
$$O_2 + \text{organics} \longrightarrow CO_2 + H_2O + \text{energy}$$

For the purposes of life support bacterial methods of waste degradation may be too slow or inefficient. For this reason several physical chemical methods of waste oxidation are being explored.

RECYCLING: WASTE OXIDATION

The efficiency of a regenerative life support system, and thereby, its overall cost, depends on the <u>extent</u> to which consumed materials can be recycled through the system and regenerated back into consumables. The path of recycling from waste to consumables is sometimes called a "loop", and the extent of recycling is termed the degree of "loop closure".

The waste materials generated by a crew include CO_2, feces, urine, volatiles, wash water and trash. The waste materials generated during the growth and preparation of food from plants or algae include inedible biomass (mostly cellulose and lignin), volatile and soluble organics and minerals from the plant nutrient solution. It is useful to consider

Table 3

Waste Streams in a Bioregenerative Life Support System

Waste Stream	Source	Content	Fate
1. Solid waste	Crew activities	Paper, plastics, feces, undefined trash	Oxidation
2. Urine	Crew	Urine, urinal flush water	NH_3 recovery, then oxidation
3. Domestic	Crew	Shower water, dish washing, clothes washing, oral hygiene	Removal of organics, then nutrient soln. makeup
4. Nutrient	Plant growth	Spent plant and algae nutrient solutions	Oxidation
5. Inedible biomass	Plant food processing	Roots, shoots, food process waste, water	Digest cellulose, remove sugars, then oxidize
6. Volatiles	Crew and plant volatiles, materials outgassing	Small organic molecules	Oxidation

several different waste streams in a BLS system, and to treat each separately. The waste streams, their contents and fates are summarized in Table 3.

The goal of BLS waste processing activities is 1) to recover as much usable material as possible from the waste streams, and 2) to convert wastes into forms that are usable by the crew or by growing plants. Potentially usable material includes sugars and fixed (organic) nitrogen. Cellulose from inedible plant biomass may be recovered by digestion to sugars for subsequent use.[6] Because of the energetic costs of converting atmospheric N_2 into organic nitrogen, N contained in plant biomass and NH_3 in urine may be extracted for subsequent use. The form of carbon most usable in a BLS system is CO_2; therefore, the most appropriate processes for recycling extracted wastes is chemical or biological oxidation. Several waste treatment or oxidation processes have characteristics that may be appropriate for use in a BLS system.

Candidate physical/chemical technologies for processing solid wastes in a bioregenerative life support system include wet oxidation,[7,8] super-critical water oxidation

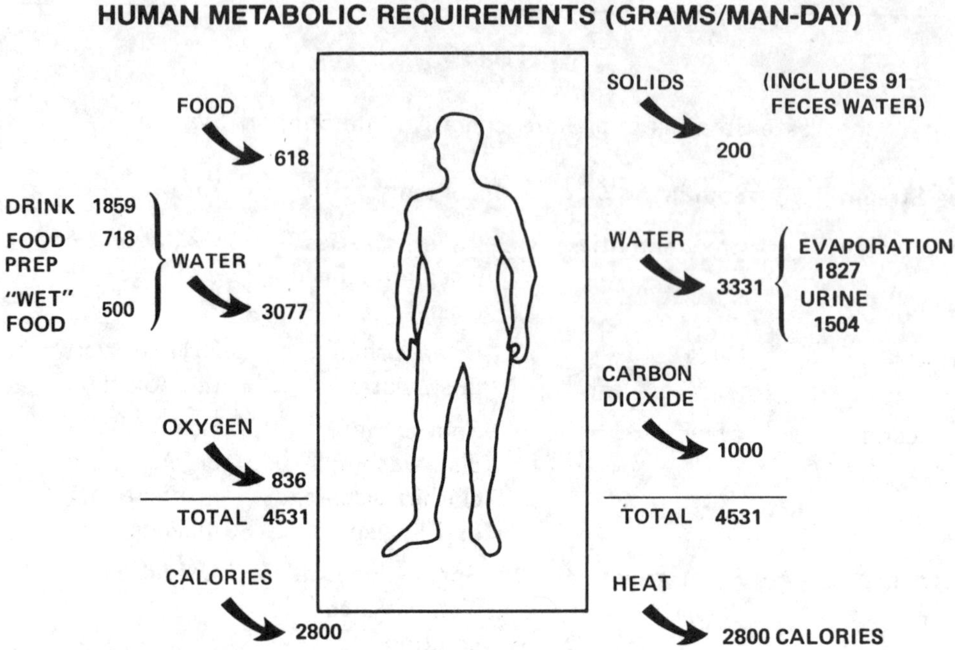

Fig. 1 Daily input/output balance of a 70 kg person performing moderate work (adapted from reference 1).

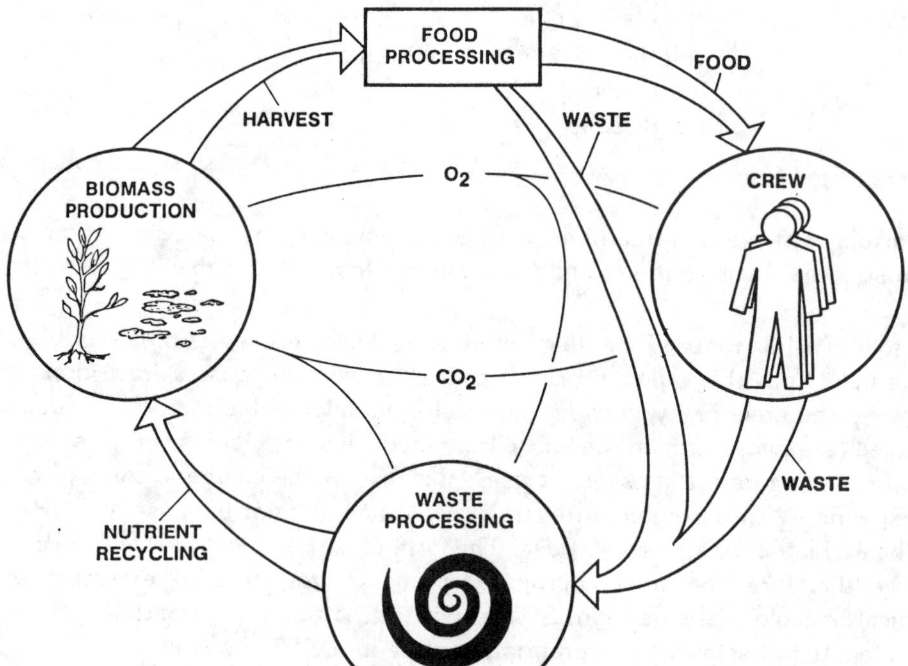

Fig. 2 Cycling of major materials in a bioregenerative life support system.

(SCWO)[9] and incineration.[10] Wet oxidation and SCWO can be used to treat dilute wastes and therefore do not require an energy intensive drying step. Incineration, on the other hand, requires a high concentration of solids in order for combustion to occur.

In wet oxidation and SCWO a dilute slurry of the process stream is contacted with air or oxygen at high pressure and temperature for a sufficient time to convert the organic waste to CO_2, water and soluble and insoluble mineral salts. Wet oxidation requires temperatures of about 250 C and pressures of about 1500 psi (105 kg/cm^2),[7,8] whereas SCWO utilizes temperatures between 500 C and 700 C and pressures of from 3000 to 4000 psi (210 to 280 kg/cm^2).[9] Assuming that the oxidation of organics is complete, the product CO_2 can be used for plant growth without further treatment. The mineral salts can also be used for plant growth,[11] however it is likely that ions which normally form insoluble salts, for example Ca^{2+}, PO_4^{3+}, Mg^{2+}, will not be recycled. Soluble minerals, such as Na^+, K^+, Cl^- and SO_4^{2-}, can probably be recovered,[12] and after concentration adjustments, be used to regenerate the plant nutrient solution.

Combustion or incineration can be used on relatively dry solid wastes to convert organic matter to CO_2, water and dry ash. Particulates produced during combustion may pose a problem which requires further assessment before this method is used to treat solid wastes in a BLS system. In addition, the fate of organic sulfur and nitrogen during incineration has not been studied in detail yet. As mentioned previously, biological waste processing, if used alone for solid waste processing may be too slow and inefficient. However, such methods may be appropriate for recovering fixed nitrogen (NH_3^+ or NO_3^-) from solid organic materials. It is anticipated that waste recycling in a BLS system will be accomplished with several processes that will be chosen by their capabilities to reduce the mass, volume and power demands of the system.

There are similarities between bio-regenerative life support systems and a natural ecological systems (Figure 2). Some comparisons have been helpful because they have highlighted processes, such as ecological interactions, that will inevitably be a part of a BLS system. However, a BLS system cannot be allowed to operate in response to natural physiological or genetic control mechanisms. One reason is that a BLS system will be very small in comparison to an ecological system, hence the transport of materials through its cycles will need to be very much faster. A second reason is that the goal of a BLS system will be, very specifically, to support human life, rather than to place humans in competition with other life forms. As a consequence, the operation of the system will depend upon the intelligent control of the location and flow of materials in the system with the intention of making the system operate efficiently and in a stable mode over a long period of time.

THE ECONOMICS OF BLS SYSTEMS

The benefits of bio-regenerative life support systems for future missions will depend on the system's economics. The primary costs of a BLS system are associated with system development, deployment for specific missions, and long-term operation, including repair and replacement costs. In each of these areas a BLS system must compete with alternative approaches, such as storage of expendables and physical/chemical regeneration methods, and for final decisions to be made the sum of all costs must be compared. While the advantages of regeneration of life support materials are primarily economic; the advantages of a BLS system include both psychological and economic factors. It is difficult to quantify human reactions to the presence of "fresh food" in space, yet such responses, which can promote a sense of well being among a crew, will undoubtedly be significant factors in choosing a life support system for extended missions.

At the present time there are many uncertainties regarding the specifics of potential missions, and insufficient information is at hand to make reliable or realistic cost estimates for life support systems. For this reason, research is proceeding to explore the development of bio-regenerative life support systems under the aegis of the NASA Headquarters Life Science Division's Controlled Ecological Life Support System (CELSS) program. As research progresses, and missions become more clearly defined, cost comparisons will become increasingly accurate. However, even at this time cost can be estimated sufficiently well to strongly suggest that a BLS system is well worth consideration.

RESEARCH DIRECTIONS AND MISSION SCENARIOS

To a significant degree the directions of research in the CELSS program are dependent upon the kinds of space missions that are projected. A transportation analysis,[1] completed in 1982 provides a basis for identifying research priorities. The transportation analysis allows a comparison to be made among life support systems that a) depend on resupply, b) rely on physical/chemical methods of regeneration, and c) utilize bio-regenerative approaches. The 1982 study resulted in a series of break-even analyses which serve to illustrate the logic used in identifying research directions and priorities.

A space mission that relies totally on resupply, without any recycling, (for example, in low Earth orbit) requires an initial launch mass of 17,895 kg, and has a re-supply demand of 13,552 kg every 90 days. To evaluate the impact of regeneration techniques scenarios for physical/chemical and bio-regeneration were projected. Regeneration techniques were projected by calculating the mass that must be launched into orbit solely for life support purposes. These data were extrapolated for a period of several years for 3 different life support scenarios (Figure 3). The scenarios were applied to a low Earth orbit mission (e.g. Space Station), and assumed a 4 person crew.

In one case ("water and air closed") in which air and water were regenerated by physical/chemical methods the initial launch mass required for a crew of 4 is 4,153 kg, and resupply at intervals of 90 days requires 1,069 kg each time. In contrast, a bio-

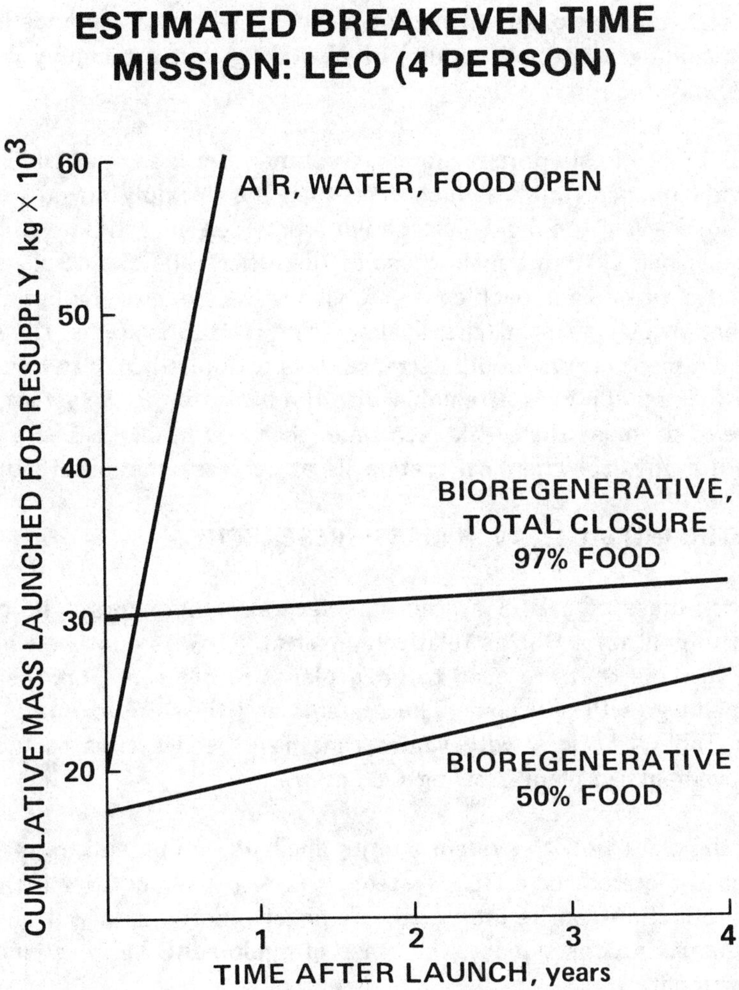

Fig. 3 Breakeven analysis for bioregenerative life support techniques compared to physical-chemical life support methods based on data gathered in 1982. The time, in years, at which the curves intersect represent the time at which the mass of materials brought into orbit for the systems or scenarios become equal. The CELSS research program has focussed on methods of reducing the initial launch mass; the slopes of the curves remain constant based on the assumption that the extent of regeneration remains the same.

regenerative system capable of supplying 97% of the crew's food demands has an initial launch mass (including a special module) of 27,000 kg, but a resupply mass of only 237 kg every 90 days.

In comparing different life support scenarios, the time after launch at which the cumulative total mass launched (initial launch mass plus the resupply mass) become equal is called the "break-even" point. This technique provides an estimate of the time at which the costs of one system equals those of the other. The extent of recycling and regeneration that is possible in each case establishes the slope of the lines, and during an evolution from strictly physical/chemical systems to BLS systems, the slope of the life support requirement curves would decrease. It is estimated that an efficiency of recycling of 97% (3% resupply) is attainable with the projected BLS system. The single option available to decrease the break-even time, that is, to make a BLS system more competitive with a physical/chemical system, is to decrease the initial launch mass.

DECREASING THE INITIAL LAUNCH MASS: RESEARCH

The initial launch mass of a BLS system has several components. The mass of the module needed to contain a BLS is relatively insensitive to its contents; however, the masses of the systems that are used to grow plants or algae respond nearly linearly to changes in plant growth efficiency. The systems and their approximate weights are summarized in Table 4,[1] along with values that have been estimated more recently based on improvements in plant growing techniques.*

It is clear that the plant growth system and its ancillary equipment are major contributors to the mass required for a BLS system. It is also obvious that if the efficiency of plant food production can be increased, a smaller planted area will be required to support one person. In consequence, the mass of equipment, lighting, and water will decrease proportionately.

To compare the food production efficiency growth of different kinds of plants that might be needed to provide a complete human diet, scientists in the CELSS program frequently use a technique based on the caloric content of food material produced by a plant as a percentage of the daily human caloric requirement, without immediate consideration of diet content (carbohydrate, lipid, protein). It is useful briefly to expand on this comparison to provide an example of how applied biological research can alter system concepts and design.

One example of the calculations involved is based on the growth of wheat plants, and assumes a food growth scenario in which enough wheat plants are harvested each day to yield sufficient wheat berries to provide 2800 Calories. Each day the same number of

---

* Personal communication, Dr. Bruce Bugbee, Utah State University, Department of Plant Science

Table 4

Initial Launch Mass of Components of a BLS System
(assumes 25 m^2 required for food growth, per person)

Material	Mass (kg) (per person)	Mass (kg) (per person)
Year of Data	1982	1987
1. Water	1867	896
2. Water (nutrient) Tanks	584	280
3. Structures (plant growth)	180	86
4. Ancillary equipment	1336	641
5. Harvest/Process equip.	134	134 *
6. Waste process equip.	56	56 *
7. Power (lighting) (electric, 43 Kw/person, 45.3 kg/kw)	1970	945
TOTAL	6127	3038

* not affected by plant growth efficiency improvements

wheat plants are started. The amount of volume required for the growth of a "perpetual crop" depends on the time from seeding to maturity, the number of wheat berries per head, the average weight of each wheat berry, the caloric content of the berries, the number of plants per unit area, and the height of the plant.

At the time the first Transportation Analysis was done (1982) the data available indicated that 25 m^3 was sufficient to support the caloric requirement of one person. Through extensive research, done by Professors Frank Salisbury and Bruce Bugbee of Utah State University,[13] under the support of the CELSS program, the fundamental efficiency of wheat growth has changed significantly. The research has resulted in an improvement in the growth of wheat to the point that the volume required has been reduced to 12 m^3. This reduction of 52% in the required growing area and volume can be reflected in a proportionate reduction in the mass of plant growth and lighting equipment.

The success of this work has been based on careful examination of the effects of the environment in which the wheat plants are grown: the temperature, humidity, the intensity, quality and cycling of the light received by the plants, and the rate at which essential plant nutrients are supplied to the plants. Other significant factors include the strain of wheat planted, and the ratio of food weight to the weight of inedible biomass.

While fundamental work on the growth of plants and algae have had major effects on the efficiency of BLS systems, technological improvements also contribute major increases in efficiency. In fact, physical/chemical systems will constitute a major fraction of the functions of a BLS system, and development work on such sub-systems will eventually consume the largest part of the resources devoted to the development of a BLS system.

An example of increased efficiency due to technological advances is given by consideration of the use of power by a BLS system. Electrical lighting requires that a solar power array be launched capable of supporting the lighting demands of plant growth. Recent work[14] by Prof. K. Mori of Japan has allowed the development of a method of directly capturing sunlight by a collector array and transferring it into a plant growth module through optical cables and diffusing the captured light for plant growth. The efficiency claimed for the system is 60%,[14] contrasting with solar power arrays that have a sunlight to electric power conversion efficiency of between 2% and 10%,[15] coupled with electric lamps that have efficiencies of about 40%, yielding an overall efficiency of between 0.8% and 4.0%. The direct sunlight-capturing system also removes much of the undesirable solar infra-red and the ultraviolet radiation through the use of focussed, concentric Fresnel lenses. The use of such systems can substantially reduce the electric power demands of plant growth, and dramatically reduce the initial launch mass of a BLS.

CONCLUSIONS

Thus, research and development can substantially decrease the break-even point of BLS versus physical/chemical systems, primarily by decreasing the required initial launch mass. The applied and fundamental research conducted by the CELSS program is focussed on such problems as maximizing plant growth through understanding the environmental conditions required for their growth, exploring methods of fractionating algae and preparing food from algal biomass, understanding waste oxidation processes, developing the models and control strategies required to maintain a productive and stable BLS system, and developing crucial flight experiments and tests to evaluate the effects of gravity and the space environment on the biological and fluid handling machinery.[15,16] As understanding of the operation of a BLS increases it becomes increasingly possible to define the requirements of the system, and to identify the technology development that must be initiated.

In general, the CELSS program is responding to the requirements and constraints of operating in the space environment. The requirements include 1) supplying human life support needs, including dietary constituents and calories, appropriate concentrations of atmospheric gases and potable water; 2) maintaining an ability to recycle and regenerate materials; and 3) insuring system stability and resilience. The constraints of the space environment include 1) limitations on mass and volume use; 2) restrictions on power and human labor use; and 3) the effects of in-space gravity and radiation.

Bio-regeneration is potentially useful, and possibly essential, in establishing autonomous life support systems for use on the Moon. But short of system autonomy, a BLS system is capable of performing all of the functions of life support. The extent to which a BLS system can regenerate life support materials, and thus substantially reduce resupply, makes it attractive even for Earth-orbiting missions with complex logistics. The advantages of bio-regeneration include psychological as well as economic factors. For example, it is difficult to quantify a crew's reactions to the presence of "fresh food" on a mission, but an occasional salad would be a significant factor in promoting a sense of well being. Thus, for a number of reasons it is appropriate to explore the usefulness of a bio-regenerative life support system for future NASA missions, including Space Station, Lunar Bases, Transit to Mars and Mars Bases.

REFERENCES

1. E. Gustan and T. Vinopal, "CELSS Transportation Analysis", NASA CR 166420, 1982

2. P. D. Quattrone, "Extended Mission Life Support Systems", The Case for Mars, Edited by Penelope Boston, AAS Science and Technology Series Volume 57, 1984.

3. F. H. Schubert, R. A. Wynveen, and ¡P. D. Quattrone, "Advanced Regenerativve Environmental Control and Life Support System: Air and Water Regeneration," Adv. Space Res., Vol. 4, No. 12, pp. 279-288 (1984).

4. R. D. MacElroy and J. Bredt, "Current concepts and future directions of CELSS", in: Controlled Ecological Life Support Systems, R. D. MacElroy, D. T. Smernoff, H. P. Klein, eds, NASA CP 2378 (1985).

5. O. Ciferri, "Spirulina, the Edible Microorganism," Microbiol. Revs., p. 551, (Dec. 1983).

6. M. Karel, "Evaluation of Engineering Foods for Controlled Ecological Life Support Systems (CELSS)," NASA CR 166359, 1982.

7. R. B. Jagow, "Development of a Spacecraft Wet Oxidation Waste Processing System", The Society of Mechanical Engineers for presentation at the Environmental Control and Life Support Systems Conference, San Francisco, CA, August 14-16, 1972.

8. C. C. Johnson and T. Wydeven, "Wet Oxidation of a Spacecraft Model Waste," SAE Technical Paper Series 851372, Fifteenth Intersociety Conference on Environmental Systems, San Francisco, CA, July 15-17, 1985.

9. G. T. Hong, Patrick K. Fowler, William R. Killilea, and Kathleen C. Swallow, "Supercritical Water Oxidation: Treatment of Human Waste and System Configuration Tradeoff Study," SAE Technical Paper Series 87144, 17th Intersociety Conference on Environmental Systems, Seattle, Washington, July 13-15, 1987.

10. H. P. Meissner and Michael Modell, "Recycling Plant, Human and Animal Wastes to Plant Nutrients in a Closed Ecological System," The American Society of Mechanical Engineering for presentation at the 9th Intersociety Conference on Environmental Systems, San Francisco, CA, July 16-19, 1979.

11. B. L. Onisko and T. Wydeven, "Wet Oxidation as a Waste Treatment in Closed Systems", The American Society of Mechanical Engineering for presentation at the Intersociety Conference on Environmental Systems, July 13-15, 1981.

12. E. Ballou, L. A. Spitze, F. W. Wong, T. Wydeven, and C. C. Johnson, "Ion-Exchange Chromatography Separation Applied to Mineral Recycle in Closed Systems," The American Society of Mechanical Engineering for presentation at the Intersociety Conference on Environmental Systems, San Francisco, CA July 13-15, 1981.

13. F. B. Salisbury and B. G. Bugbee, "Wheat Farming in a Lunar Base," in Lunar Bases and Space Activities of the 21st Century, W. W. Mendell, Ed., Lunar and Planetary Institute, Houston, pp. 135, 1985.

14. K. Mori, H. Ohyo, K. Matsumoto, and H. Furume, "Sunlight Supply and Gas Exchange Systems in Microalgal Bioreactor," Adv. Space Res., Vol. 7, No. 4, pp. 47-52 (1987).

15. M. Oleson and R. Olson, "Controlleed Ecological Life Support Systems (CELSS): Conceptual Design Study," NASA CR 177421, 1986.

16. R. D. MacElroy, J. Tremor, D. T. Smernoff, W. Knott, and R. P. Prince, "A Review of Recent Activities in the NASA CELSS Program," Adv. Space Res., Vol. 7, No. 4, pp. 53-57 (1987).

QUESTIONS AND ANSWERS - LIFE SCIENCES

Dr. Arnauld E. Nicogossian: We have ten minutes left. I hope that we have some written questions. The host chairwoman, Dr. Huntoon, will be distributing the questions.

Dr. Carolyn L. Huntoon: The first question is for Dr. Stewart.

Question: With no rapid return capability from the Space Station, how will you monitor and treat radiation dosages, say, from a solar event?

Dr. Donald F. Stewart: I'm not an expert in the area of radiation biology but presumably, if that happened, if we had some sort of solar event at a time when there was no rapid return capability, we would just have one or two options. Hopefully, there's enough shielding at the orbit of the Space Station so that the doses would not be lethal. Above and beyond that, I think you would have enough lead time so that we could mount a rescue mission. Once again, I'm not a solar physicist but there is usually some kind of a warning of the solar activity before you see any kind of event that would be of biomedical significance. So hopefully, we could either mount a rescue or the doses would not be biomedically significant.

Dr. Carolyn L. Huntoon: The second question is from Pat Jackson, who is a freelance space writer and video maker. And she asks Dr. Nicogossian.

Question: How helpful have the Soviets been in sharing life sciences information and data? Have they been completely open? Can we or should we be doing experiments they've done and build on what they've learned?

Dr. Arnauld E. Nicogossian: The answer is very simple. The Soviets have been very open with us and shared whatever they had to share. It was useful in some cases; in other cases, it left us with more questions than answers. From my experience, I think that when we asked a question, either we were told that they didn't know or they knew and gave us an answer. I suspect that Dr. Huntoon will support what I'm saying.

Dr. Carolyn L. Huntoon: Are there any other questions? Let me thank you all for coming. I can tell you that life sciences is facing a great challenge. It certainly is going to be a challenge for all of us, the Government and private sectors. Join with us and help us, as I know most of you have been doing and will continue to do, because we have got a lot of exciting work before us. Thank you for coming.

TRACKING AND DATA SYSTEMS

PANEL SEVEN

Chairman:	Charles T. Force Deputy Associate Administrator for Space Operations NASA Headquarters
Host Chairman:	Phillip M. Hopkins Manager Tracking & Communications Management Lockheed-EMSCO

AAS 87-648 to -649

TRACKING AND DATA SYSTEMS - INTRODUCTION

Phillip Hopkins* and Charles T. Force†

Dr. Phillip M. Hopkins: Welcome to Panel 7. My name is Phil Hopkins. I'm the host chairman and I have the easiest job today, because I don't have to do very much.

It's my pleasant chore to introduce Charles T. Force, who is the chairman of this panel. Charles Force is presently the Deputy Associate Administrator for Space Tracking and Data Systems at NASA Headquarters. He was appointed to that position in September, 1986, and that office was absorbed into the newly formed Office of Space Operations, OSO, in January, 1987. OSO is responsible for the development of a comprehensive plan for managing NASA space operations to be implemented when the Shuttle recovery is complete, and by the way, to continue the overall management of NASA's Tracking and Data Systems.

Prior to that, going back to 1965, when Charles joined NASA, he was the Director of the Guam Tracking Station. Subsequent to that, he was Director of the Quito Tracking Station. So he does have experience in how the data system really works. He was the Deputy Director of the TDRSS Project, he was the Chief of the Network Operations Division at the Goddard Space Flight Center. In 1980 he left for a couple of years to work in industry. He returned in 1982 as Program Manager for the Deep Space Network in the Office of Space Tracking and Data Systems in NASA Headquarters. He was Director of the Ground Networks Division, responsible for STDN, Deep Space Network, the ground tracking and data support for the Aeronautics Balloon and Sounding Rocket Program for the Laser Tracking Network. And another responsibility, by the way, was the construction of facilities and the safety program for the OSTDS. So certainly he is well qualified to speak to us today on the subject of Tracking and Data Systems.

He has received the NASA Exceptional Service Medal for his work in the Deep Space Network, the Department of State Meritorious Honor Award, for his work in Quito, Ecuador. Charles is a native of Indiana and graduated in Science and Aeronautical Engineering from Purdue University in 1957, *cum laude*, he lives with his wife, Marilyn, and five children in southern Anne Arundel County in Maryland. Join me in welcoming Charles T. Force.

---

* Dr. Hopkins is Manager, Tracking and Communications Management, Lockheed Engineering and Management Services Company, 2400 NASA Road One, Houston, Texas 77058.

† Deputy Associate Administrator for Space Operations, NASA Headquarters, 600 Independence Avenue, S.W., Washington, D.C. 20546.

Charles T. Force: Phil said he had the easiest job in that he didn't have much to do. I think he misled you, for he's done a great deal already in preparing for this session.

I would like to welcome you all, on behalf of the panelists, to the session on Tracking and Data Systems.

To provide a perspective I'd like to characterize, what we mean generally by tracking and data systems, what the characteristics are, particularly as they pertain to the theme of the session of this symposium, and how we can all be of best service to the new community which is emerging to use space. The primary characteristic of tracking and data that dominates what we do is the fact that it's a service element. It's a means to an end, it's not an end in itself. This has some rather significant repercussions in the way we operate, in the sense that ultimately most of what we do is kind of a zero sum game. I think that the more efficiently we can handle the tracking and data element, then the more resources are available for the more front line activities of the aerospace program. If we do our job well, it's generally invisible to our customers, which makes us feel sometimes as if we're the 'unsung people', and on occasions even 'unsung heroes', when we do something heroic regarding the things that occur. Unfortunately, when we 'goof', then everybody knows immediately.

Another characteristic of tracking and data is that it has evolved over the years, which influences the way we look at the future. In the early days, we had, say, a radar out in the field which tracked the satellite; we had a transmitter to send a command to it and a receiver to receive telemetry. Normally, in the very early days, we had a completely separate system to send voice, if it was a manned mission, another one to receive it. A lot of data was recorded. On the manned missions the Johnson Space Center used to send flight controllers out to the remote sites to take actions there. Over the years, technology has let us operate much differently, much more efficiently. But it gives us quite a bit more challenge.

What has happened is that we have integrated to the point where we now usually have just one radio link up. We use it for tracking, for transmitting instructions up, for transmitting ranging information. We have one link down and we use it in reverse. The whole thing now has to tie together. It's kind of a dynamic interactive network. We operate with real time data and we'll hear more about that from one of our panelists this afternoon as to what is involved, in particular, with the OMV, and trying to operate it remotely from the ground. There is another characteristic which is serendipitous, and which has led to some of the international activities in the space programs, and that is that tracking and data have been internationalized. The reason was simply geographical. We had to go rent somebody else's turf in order to operate around the world. That gave us entrees, which then served as a base for a lot of the international cooperation, which made sense.

Another characteristic I want to point out is the fact that tracking and data elements are usually separate. There are a couple of reasons why that happened but, generally, an element has a tracking and data activity and is not handled as part of the main line program; let's say it's a flight project. It's a support element, such as NASA's Office of Space Tracking and Data, now Space Operations. One of the reasons, I think, it has evolved this way is the fact that the expertise is fairly specialized. There's a fairly small community of those of us working in it. So it's simply a lack of resources that has made that an economical way to go. The other thing is that it's an infrastructure element, a service element. Any particular organization needing tracking and data usually has a number of related activities and can support all of them, say, with a common tracking and data element. It therefore makes economic sense also to do it that way.

Our primary objective this afternoon is not to provide a forum for those of us here who are spending our lives in tracking and data, trying to communicate among ourselves about what our problems are and what our solutions are. It's rather a forum for us to try to interact with what we view as the broader audience that we have with us this afternoon. And that is, how can tracking and data help in looking at the priorities as we see them, the needs as we see them, the needs as you see them, the priorities as you see them? Looking at the limitations, technology, for example, gives us limits on what we can do. And these are not always well understood by those who are designing the missions which require support from tracking and data. This afternoon we're looking primarily to share with you what tracking and data mean to us and to get an input, hopefully, in the form of questions, as to how we can be more responsive to the needs as you see them in the future.

The panelists that we have with us this afternoon cover a rather diverse area and a wide range. The tracking and data function like most fields is quite broad within itself. The first paper covers tracking with respect to Space Station. We're going to be focusing on something as far out, literally, as trying to navigate, say, the Voyager II spacecraft through the Neptune System. That spacecraft, by the way, is so far out that it takes a five-hour one-way light time just for a signal to get to it. So trying to navigate with precision at that distance without running into the planets or the Moon, but yet come close enough to get the needed scientific data, is quite a challenge.

The other end of our session and the last paper will be looking at something very close to Earth in the future, and that's your Orbital Maneuvering Vehicle, the OMV, and there we're faced with challenges in the sense that we need a lot of data and need it very fast. Even transmission time has become a factor because if someone on the Earth is trying to fly a satellite to rendezvous with another one, which is possibly even spinning or tumbling, time delays become quite critical.

So the panelists this afternoon will try to cover the whole spectrum, not with the intent of covering each one of them in any depth but with the intent of bringing out the issues which span the whole spectrum. One of the issues, just to illustrate, which I think we would appreciate a broader understanding of, is a need for standards. There are many ways that we can do a tracking and data job, and a lot of them are quite arbitrary. If we choose a certain set of frequencies, a certain set of modulation characteristics, and somebody else chooses a different set for the infrastructure or support system, we need two of them to do a job that one, in general, could accomplish. So

I'd like you to give thought to what kind of issues underlie the tracking and data job as we go through the session this afternoon. The executive chairman suggested that we try to introduce each session with a video tape describing our particular area. Through the cooperation of the Jet Propulsion Laboratory, we have one for this session which I'd like to show at this time.

(SHOWING VIDEO TAPE HERE)

INTRODUCTIONS OF PANELISTS

Our first panelist is Dr. J. Frank Jordan from the Jet Propulsion Laboratory. I mentioned to you that Frank was going to talk about how he was going to miss hitting Neptune. However, he has told me that this is not the case. Frank bears some responsibility for this subject, so I don't know if there's a message in this or not. He's going to talk on a subject a little nearer to Earth about precision tracking. Frank has a Ph.D. in engineering mechanics from the University of Texas at Austin. He has spent some 22 years at the Jet Propulsion Laboratory working in deep space tracking and navigation. He has received several professional awards and is the author of quite a number of technical papers.

Most of you have seen a lot of pictures of astronauts today. There's a gentleman named Charlie Walker who has flown three times as a mission specialist, and virtually all the pictures you've seen of Charlie are when he is buried inside the equipment. One of the advantages of sending men and women into space is the fact that when things don't work right, then they've got the expertise to fix them.

Our second speaker is a gentleman who was on the ground at the end of those conversations and trying to help Charlie work with those problems and has quite an insight into what tracking and data did and where it fell short, in helping Charlie troubleshoot the problems on those flights. He is David W. Richman, Deputy Manager of McDonnell Douglas's Electrophoresis Operations in Space Program. He has been involved in Gemini programs, Space Shuttle, and Skylab. And he has a number of patents for electrophoresis hardware and process innovations. His degrees were an undergraduate degree from the University of Missouri, and M.A. from Purdue. I hope I don't get accused of overburdening the panel with Purdue graduates, because it turns out there are several besides myself here. Mr. Richman also has written a number of papers and has received a number of awards.

Following up on where technology is taking us, our third panelist is Dr. Kumar Krishen. Dr. Krishen is presently manager of the Advanced Programs in the Tracking Communications Division at NASA-Johnson Space Center. He received his advanced degrees including a Ph.D. from Kansas State University and was an Assistant Professor in Electrical Engineering at Kansas State before he joined Lockheed in 1969. He joined JSC then in 1976. He also has won a number of awards and written several papers.

The next speaker is going to be talking about Space Station Communications. He is Dr. Albert W. Weinrich, Division Vice President of Transmission Systems at RCA Government Systems Division. Dr. Weinrich has extensive experience in space-related programs. At RCA's Electro Space Division, he was manager of the Series 4000 Satellite Program which was responsible for the design, development, and production of the Series 4000 Ku Band Spacecraft. Dr. Weinrich has received his bachelor's in civil engineering from Missouri School of Mines, a master of electrical engineering at Drexel, and a Ph.D., I hesitate to mention, from Purdue University.

The last panelist is Arthur G. Stephenson from TRW. He is going to talk about the Orbital Maneuvering Vehicle, OMV, which is being developed by TRW under contract to the Marshall Spaceflight Center. It's now one year into the development, with a launch scheduled in 1993. The vehicle is unmanned and teleoperated from the ground by a TDRSS. As you can anticipate, the tracking and data challenges are quite significant for that program. Art has been with TRW for 23 years. During that time he has managed the development of the Pioneer Jupiter spacecraft receiver and the Shuttle orbital S-band communications transponder, among other things.

AAS 87-650

TRACKING REQUIREMENTS TO SUPPORT SPACE - VERY LONG BASELINE INTERFEROMETRY

J. Frank Jordan*

Thank you, Charles. It's always tough to follow a good video tape. I hope that my words dovetail with the tape that Charles showed. I'm going to attempt to show a single future mission example. and how it has a very heavy dependency on space tracking. My example is the placing into space of a radio telescope. The reason we place a radio telescope into space is to increase the resolution with which we can image the observed radio sources. Here I show a cartoon of high image resolution of a very distant source (Figure 1). The mechanism by which we get high resolution is called Very Long Baseline Interferometry, VLBI, and it works on the following principle. Multiple antennas observe the source simultaneously, and then the data from each source, the data from each pair of antennas, are correlated at a common site.

Figure 1 QUASAT: High Resolution Radio Interferometry from Space

Interferometric tracking is conducted from the ground today at the VLA, the Very Large Array, that George Newton spoke of this morning, in Socorro, New Mexico. It's in an array of 27 antennas across 30 kilometers in central New Mexico.

* Dr. Jordan is Manager, Systems Division, Office of Technical Divisions, Jet Propulsion Laboratory, 4800 Oak Grove Drive, Pasadena, California 91109.

Each antenna observes a common radio source. As the Earth turns, these baselines sweep out and cover a quasicircular type of an aperture. VLBI is also the technology which is being implemented in what we call the VLBA, the Very Long Baseline Array, implemented by the National Radio Astronomy Observatory, and will be operating as a continental array of ten telescopes starting in 1992. The advantage of a space radio telescope is that whereas a continental array synthesizes an area of about the size of a continent, a space telescope adds to an array synthesis more like the size of the orbit, very much larger than the Earth indeed.

Now I have a cartoon to illustrate what I mean (Figure 2). Here we have the continental array and in cartoon fashion the synthesized radio telescope, and we have a space orbiter acting in concert with the ground orbiters synthesizing a very large aperture. Now, I hasten to add that when I talk about a large aperture, I mean an aperture in the resolution sense and not in the sensitivity sense. The sensitivity is, of course, determined by the actual areas of the particular elements. The resolution is determined by the baselines which are correlated over, and those baselines now, instead of being continental links, are much larger. In this fashion we can construct the largest camera ever, which can produce resolution (if I define resolution the way one does in college physics textbooks) the wavelength divided by the diameter of the aperture. This resolution is three orders of magnitude higher than the best achieved with the Hubble Space Telescope, three orders of magnitude higher then the best achieved with the VLA, and an order of magnitude higher then can be achieved with the VLBA.

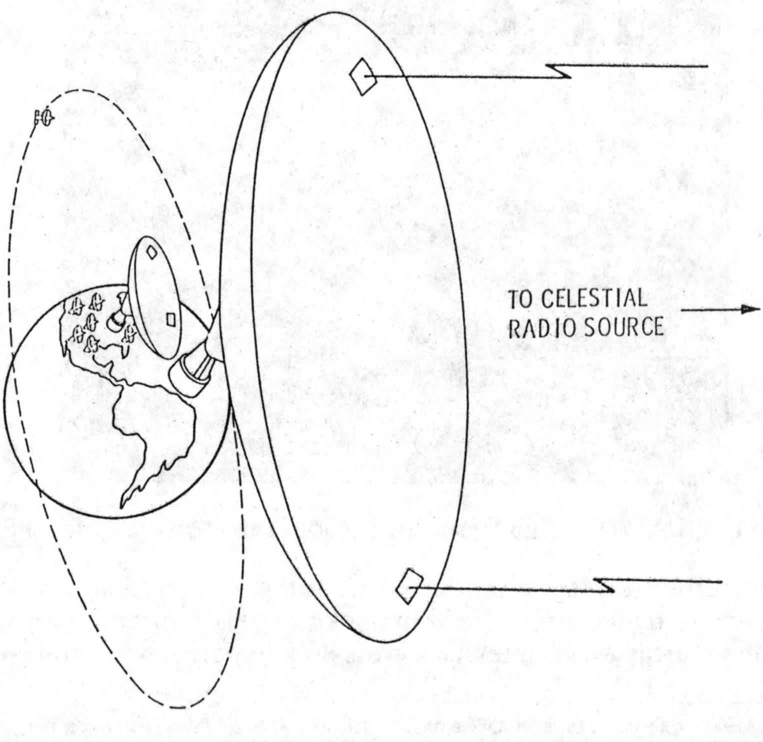

Figure 2 Space VLBI: Synthesizing Antennas Larger Than the Earth

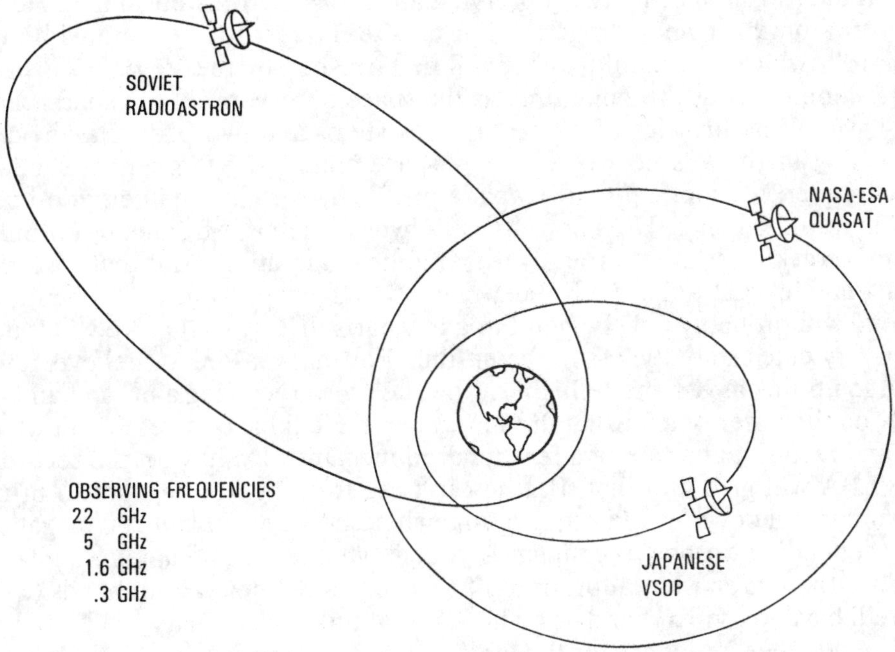

Figure 3 Space VLBI Missions in the 1990s

There are three separate studies for space VLBI observatories that are currently being conducted among the space agencies, all of which have the potential for launch in the mid-1990s (Figure 3). The first is the Soviet concept, RADIOASTRON, that proposes a ten-meter dish in a 24-hour elliptical/geosynchronous orbit, so that for the bulk of the orbit, it can be timed to be placed over the Soviet ground stations. ESA and NASA have studied over the past few years a concept called Quasat, perhaps it will eventually be a joint agency mission. Quasat proposes a 15-meter antenna and subsynchronous orbit. Finally, the Japanese, at the Institute for Space and Aeronautical Sciences in Tokyo, have been studying a mission they call the VLBI Space Observatory Program, VSOP, also scheduled for possible launch in the mid-1990s. All of these possible space VLBI observatories could be launched in the mid-1990s and they will all operate at standard radio astronomy frequencies between a third of a gigahertz and 22 gigahertz.

The point of this discussion is that there's quite a price to pay for radio astronomy in space. In addition to the price of implementing a spacecraft and launching it, there's the price of the tracking, which is in heavy demand. There are two key requirements that influence the stringent tracking requirements in space VLBI observatories. Thus, a radio telescope is, in essence, a directed antenna, a receiver, a master clock which must function as a very stable frequency reference to drive the receiver, and then a data recording system. The requirements on both the frequency stability for space VLBI and the bandwidth and recording data rates are so stringent that you can't carry everything you need on board, so you're forced into a tracking space system that looks like the following schematic (Figure 4). It shows an independent radio astronomer merrily observing an extragalactic radio source. He drives his receiver from his own hydrogen maser on site; he records his data with his own data recorder which, in the mid-1990s, will be furnished by the VLBA as a standard item. He's doing this at the same time as the space observer, but the space observer doesn't have all the luxuries of big recording systems and hydrogen maser on hand. Very possibly in the 90s he won't have a space-qualified hydrogen maser able to produce frequencies good up to a few parts in 10^{14} over common integration periods of a few hundred seconds. He'll have to use a hydrogen maser frequency up-linked to him from a tracking station for all of the time he's operating. And that's a stringent requirement. He's got to have a frequency good to a few parts in 10^{14}. The maser on the ground will probably only be good to a few parts in10^{15}, so that we won't be able to suffer very much stability loss in the up-link. For that reason, we believe, it will be required to up-link at X-band or higher. Now, he's also receiving a broad band signal, probably on the order of a few hundred megahertz. He'll have to get that down to the ground and record it at a rate of a few hundred megabits. He has got the recorder for it; the VLBA will give him that. But he can't use it on board; he's got to bring the broad band signal down. It will require somewhere between 50 and 200 megahertz to do that, depending on how the signal is coded, whether it's analogue, digital or so forth. Since there aren't any allocations of that size in the near Earth bands (S- or X-band) we'll have to use Ku-band or higher for that link. In summary, we'll have to use X-band, or a higher frequency for the up-link, and Ku-band for the down-link.

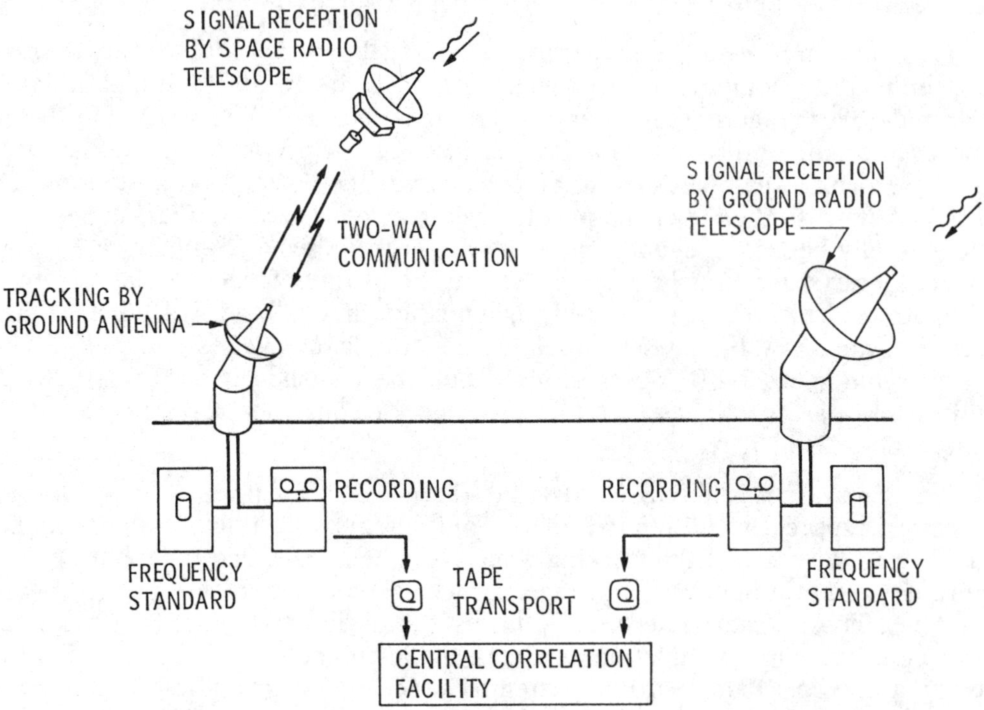

Figure 4 Diagram of a Space-Ground VLBI System

The full contingent of requirements from the tracking system also includes the spacecraft control up-link which can be at S-band and the down-link engineering telemetry of a few kilobits. And then of course, are the two-way requirements for Doppler and ranging for precision orbit determination.

Now, let me say a word about tracking coverage. Here is a map showing some of the potential ground tracking sites for high Earth orbiters (Figure 5). If I take the Quasat Mission, as an example, which has a five-hour period with the orbiter at a distance from the Earth of, say, ten thousand kilometers; the tracking passes tend to be about two or three hours in duration occurring two or three times every day. A station that commits itself to supporting a space VLBI terminal, then would be active about 25% of the time on that single mission. Now, in the video tape, the Goldstone, Madrid and Canberra sites were mentioned. If these three deep space net antennas were trained on a space VLBI orbiter, it would only cover it about 65% of the time, making it a two-thirds operable space observatory. If ESA upgraded some antennas they use in South America and in Africa, we could drive that up to 85% of the time. And with the addition of a Far Eastern terminal we'd get almost full time coverage. The major point here is that the tracking coverage needs to be continual. It may involve many ground stations with international cooperation. The tracking will have to be at higher frequencies than we're used to, X-band or even Ku-band for the up-link, and certainly Ku-band for the down-link. And all the links will be required because of the frequency stability requirements, and the broad band data will have to be secured through an allocation of a few hundred megahertz, perhaps at Ku-band.

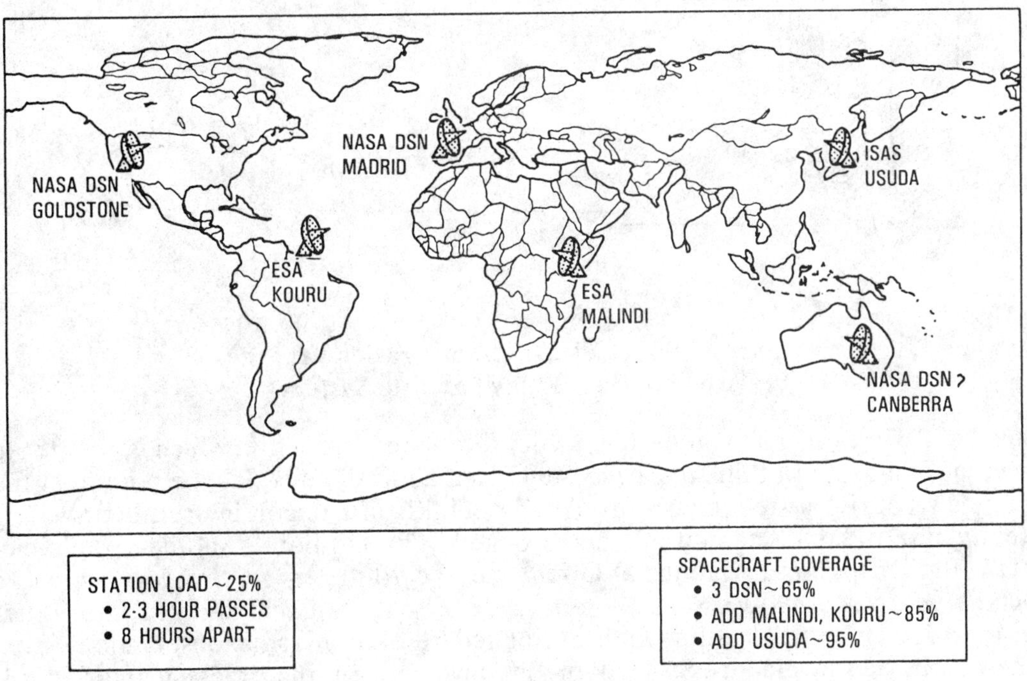

Figure 5 Space-VLBI Tracking Coverage

Now, I want to make one more point in closing, and that is that we have demonstrated the technologies for space VLBI with a series of experiments conducted last January with the Tracking Data Relay Satellite System which also appeared in the video tape (Figure 6). We operated it in conjunction with the DSS-43 in Canberra and a couple of Japanese stations. Those three stations and the space antenna recorded simultaneous data for a series of Southern Hemisphere radio sources. We were successful in observing all of them and correlating all of the data at S-band, and we'll repeat this experiment again next spring at Ku-band. I think this successful experiment demonstrates two important things: (1) That the space VLBI technique works and that the technologies are now available, and (2) that international partnerships can be successfully brought to bear to coordinate, the observational scheduling of a space antenna along with ground antennas on three separate continents. Thank you.

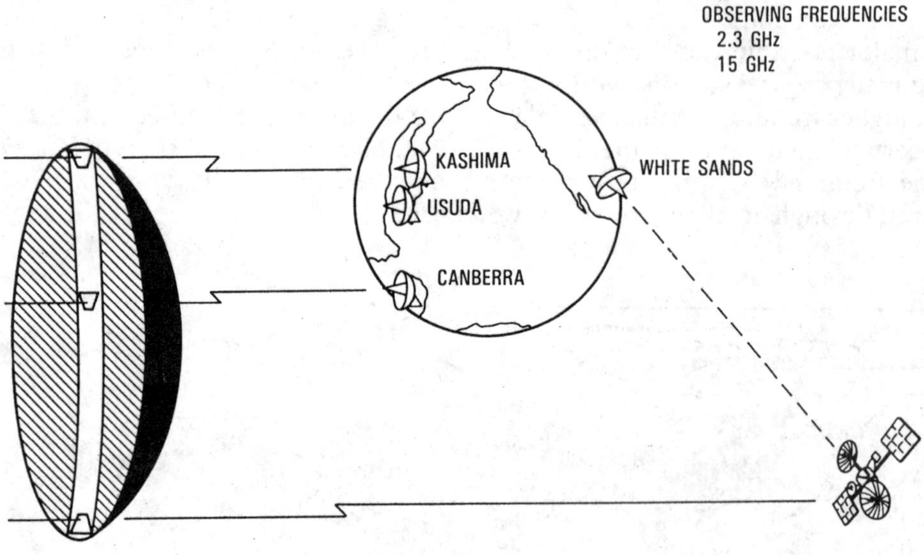

Figure 6 Space VLBI Demonstration with the Tracking Data Relay Satellite System

Charles T. Force: Thank you, Frank. I think Frank makes a point which is counter to one that I made, but I think it's a very important and valid one. It's the point that the antenna's tracking system may be used as a scientific instrument in itself in providing useful data. And that's a point that I overlooked. It's a key point. Another example of that is the Deep Space Antenna at Goldstone, the 70-meter antenna you saw in the picture, which is used as a Solar System radar of sorts, as a 400 kilowatt transmitter on it which is actually getting rather detailed resolution radar observations from some of the nearby planets. So I will modify my comment that tracking and data are only a means to an end because, in some cases, it's the scientific instrument itself.

AAS 87-651

ELECTROPHORESIS EXPERIMENT COMMAND AND DATA HANDLING

David W. Richman*

The Continuous Flow Electrophoresis System (CFES) flew on the Space Shuttle seven times between June 1982 and November 1985. As the flight program evolved, data handling requirements increased along with the complexity and length of operations. This paper discusses these operations and the approach taken to experiment command and data.

FLIGHT OPERATIONS

The Continuous Flow Electrophoresis System as it first flew on STS-4 in June 1982 consisted of three modules. First, there was a fluids systems module containing the electrophoresis chamber. This module is approximately six and one-half feet high, two feet wide, one and one-half feet deep, and weighs about 500 pounds and is mounted in the Middeck of the Shuttle just inside the hatch.

The other two modules were mounted on the forward avionics bulkhead. One is the experiment control and monitoring module, which is essentially a single 8-bit microcomputer and power supply. It weighs 80 pounds and is approximately two feet high, one and one-half feet wide, and one foot deep. It controls the separation process going on in the fluid systems module and alerts astronauts when it is time to insert new samples or remove separated samples. It also alerts them when important experiment events are taking place in the chamber so they can take photographs. The other is the sample storage module. It is a small thermoelectric refrigerator in which sample materials are stored before and after separation.

Astronaut Henry Hartsfield operated the CFES unit on STS-4. He put the syringes containing the sample materials into the unit, started the flow of carrier fluid, and put in the collection cassettes. After the flow conditions were constant, he punched a continue code into the computer. The system then automatically injected samples into the chamber, starting the separation process. When the process was complete, he removed the collection cassettes and placed them in the refrigerated sample storage module.

Hartsfield also took photographs of the process during various stages of the separation. These photographs and those on subsequent missions constituted an im-

---

* Branch Chief Technology, McDonnell Douglas Astronautics Company, Box 516, St. Louis, Missouri 63166.

portant part of the experiment data. During the STS-4 mission six samples were separated during two six-hour periods of operation on two successive days.

Experiment operations were monitored by thirty-nine sensors that interfaced with the microcomputer and the experiment data was recorded in 64K of battery backed random access memory. The continuous nature of the continuous flow electrophoresis process where parameters are relatively constant for long periods of time resulted in the ability to use this small amount of data storage. Internal data storage was a real advantage in simplifying integration into the Shuttle, allowing the experiment to go from concept to flight in less than two years. The lack of downlinked data to the mission support team in the Mission Operations Control Center was overcome by voice downlink of the fifteen most significant parameters several times during the operations. This operation was reminiscent of early manned space missions where spacecraft systems status was evaluated using periodic quick look data telexed from remote sites during station passes. For many experiments this approach is more than adequate, especially considering the amount of computing power and data storage that can be resident in the experiment.

On STS-6 which flew in April 1983 the system was modified to include a water cooling module. This module facilitated the use of more separation power without increasing cabin temperature. Command and data provisions for this flight and for STS-7 in June 1983 and for STS-8 in August 1983 were the same as for STS-4.

The unit was then modified from separating six small individual samples to continuously separating one large sample of protein material for subsequent ground tests. These modifications included changes to the software and the addition of bubble memory for data storage. The additional data storage was required because operations were extended from twelve hours to eighty hours during the mission. The extended operations were required to process the larger sample. This unit flew in August 1984, April 1985, and November 1985 with a McDonnell Douglas Payload Specialist, Charles Walker, operating the equipment. The extended operations on these flights included periods of transient cabin pressure operation which posed a challenge to the stability of the fluid systems module which has internal fluid pressures that are referenced to the atmosphere. It was during these periods that nearly continuous down and uplink between the Mission Control Center and the Payload Specialist minimized the loss of experiment output. The purpose of the downlink communications was principally to describe the current state of the experiment to the support team so that they could be of assistance to the Payload Specialist. Consequently, it was apparent that at least some downlinked data is probably required for efficient operation of any complex experiment.

DISCUSSION

The electrophoresis experiment is indicative of a trend in the design of experiments, the ability to easily incorporate both increased data storage and data reduction computer capability within the experiment. Most experiment data processing is well within the capabilities of today's personal computers and it has been estimated that by 1997 all of those capabilities could be resident on a single chip. Similarly, data storage capability for magnetic media is about 20 million bits per square inch today and is increasing rapidly. This facilitates the use of streaming tape or hard disk drives for data storage. With the ability to store large amounts of data and to interpret the data using the experiment computer the amount of data that must be downlinked will be, in general, greatly reduced.

One reason that experimenters want real time downlinked data is to facilitate real time command of the experimental process from the ground. Direct command uplink does pose a safety issue and requires that the process and its implementation are both well enough understood to predict that the outcome of any particular command given at any time will be a safe one. The need to test commands in software in the electrophoresis experiment was satisfied by parallel operation of a flight experiment computer on the ground. In general, this requirement will increase the use of higher level languages and the incorporation of expert systems in software. In turn, the incorporation of expert systems in software will reduce the need for real time ground intervention in the process. So while some experimenters will probably still want real time video and real time control of all process parameters, they will probably be in a minority.

SUMMARY

In summary, the increased computing power and data storage capability of future experiments will yield better scientific results while minimizing command and data handling requirements.

AAS 87-652

ADVANCED TECHNOLOGY FOR SPACE COMMUNICATIONS, TRACKING, AND ROBOTIC SENSORS

Kumar Krishen*

Technology advancements in the areas of tracking, communications, and robotic vision sensors being pursued within NASA, as applicable to future missions and associated space operations, are presented. Systems concepts for efficient and reliable communications, tracking, and robotic vision are discussed. Communications systems for multiple access, broadband, high data rate, and efficient operations are given. Optical-, laser-, and millimeter-wave-based implementations of both communications and tracking systems are discussed. The incorporation of NASA's anticipated space communications and tracking assets, which includes Tracking and Data Relay Satellite Systems, Shuttle(s), Space Station, Global Positioning System, Advanced Communication Satellites, Space Communications Center(s), and Orbital Transfer Vehicles, in the overall scenario is presented. Current NASA efforts at Ka- and W-bands are summrized. The use of smart television, laser, and microwave sensors for robotics operations is discussed. Optical processing for autonomous rendezvous and docking operations is presented. The fusion of systems ranging from microwave to optical wavelengths and high-level data processing techniques for remote control, station keeping, tracking, inspection, repair, and assembly are detailed.

INTRODUCTION

The communications environment for the Space Station encompasses communications within its own elements and a large number of external elements (Fig. 1). Communications for initial configuration are anticipated to be at S- and Ku-bands, the frequencies in use for the Shuttle and Tracking and Data Relay Satellite Systems (TDRSS). Several zones of communications coverage have been identified for specific purposes. Zone 1 is the Proximity Operations Zone and extends to 1Km radius from the Space Station. This zone is for fly-arounds, station keeping, final approaches, and docking with the Space Station. Zone 2 extends to 37 Km on all sides of the Space Station and is for

* Ph.D., Tracking and Communications Division, NASA Johnson Space Center, Houston, Texas 77058.

command and control purposes. Zone 3 is designated for Departures, and extends from Zone 2 up to 185 Km forward of the Station. Zone 4 is the Rendezvous Zone where all normal rendezvous with the Space Station will be planned/targeted to arrive. It begins at the end of Zone 3 and extends to 185 Km behind the Space Station. Zones 5 and 6 are for co-orbiting, satellite/free-flyer operations. Zone 5 is the leading, and Zone 6 is the trailing co-orbiting zone. Both begin at the end of the rendezvous zones and extended to 2000 Km. The space-to-space service requirements are anticipated to increase for the growth configuration. An estimate of this growth is displayed in Table 1.

The communications needs for the Growth Space Station (GSS) are envisioned to drive NASA to seek unique concepts and capability to establish this permanent presence in space. Furthermore, it will provide a facility to assemble, test, and deploy rather large and unique communications systems/subsystems. GSS is envisioned to need or desire the capability to communicate with many more satellites and spacecraft than the Initial Operating Capability (IOC). The increased interconnectivity will include links with numerous NASA and other U.S. Government satellites (perhaps including military satellites), commercial satellites, foreign spacecraft, and deep space missions. This implies the need for significantly improved switching/processing capability with the ability to receive/transmit a variety of signals (different frequencies, bandwidths, modulation formats, etc.) and to dynamically reconfigure the onboard switching as required. In parallel, the payloads/experiments on Space Station are expected to increase in numbers and in terms of data gathering capabilities. The use of automation and robotics will require high data rate and extremely reliable links.

The GSS will need to accommodate continually evolving and largely unknown future requirements for coverage, data rates, number of users, etc. This requirement for flexibility over a long term will provide a unique challenge to develop systems which are user transparent and which are quickly reconfigurable. The overall thrust is to maximize the cost effectiveness of both the GSS and user space systems.

Deep space communications are driven by the relatively near-term envisioned missions to the Moon and Mars (Ref. 1). The Mars missions comprise the unmanned sample return and manned explorations. For the Moon, a Lunar Base is proposed. In addition to these, projected NASA missions include Saturn, Uranus, Neptune, and Comet/Asteroid probes. These future deep space missions will require highly reliable, long life, and very efficient communications to ensure success. A significant element of the Moon and Mars missions is exploration and utilization of the surface of these planets. The communications for these missions could be planned to have an evolutionary growth with the communications and tracking initially controlled by Earth or Space Station then by the near-side base on the planet; and, finally, there could be a planet orbiting service station and a far-side base. One such concept is illustrated in Figure 2 for the Lunar Base mission.

Another significant element in the overall discussion of the future NASA space communications is the TDRSS. TDRSS is currently planned to have an operational life, including the current configuration, plus the augmentations, until the year 2000. An advanced system to be implemented after TDRSS is called the Tracking and Data Acquisition System (TDAS). The TDAS performance goals include: (1) improving TDRSS capabilities, (2) elimination of blind spots (backside satellites and cross linking), (3) increased number of direct links to ground with Ka-band implementation, (4) increased number of single access S-band, (5) addition of K-band, 60 GHz, and laser links, (6) Geostationary Orbit (GEO) to GEO 1 Gb/s laser links, and (7) Low Earth Orbit (LEO) to GEO 1 Gb/s laser links. These advancements for TDAS have been envisioned to satisfy the needs for increased data rates as a result of many payloads and experiments that will evolve and become operational in the Space Station era.

Future NASA space proximity operations will require systems capable of supporting rendezvous, station keeping, and soft-docking between various vehicles, Shuttle, satellites, unknown objects, and Space Stations. Representative activities and space vehicles are given in Table 2, and capabilities required for various missions are shown in Table 3. The advantages of accurate range, range rate, bearing information, and attitude data for proximity operations include: (1) enabling soft-docking using orbital mechanics rather than breaking thrust, (2) saving fuel by minimizing or eliminating wasteful off-axis thrusting, which is presently required to minimize plume impingement, (3) minimizing plume contamination of the environment and surface of the target vehicle, (4) avoiding translational and rotational perturbations of the target vehicle by plume impingement and contact forces, (5) minimizing time required for docking maneuver (especially important for the Space Station because docking port is not at the same orbital attitude as center of gravity/mass), (6) providing early warning to crew for corrective action for missed approach, (7) enabling automatic soft-docking, and (8) providing data inputs which will enable implementation of robotics and automation operations.

The Space Station will be a multipurpose facility with missions which will include science and applications, observations, technology developments and demonstrations, commercial laboratories and production facilities, and operational activities such as servicing/maintenance, repair of satellites, support of unmanned platforms, assembly of large-scale systems, and as a transportation node for transfer to other orbits and planetary missions. The overall thrust of automation and robotics for space missions and operations is to increase the productivity through autonomy, increased operational capability, and flexibility. The evolution of vision systems seems to have been dictated by the need for quality inspections and improvements in productivity. Early vision systems were cameras whose data was fed to a computer. As the role of robotics began to be widespread, the need for vision systems which are smaller, faster, more flexible, provide accurate images, have high resolution, sense color, and provide location and orientation data became obvious. More recently, intelligence has been identified as another desired feature for these systems (Refs. 2 and 3). Associated with the sensor capability for teleoperated or autonomous robotic systems is the need for advanced communications to handle the transfer of vision and other command, control, and sensor data.

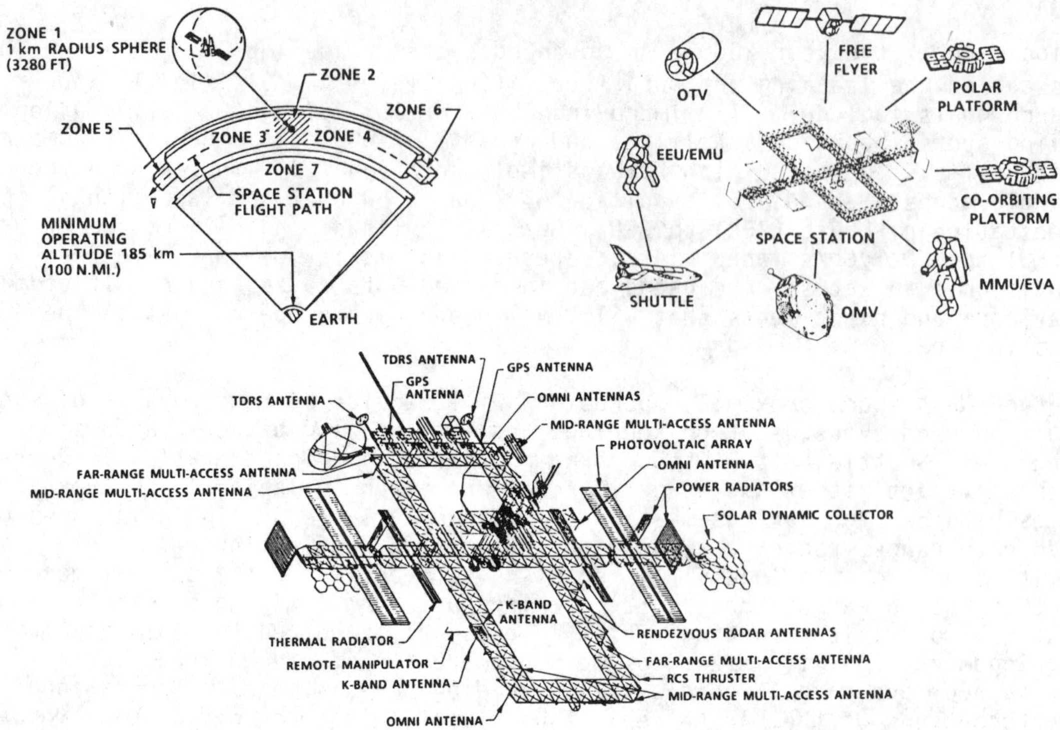

FIGURE 1.- SPACE STATION COMMUNICATION ENVIRONMENT

USER	INITIAL CONFIGURATION			GROWTH CONFIGURATION		
	QUANTITY	DATA REQUIREMENTS		QUANTITY	DATA REQUIREMENTS	
		FWD	RET		FWD	RET
EVA	2	A	A / V / T	4	A / G	A / V / T
SSO	1	A	A	2	A / C	A / T
OMV	1	C	V / T	2	C	V / T
OVT	–	–	–	1	C	V / T
FF	–	–	–	8	C	V / T
COP	–	–	–	1	C	V / T
MSC	1	C	3 V / T	1	C	3 V / T

FWD – FORWARD LINK (TO USER)
RET – RETURN LINK (TO SPACE STATION)
C – COMMANDS
T – TELEMETRY

A – AUDIO
V – VIDEO
G – GRAPHICAL DATA

TABLE 1 - SPACE STATION SPACE-TO-SPACE SERVICE REQUIREMENTS

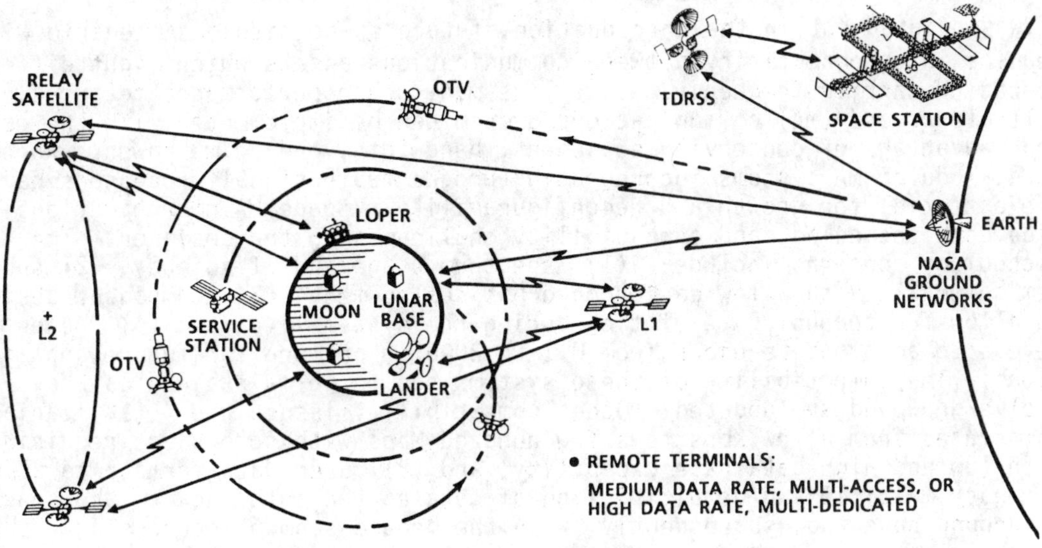

FIGURE 2.- LUNAR BASE COMMUNICATIONS/TRACKING CONCEPT

	SPACE SATELLITES	SPACE STATION	SHUTTLE / VEHICLES	UNKNOWN OBJECTS
DEPLOYMENT	X	X	X	
CONSTRUCTION		X	X	
OPERATION	X	X	X	
INSPECTION	X	X	X	X
REPAIR	X	X	X	
RETRIEVAL	X	X	X	X

TABLE 2 - FUTURE SPACE ACTIVITIES AND ENTITIES

	RENDEZVOUS	STATIONKEEPING	DOCKING
DEPLOYMENT		X	
CONSTRUCTION	X	X	X
OPERATION	X	X	X
INSPECTION	X	X	
REPAIR	X	X	X
RETRIEVAL	X	X	X

TABLE 3 - CAPABILITIES FOR FUTURE SPACE MISSIONS

TECHNOLOGY/TECHNIQUE ADVANCEMENTS

As was pointed out in the Introduction, future space missions scenario will require interconnectivity between communications assets which might differ in operating carrier frequency, as well as many other performance features. A multiple access (MA) communications system can be implemented with the resultant advantage of conserving equipment, bandwidth, and prime power. Furthermore, modular MA systems incorporating programmable signal processors have the potential for growth and reconfigurability. Space MA communications provide both technology and compatibility challenges to the designer. The technology concerns include: (1) spectrum usage and efficiency, (2) subsystem commonality to allow easier on-orbit replacement, (3) component design to allow for technology evolution during the system life, and (4) power levels to accommodate users from 0.1 to 2000 Km or more for future applications. The compatibility of these systems has to be maintained as they evolve/grow and are updated. Other compatibility issues are: (1) variable data rates from a few Kbps to a few hundred Mbps with some users requiring both low and high data rate capability, and (2) user data formats to allow for easy multiplexing/demultiplexing of data at the interface to the space-to-ground node and, subsequently, with the ground-segment processing.

Typical MA systems include Frequency Division (FD), Code Division (CD), Time Division (TD), Frequency Hop (FH), Space Division (SD), and any combination of these. Recently, the relative performance of three modulation techniques in a FDMA with nonlinear transmit power amplification was compared for the Space Station application (Ref. 4). The comparison for the three techniques, Quadrature Phase Shift Keying (QPSK), Staggered QPSK (SQPSK), and Minimum Shift Keying (MSK) was accomplished in terms of their spectrums, effects of filtering the post modulated spectrum, and optimum selection of detection filter bandwidth. The simulation block diagram is shown in Figure 3. From these studies, the FDMA, SQPSK modulation was determined as the superior one for nonlinear channel in overall performance based on bit error rates, spectrum width, channel interference, and receiver complexity for the Space Station high data rate links (Figure 4).

The potential advantages offered by a particular MA technique for certain space links make it possible for a space communications scenario to have interconnectivity to several MA systems. The carrier frequencies, data rates, coding, time/space coverage, and system signal margins on these MA links make it a complex problem for interconnectivity. In the case of the GSS antenna, systems for MA communications/tracking are needed to provide: (a) spherical antenna coverage in the near zone, (b) several simultaneous links to near and far satellites/targets, and (c) links which cover several frequency bands (for example S-, X-, Ku-, Ka-, W-bands). For making the antennas reliable and agile beam, distributed arrays are needed. In view of these requirements, advancements in antenna technology are currently in progress to provide distributed array antennas with multiple beam, wide scan capabilities (Figure 5). In the higher bands, microwave integrated circuits (MICs) and monolithic MICs (MMICs) are needed for achieving higher reliability, smaller size, and better signal performance (Ref. 5). The multi-beam antennas can be used to increase communications capability through frequency reuse and scanning beam capabilities. These antennas are efficient in handling nonuniform traffic distributions. Microstrip feed arrays or other approaches to adaptive beam-forming networks are needed. Some applications call for extremely large infurlable reflector antennas ranging in diameter

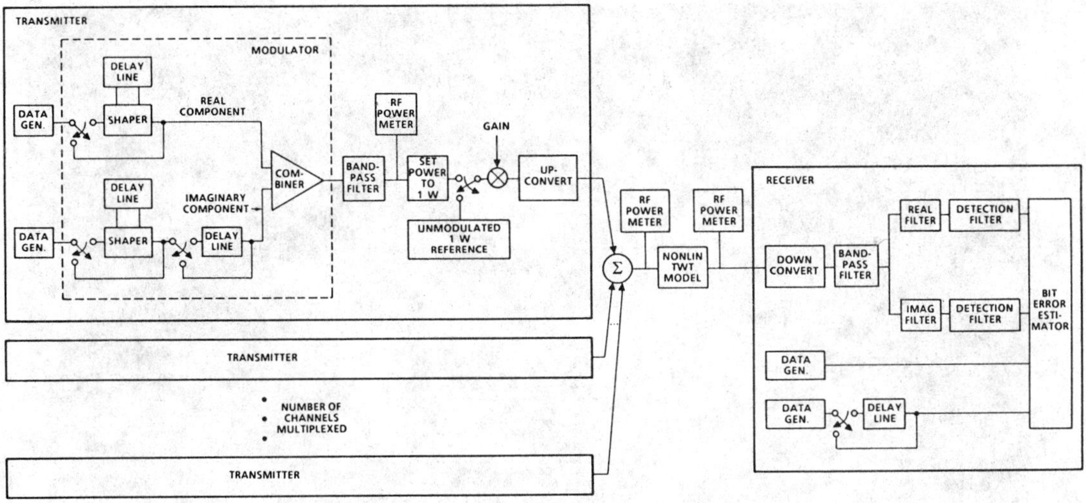

FIGURE 3.- SYSTID SIMULATION BLOCK DIAGRAM FOR 5 CHANNEL FDMA BER ESTIMATION

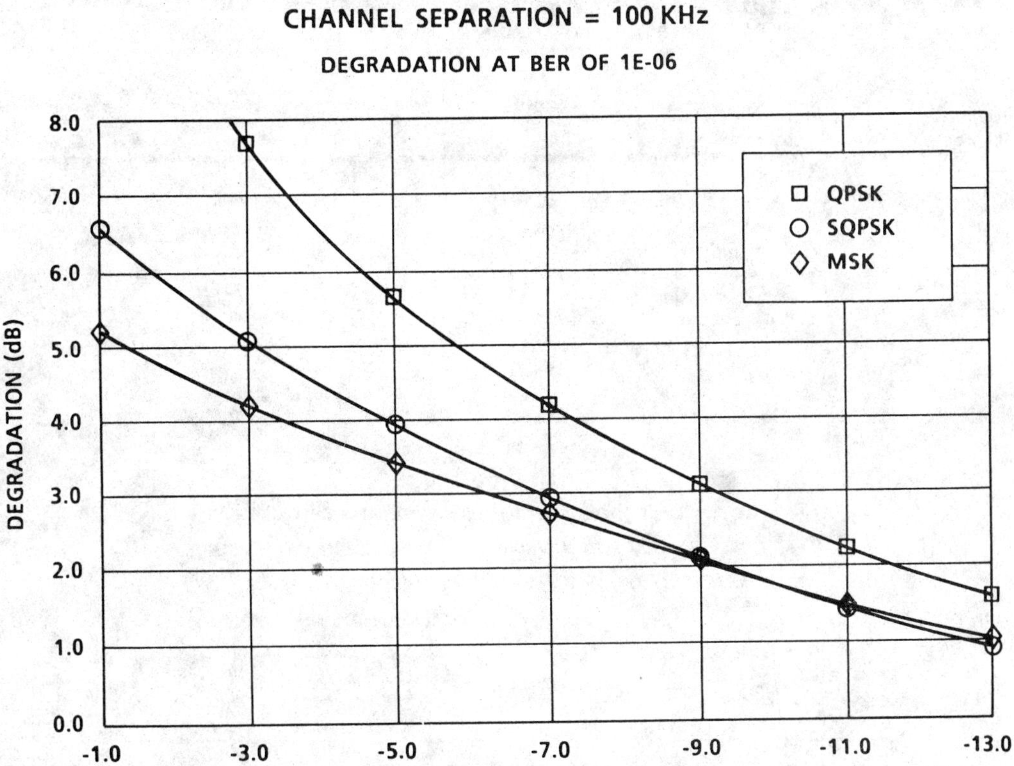

FIGURE 4.- COMPARISON OF FDMA SCHEMES

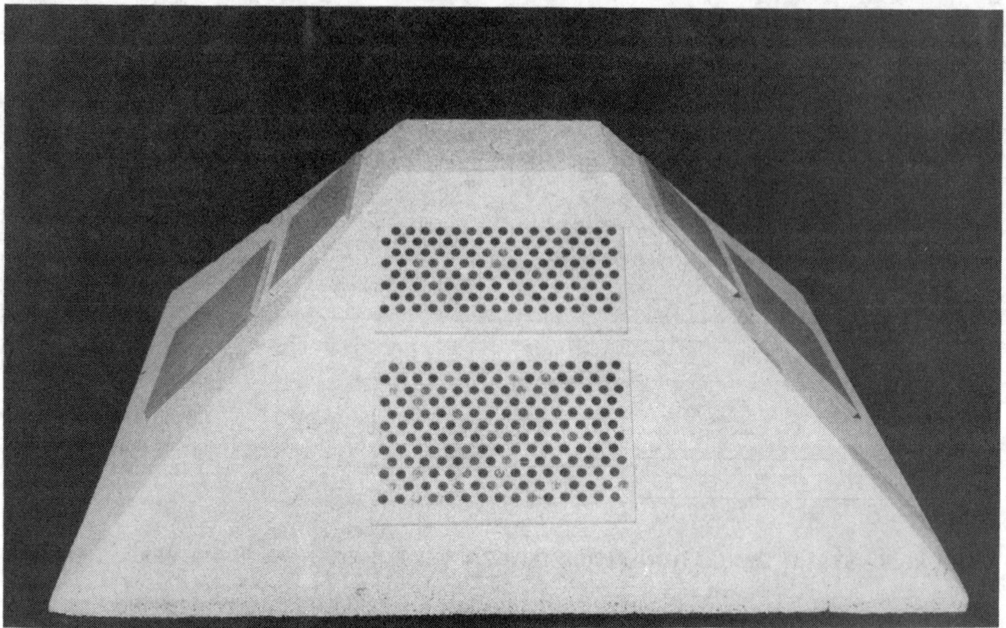

FIGURE 5.- MULTI-BEAM KU-BAND PHASED ARRAY USING MMIC TECHNOLOGY

ADVANCED SIGNAL PROCESSOR DESIGN CONCEPT

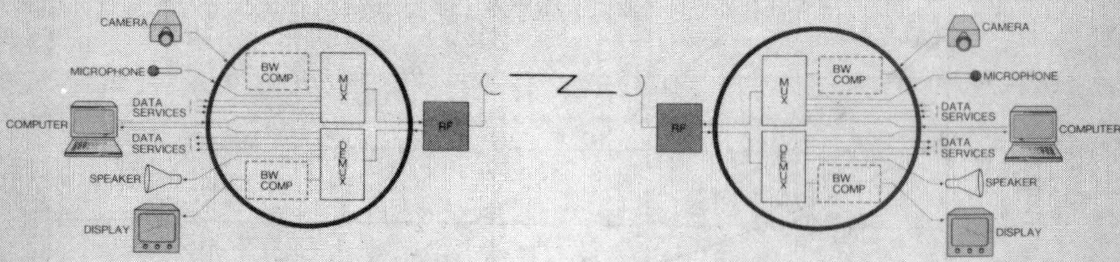

- **SINGLE STANDARDIZED DESIGN TO SUPPORT FREE-FLYERS, OTV'S, EVA'S, ETC.**
- **FULL DUPLEX OPERATION SUPPORTING A WIDE VARIETY OF DATA SERVICES**
- **COMPLEMENTARY TRANSMIT AND RECEIVE FUNCTIONS:**

TRANSMIT FUNCTIONS

- PROCESS HIGH QUALITY AUDIO
- PROCESS HIGH QUALITY VIDEO
- PROCESS MULTIPLE DATA SERVICES
- PROVIDE BANDWIDTH COMPRESSION FOR AUDIO/VIDEO CHANNELS (IF REQUIRED)
- MULTIPLEX DIGITAL AUDIO, MULTI-FRAME RATE DIGITAL VIDEO, AND DATA SERVICES
- GENERATE TIMING FOR DATA SOURCES
- PROVIDE CHANNEL ENCODING AS REQUIRED
- PROVIDE ENCRYPTION AS REQUIRED

RECEIVE FUNCTIONS

- PROCESS AUDIO
- PROCESS VIDEO
- PROCESS MULTIPLE DATA SERVICES
- RECONSTRUCT AUDIO/VIDEO CHANNELS FROM COMPRESSED DATA (IF REQUIRED)
- DEMULTIPLEX DIGITAL AUDIO, MULTI-FRAME RATE DIGITAL VIDEO, AND DATA SERVICES
- REGENERATE TIMING
- PROVIDE CHANNEL DECODING AS REQUIRED
- PROVIDE DECRYPTION AS REQUIRED

FIGURE 6.- ADVANCED SIGNAL PROCESSOR DESIGN CONCEPT

from 20 to 50 meters (Ref. 6). Required surface tolerance of these antennas calls for development of materials and construction techniques to reduce distortion due to thermal effects and/or adaptive beam forming networks to compensate for distortion (Ref. 7). Unfurling techniques are also needed which would maintain the required tolerances.

In many space communications systems, multiple scanning beams require rapid switching between any beam and any transponder. For these systems, advances in onboard processing and switching are needed. For efficient combining/multiplexing of a wide variety of asychronous data, baseband signal processing has potentially higher flexibility and link efficiency. This dictates a need for innovative and advanced designs in processor modulator/demodulator to significantly increase speed and reliability while reducing mass, size, and power consumption. The key drivers to the design of these systems are the high data rate links ranging from a few hundred Mbps to one Gbps. These data rates are encountered in packetized video, audio, robotic vision, and experiment systems (such as synthetic Aperture Radar imagers) data handling (Ref. 8). In addition to multiplexing and demultiplexing algorithms and multiplexing formats, high performance programmable signal processors are needed for increased efficiency. These signal processors should be able to handle various carrier frequencies, variable data rates, compression algorithms, flexible protocols, and have coding/modulation diversity. Hardware implementations for high data rate include VHSIC technology, FIFO buffers, LSI encoders/decoders, retiming recovery circuitry, retransmission protocol subsystem, and radiation immunity. Figure 6 shows advanced signal processor design concept being developed at JSC. Coding/decoding has been responsible for immense gains in communications systems. With recent advances in high-speed computers, miniaturization of electronics, and advances in photonics, new coding techniques can be expected to make greater contributions to the increased efficiency and reliability of advanced communication systems. As shown in Figure 7, the projected coding gains over an uncoded system in the year 2000 exceed 9.0 (Ref. 9).

Intersatellite links between platforms and satellites can eliminate double (or more) hops between ground and GEO that increase transmission time and path losses. In addition, these links can be implemented to transfer data between orbits and between orbits and deep space missions. These links can be implemented at very high microwave frequencies; for example, 44 and 60 GHz and laser bands. Although, presently, systems can be developed at the microwave bands, improvements in several areas of component technology are needed. In particular, solid state amplifiers and low noise receivers are needed for increased reliability. Low noise receiver development programs at 44 and 60 GHz are being pursued by both DOD and NASA. The DOD hybrid preamplifiers are being integrated with monolithic mixers and IF amplifiers. Also being pursued is a completely monolithic 44 GHz pre- amplifier with 5 bit phase shifter. Such a preamplifier would be suitable for active aperture receive antennas. Similar development is in progress at the 60 GHz at NASA for implementation into the TDAS system. The immediate goal is to produce 2 watt and 20 percent efficient GaAs diodes at 60 GHz (Ref. 5). Amplifier development is continuing with emphasis on achieving 10 percent power added efficiency for 5 and 10 watt output levels (Ref. 8). Bandwidths in excess of 1 GHz are aimed at in the NASA development efforts at 60 GHz.

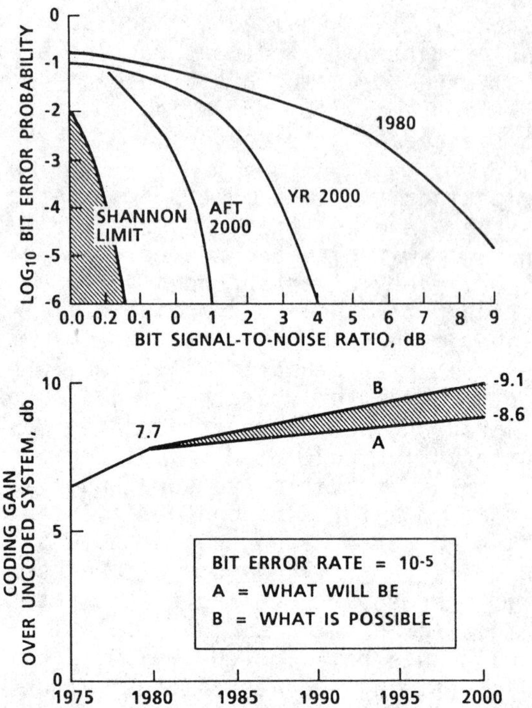

FIGURE 7.- U.S. AIR FORCE PROJECTIONS OF BIT ERROR PROBABILITY AND CODING GAIN

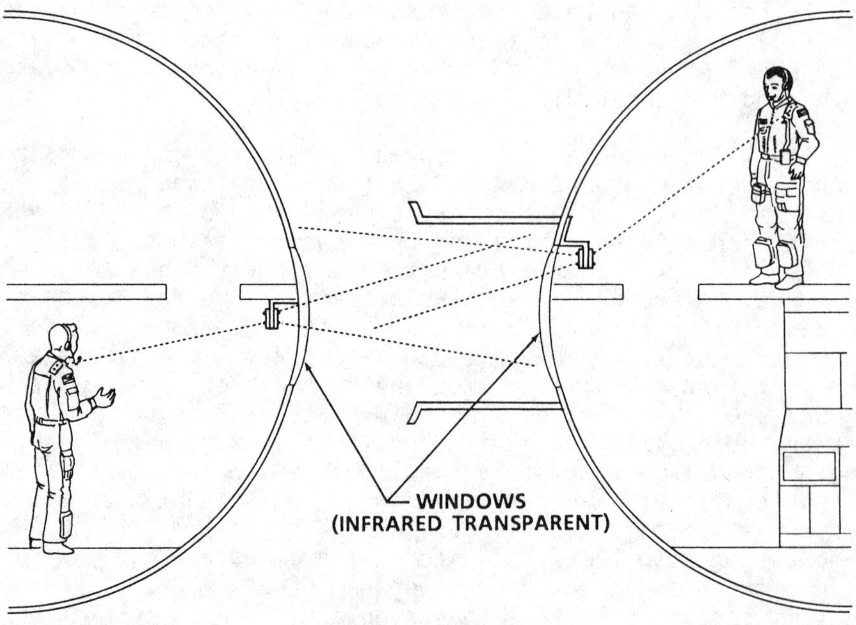

FIGURE 8.- MODULE-TO-MODULE DIGITAL INFRARED/LASER COMMUNICATIONS

There are two basic classes of laser communications (LASERCOMM) systems; coherent and incoherent. Coherent systems can use either heterodyne or homodyne techniques, while incoherent uses direct detection circuitry. The technology issues in LASERCOMM center around high power laser diodes (1 watt single device), heterodyne/homodyne detection, and reliability and long life of laser diodes. Heterodyne systems are sensitive to the frequency and phase of incoming optical signal. However, these systems can be as much as 20 dB or more sensitive than comparable direct detection systems. For higher data rates (between 1 to 100 MB/s), the aperture diameters are very small compared to direct detection systems which respond to the magnitude of the incoming optical energy. The technology for heterodyne systems is emerging. Currently extensive studies are being conducted by the Air Force in conjunction with MIT Lincoln Laboratories (Ref. 9).

In internal point-to-point communications on spacecraft, electro-optics provide a viable technology with attendant advantages which include: (1) size and weight savings versus copper cables, (2) noise immunity, (3) potential without rewiring to accommodate growing number of users and data rates, and changing requirements, and (4) modular development of the systems. Several development activities are aimed at 200 MB/s devices, local area networks, and wavelength multiplexing for future growth. NASA/LaRC has developed a 500 MB/s fiber optics transceiver (Ref. 5). This transmitter and receiver pair utilizes preamplifier and laser device modules to achieve a size reduction of 3:1 over commercially available devices. The long-term objective of NASA is to develop 1 GB/s transceiver which could satisfy multifunction requirements of many video, control, and data systems for the GSS data applications. Further technology implementation has to be incorporated in terms of space-qualifying these advanced systems. Diffused infrared (IR) communications have shown many advantages for internal communications for manned spacecraft such as the Shuttle, and the Space Station. Module-to-module communications on the Space Station could also be implemented with Laser or IR links (Figure 8). The advancements in these systems are aimed at increasing the number of duplex channels (24 or more) with a capacity of 20 KB/s and multi-access protocols.

As the communications assets increase on any particular spacecraft(s), the advantages of automatic control and monitoring (CAM) using expert systems become apparent. Such a subsystem is under design and development for the Space Station. The functions which are automatically controlled include: (1) planning and scheduling of communications missions, (2) traffic control and routing, (3) subsystem and system performance verification testing, fault detection, and isolation, (4) configuration and performance control, and (5) antenna and television pointing and performance control. As the communications resources evolve in space, the need for reuseable, modularized, expert system implementation becomes advantageous. The CAM modules (includes software and hardware) can presently be configured as intelligent subsystems based on decision logic algorithms and can readily be adapted to various space platforms/spacecraft. The language desired is ADA for its flexibility and interconnectivity with other control and monitoring subsystems. The conceptual control and monitoring subsystem for Space Station is shown in Figure 9. In addition to the software developments, the hardware implementation of CAM calls for technology developments in the three key areas: (1) parallel

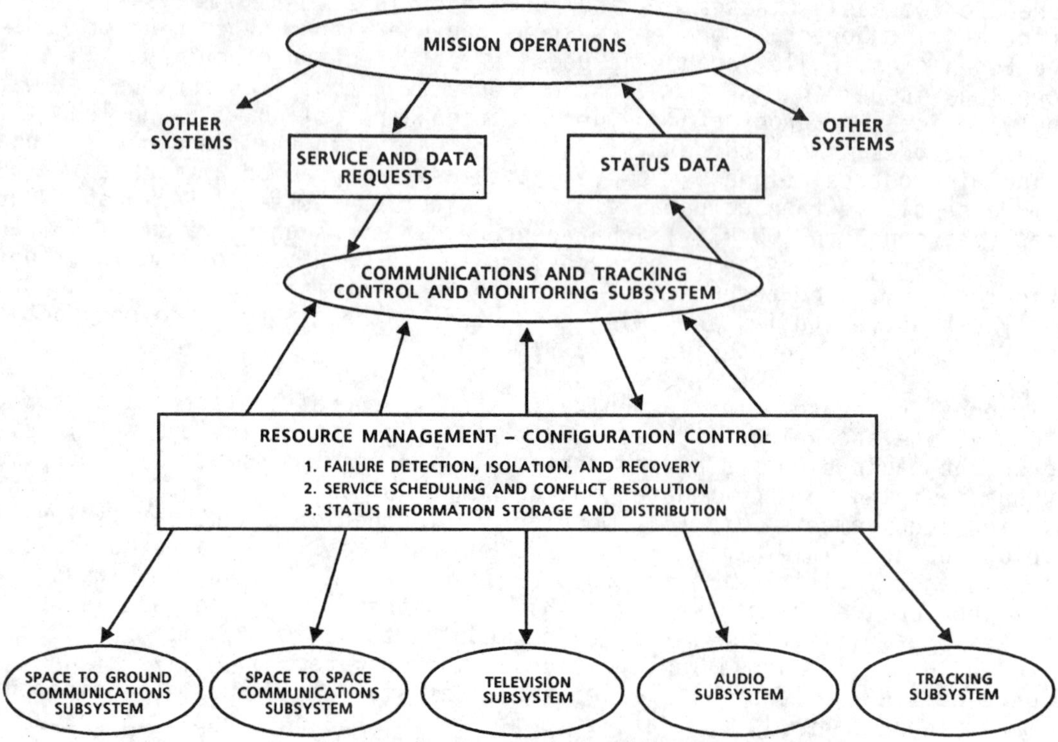

FIGURE 9.- CONTROL AND MONITORING FOR COMMUNICATIONS AND TRACKING

LASER SYSTEM TECHNOLOGY	
• BEAM STEERING • TORQUE MOTORS • LASER DIODE PHASED ARRAYS	• APD DETECTOR WITH ROTATING SLIT • TONE RANGING THROUGH APD DETECTOR • MEASURE RADIAL DISTANCE AND ANGULAR POSITION OF SPOT THROUGH SLIT
• CID DETECTOR ARRAYS • 256 × 256 ARRAY • SOFTWARE TECHNIQUES TO GIVE RESOLUTION OF 1 PART IN 25,000	• IMAGE DISSECTOR RECEIVER • ENHANCED RESPONSE OF IMAGE DISSECTOR TUBE IN THE INFRARED SPECTRUM TO ALLOW USE OF SEMICONDUCTOR LASER SOURCE
MM WAVE COMPONENTS / SUBSYSTEMS	
• 40 GHZ TO 200 GHZ • DISTRIBUTED PHASED ARRAY • SOLID STATE SOURCES AND OTHER DEVICES	• MMIC MIXERS • LOW NOISE RECEIVERS / AMPLIFIERS

TABLE 4 - TECHNOLOGY DEVELOPMENT AREAS

processors, (2) rapid access reliable mass storage devices, and (3) fiber optic networks. Advanced CAM systems are expected to become the heart of communications satellites/systems providing autonomous operation, stabilization, and performance selection for each component/subsystem such as the agile demodulator, the matrix switch, and onboard signal processors/switching.

Future systems and techniques for the tracking, stationkeeping, rendezvous, and docking will include innovative use of the Global Positioning System (GPS), multi-target microwave systems, millimeter wave radars, and laser systems. A detailed discussion of these systems is given in Ref. 10. The technology development areas for these systems are given in Table 4. The general implementation of sensors/systems capable of tracking and proximity operation requires sensors in various electromagnetic bands with unique performance characteristics for different operational scenarios. For large spacecraft such as the Space Station, a computer/controller could be envisioned (similar to the CAM subsystem discussed in the previous paragraph) which could control the selection and operational parameters of various subsystems (Figure 10). The overall multifunction system could incorporate a high level of autonomy. Spacecraft/object recognition and aspect determination are considered a requirement for many proximity operations in the Space Station program. For this requirement, video and IR cameras are deemed essential. These sensors can also be used in robotic vision applications.

For robotic applications, there is a need for advanced technology in several areas. The initial vision systems will depend largely on television-based technology/techniques (Ref. 11). In this area, there is a need for solid-state cameras, based on charge coupled device (CCD) or charge injection device (CID) technology with variable spatial resolution, with panels of 1024 x 1024 or greater. The features should include automtic zoom and ability to detect objects as small as 2 millimeters at a distance of one meter. Furthermore, the density of the sensing elements can be patterned after the eye with high resolution in the center and low resolution in the peripheral vision (Figure 11). Brightness filters, automtic iris, and automatic gain control can be incorporated to allow handling of high intensity sunlight. Intensities of sun can range up to 13,000-foot candles necessitating bloom protection, large dynamic range, and protection from burn-in (Ref. 12). Similarly, sensitive night/drak vision modules with nonblooming characteristics are needed. In particular, very high quality and resolution imagery is needed to meet robotic vision requirements. In certain applications, large format cameras capable of precise image mensuration at the pixel level may be required. Innovative approaches to controlled illumination are required to complement natural illumination for scene definition (Ref. 3). In some applications, design features favorable to telerobotic vision may be incorporated to identify and determine aspect/attitude of various objects. Object shapes, color, and markings are some of the parameters useful for this purpose. Color vision offers an additional capability for the recognition of objects. The levels/shades of colors used in robotic vision can be substantially higher than can be distinguished by the human eye. Its immediate effect on the design of the video system is added complexity, data processing and resultant cost. Furthermore, the rate of data transmisson increases significantly. Infrared cameras may be extremely beneficial in the location

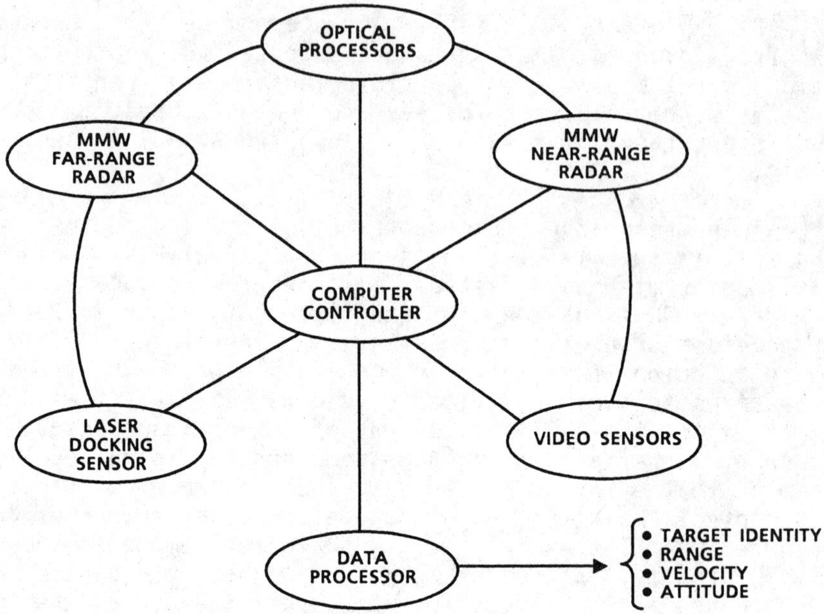

FIGURE 10.- ADVANCED MULTIMODE RENDEZVOUS AND DOCKING SYSTEM CONCEPT

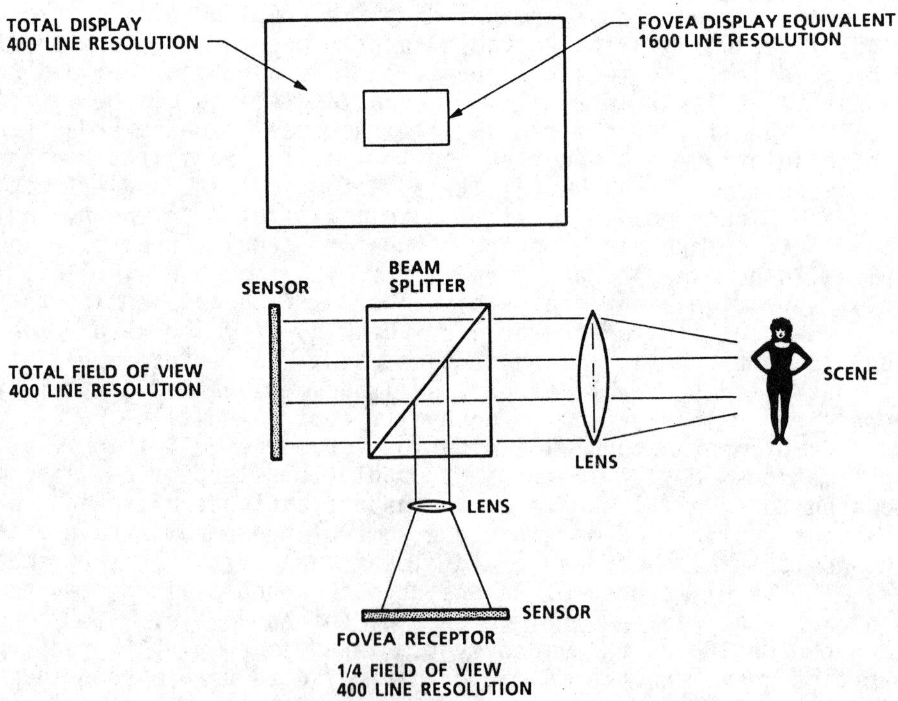

FIGURE 11.- CONCEPTUAL SIMULATED EYE SENSOR

of objects in the dark and occluded areas. Although inherently low resolution, infrared imagery can provide gross identification of the objects.

The need for video data to be able to interface with digital processors/computers has given impetus to digital TV technology. For the solid state implementation using VLSI/VHSIC, recognition/preprocessing algorithms can be implemented on the same electronic chips, making the size of these video imagers small. As this technology is rapidly moving forward, the need for handling and transmission of high data rates is becoming obvious. For a video system at 5 MHz baseband, an 8-bit digitization would generate 80 MB/s data stream. For color TV implementations and multiple systems, this bit rate can multiply significantly. For real-time processing of this data, compression techniques have been proposed which can also be implemented by innovative chip designs. The compression algorithms should be automatic and transparent to users, not destroy or discard any relevant information, and compress/decompress data at speeds significantly higher than the associated device data transfer speeds.

For 3-D vision, range information to each pixel can be added to the video imagery. Laser scanning devices capable of giving angles/position and ranges to each pixel within the field of view provide the depth/height profile of the object. Included in these measurements can be the laser reflectance of the pixel. Two technologies of the solid-state laser vision devices currently available are those using mechanical motion of mirrored surfaces, and those that involve an inertialess change in the optical properties of the transparent medium. The latter class includes diffraction of light from an acoustically generated periodic structure. Phased-array solid-state scanning devices are also currently under development. These devices have the promise of providing fast, accurate, and lightweight laser vision.

Microwave systems have been used to detect relative speed of objects and their range in many applications. Their use in space robotics applications is being studied at NASA/JSC. In particular, millimeter wave radars provide attractive performance parameters in addition to their small size. The possibility of broader beam than laser systems makes these sensors attractive for initial acquisition of moving objects. A radar at the frequency of 100 GHz has been developed at NASA/JSC (Ref. 13). This particular system is for use on a Man Maneuverable unit (MMU) to provide relative range of speeds from 0.1 to 2.0 fps. This type of radar operating at several carrier frequencies can be used to measure backscattering coefficients for various polarization combinations. The coefficients are object structure dependent. There is also possibility of penetration through thermal protection and obscuration caused by nonmetallic objects. This data can be utilized in an interactive manner with that of the video systems to provide scene definition/parameters in certain situations/scenarios.

Another promising microwave technique is the time-domain imaging, the synthesis of the scattered electromagnetic field distribution over an object plane. The transmitted pulse is an impulse source offering higher instantaneous signal-to-noise ratio, higher resolution, and option for echo time gating. Technology advances in the fast pulse generators and sampling devices allow the fabrication of time-domain imagers in picosecond region with measurements and recording of both phase and amplitude of the returned/transmitted signal. The scattered signal can be formulated as the convolution of the source with the transmitter, receiver, and scatterer responses (Ref. 14). From a set of time-domain responses obtained from different viewing directions, the two-dimensional field distribution is synthesized using a technique similar to the one in tomography (Ref. 15). The resultant image closely resembles the object geometry (Ref. 14). Thus, time-domain impulse imaging is another tool in extracting physical information about the object.

CONCLUSIONS

The initiation of the Advanced Communications Technology Satellite (ACTS) project in 1984 marked a significant milestone in space communications research and development. The key technologies being addressed through this project are: (1) multibeam antenna, which provides for frequency reuse and enables the five to 10 times increase in capacity of a single satellite in orbit, (2) satellite switching/processing system interconnecting the spot beams, and (3) Ka-band (20/30 GHz) technologies and laser communications for intersatellite links. A TDMA technique is used. The satellite communications signals are divided into short, compressed bursts, and several users can transmit and receive at the same frequency by taking turns or time sharing.

Use of TDMA is essential to the effective utilization of scanning spot beams (Figure 12). Onboard switching systems will enable interconnecting the beams and routing signals to their appropriate destinations. Two distinct switching methods will be tested on ACTS: (1) Intermediate frequency (IF) matrix switching and baseband processing, and (2) high-performance, solid-state switching devices for high speed switching between a large number of wideband inputs and outputs with low crosstalk at all output ports. In addition, dual control switch memories are needed to allow changes in the switching plan without loss of service (Ref. 16).

The ACTS baseband processor is the heart of the scanning spot-beam system and provides efficient routing of low-volume communications from small Earth terminals (such as those located directly on the customer's premises). The signals received are converted to baseband content (digital data with zeros and ones), stored, and then routed to their correct destination. The processor incorporates signal error detection and correction, and in essence, moves the switchboard from the ground to the sky.

A laser communications system using gallium arsenide diodes operating at 870 nanometers will be tested between the Shuttle and ACTS (Figure 13). This new space technology experiment will test the link at a data rate of 220 Mb/s. A comparison of the Ku-band system at 300 Mb/s and the laser link in terms of antenna size shows the laser has a 7-inch diameter while a 12-foot antenna is required at this microwave frequency. The ACTS/Shuttle laser experiment is aimed at acquisition and tracking, as well as maintaining link between rapidly moving spacecraft simulating the operational systems of the TDAS/TDRSS and the GSS and orbiting platforms.

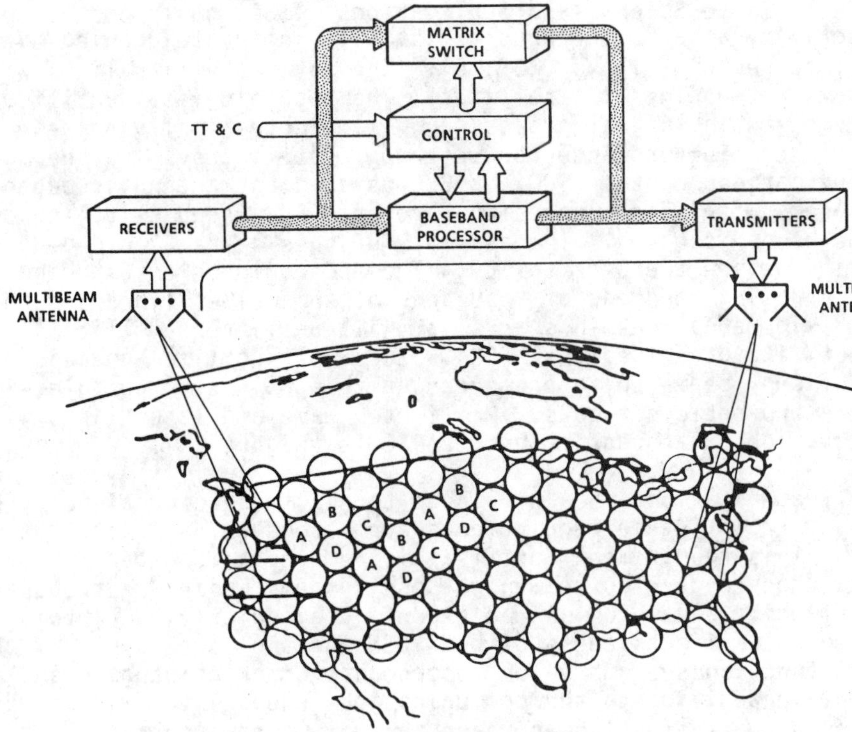

FIGURE 12.- SATELLITE COVERAGE USING SPOT CONCEPT

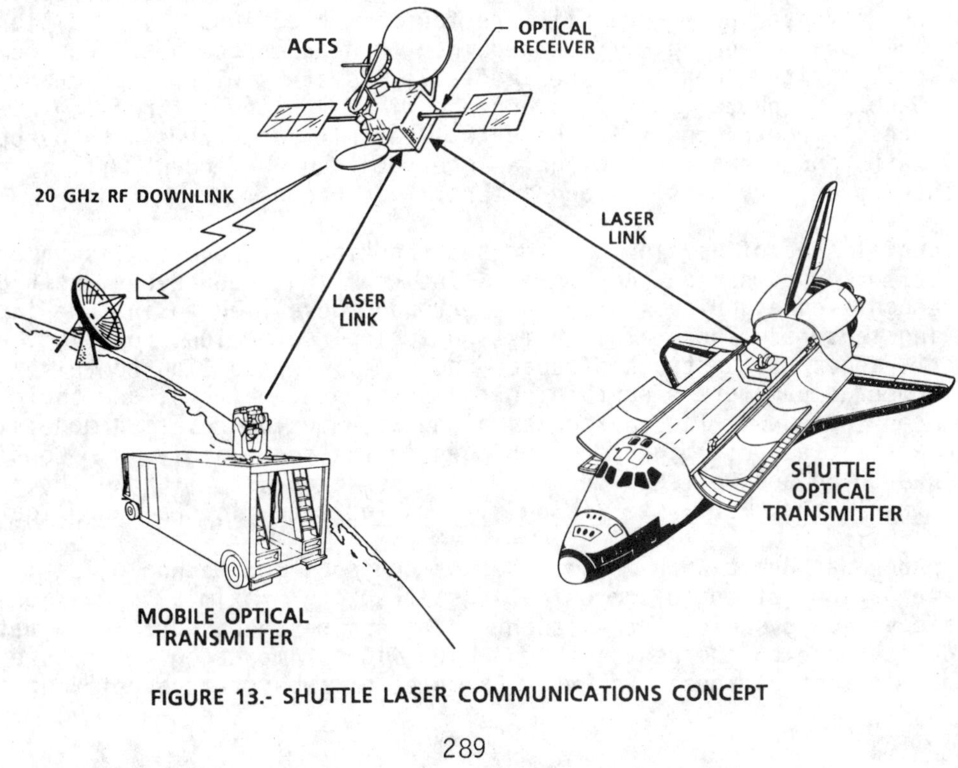

FIGURE 13.- SHUTTLE LASER COMMUNICATIONS CONCEPT

NASA/Office of Space Science and Applications (OSSA) has proposed space-earth interconnectivity as a basic thrust which will integrate growing microwave and optical terrestrial links (Ref. 6). The intersatellite links will be used as links between nodal satellites. The technologies identified include: (1) onboard switching/signal processing, (2) electronically hopped antenna beams, (3) laser links to interconnect nodal satellites. Furthermore, a Space Communications Center (SCC) in GEO using giant antennas, capable of many spot beams with interconnection through a SCC-based switching/computer complex has been discussed. The assembly of this system could be done at the Space Station before transfer to GEO. The SCC would be a switching complex capable of handling hundreds of thousands of subscriber lines and thousands of trunk or wideband lines in space. It will use intersatellite links to interconnect with other regional SCCs. SCCs will utilize hundreds or thousands of spot beams to interconnect earth stations compatible with terristrial electro-optic networks. Millimeter wave and laser links will be provided for high data transfer between SCC's (Figure 14).

The Growth Space Station (GSS) will need to accommodate continually evolving and largely unknown future requirements for coverage, data rates, number of users, etc. This requirement for flexibility over a long term will provide a unique challenge to develop communications systems which are transparent to the user community and which can be reconfigured quickly and automatically to maximize the cost effectiveness of both GSS and the users systems. Multi-mode, multi-function systems which incorporate tracking/ranging and attitude recognition, in addition to the communications capability, will need to be developed as a cost-effective approach to space communications, tracking, and robotic vision.

Deep space missions will require highly reliable, long life, and very efficient RF and laser power transmitters for communications. In addition, electronic beam steering will be required to reduce stringent requirements on spacecraft attitude control. Cost-effective systems with large diameter (from 3 to 10m) photon buckets are needed in Earth orbit for relay of optical deep space telemetry and tracking data. Laser transmitters with outputs up to 10 watts and overall efficiencies greater than 10 percent will be required for outer planetary missions and interstellar probes.

The crucial area of development for the tracking and robotic vision areas is multisensor coordination and fusion. The tracking, rendezvous, soft docking, and recognition algorithms are to be extended/developed to include interrelating sensors. These algorithms should include motion, rotation, and object changes as functions of space and time. In addition to this, the environmental parameters pertaining to the events, objects, and their status have been included. These aspects, along with motion/rotation models, integrate artificial intelligence techniques in the scene analysis. Considerable work has to be accomplished in order to provide robust, lightweight, real-time (fast), and small size vision systems for specific space applications.

This paper has dealt with a broad set of concepts and technologies for system implementations of space communications, tracking/proximity operations, and robotic vision systems. Very recent advances in superconductivity materials and devices are envisioned to have a significant impact on the future design and development of these systems. Examples of subsystems benefiting from this technology are: (1) millimeter wave high efficiency distributed antennas with broadband and large beamwidth performance (Figure 15), (2) microwave power switches, networks, and distribution circuits resulting in substantial reduction in power loss, and increase in bandwidth and sensitivity, (3) optical and infrared detector cameras for low level/dark sensing, and (4) development of programmable signal processors, neuro- networks for speech and scene recognition, and communications monitoring/ control processors.

ACKNOWLEDGEMENT

The author wishes to thank his colleagues, Messers James Ratliff, Olin Graham, Oron Schmidt, and Sidney Novosad for their contributions to this paper and to Mr. Ralph S. Sawyer, Chief, Tracking and Communications Division, for his encouragement and continued support.

REFERENCES

[1]. "Strategic Plan for NASA's Communications Applications Program, Office of Space Science and Applications," Communications Division (Code EC), Third Draft, September 30, 1985.

[2]. "Advancing Automation and Robotics Technology for the Space Station and for the U.S. Economy," NASA Advanced Technology Advisory Committee Report submitted to the U.S. Congress, NASA Tech. Memorandum 89190, Johnson Space Center, Houston, TX, October 1986.

[3]. Krishen, K., R. J. P. deFigueiredo, and O. Graham, "Robotic Vision/ Sensing for Space Applications," Proceedings of the 1987 IEEE International Conference on Robotics and Automation, Raleigh, NC, March 1987.

[4]. Ratliff, James E., "The Performance of QPSK, SQPSK, and MSK Modulation Techniques in an FDMA Nonlinear Channel Application," Technical Report, Tracking and Communications Division, NASA/Johnson Space Center, Houston, TX, January 1987.

[5] NASA-OAST Space Research and Technology, Summary Technical Report, 1986, NASA Technical Memorandum 89191, NASA/Washington, DC, March 1987.

[6] Lovell, Robert R., and C. Louis Cuccia, "Geostationary Platforms in the Space Station Era," Telecommunications, October 1984.

[7] "NASA Space Systems Technology Model, Volume II," NASA/Office of Aeronautics and Space Technology, Code RS, Third Issue, Washington, DC, September 1981.

[8] Kostiuk, Theodor, and Bill Clark, "Spaceborne Sensors (1983 - 2000 A.D.): A Forecast of Technology," NASA Technical Memorandum 86083, NASA/ Goddard Space Flight Center, Greenbelt, MD, April 1984.

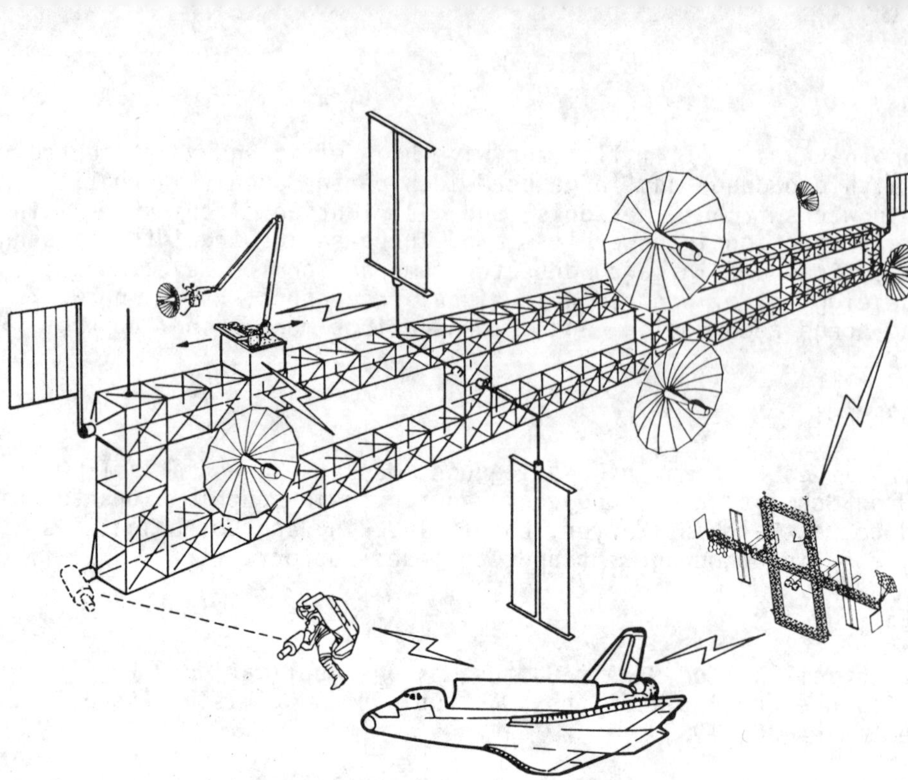

FIGURE 14.- SINGLE PLATFORM SPACE COMMUNICATIONS CENTER

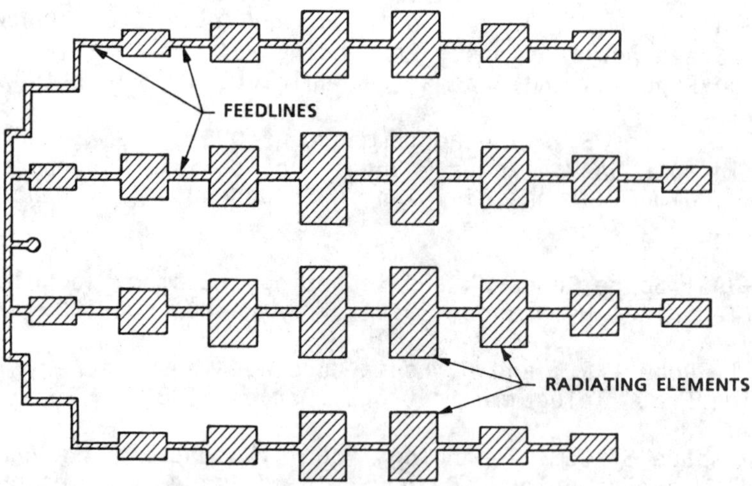

ADVANTAGES OF SUPERCONDUCTOR MICROSTRIP ANTENNA:

- HIGHER ANTENNA EFFICIENCY, HIGHER GAIN
- LOWER FEEDLINE LOSSES, LARGER ANTENNAS PRACTICAL
- LOWER FEEDLINE LOSSES AT HIGHER MILLIMETER-WAVE FREQUENCIES, MILLIMETER-WAVE ANTENNAS PRACTICAL

FIGURE 15- TYPICAL MICROSTRIP ANTENNA USING SUPERCONDUCTORS

[9] Advanced Space Communications Technology Roadmap, Space Division, Code CGX, PE63431F, U. S. Air Force, Los Angeles, CA, April 24, 1985.

[10] Krishen, K., and H. Erwin, "Review of Laser and RF Systems for Space Proximity Operations," Proceedings of the International Telemetry Conference, IFT/ISA, Las Vegas, NV, October 1985.

[11] Holcomb, L. B., R. Larsen, and M. Montemerlo, "The Automation and Robotics Research Program," AIAA/NASA Symposium on Automation, Robotics, and Advanced Computing for the National Space Program, Washington, DC, September 4-6, 1985.

[12] Bronez, Mark A., "Vision Concepts for Space-Based Telerobotics," Technical Report SSS 86-0165, Rockwell International, Space Station Systems Division, Downey, CA, January 1987.

[13] Lichtenberg, C. L., "Man Maneuverable Unit Millimeter-Wave (100 GHz) Development Radar - Design and Test Results Report," NASA/Johnson Space Center, Houston, TX, 1986.

[14] Yeung, W. K., and S. Evans, "Time-Domain Microwave Target Imaging," IEE Proceedings, Volume 132, Pt. H, No. 6, October 1985.

[15] Lewitt, Robert M., "Reconstruction Algorithm: Transform Methods," Proceedings of the IEEE, Volume 71, No. 3, March 1983.

[16] "ACTS - Advanced Communications Technology Satellite," Brochure published by NASA/Lewis Research Center, Cleveland, OH, May 1985.

AAS 87-653

SPACE STATION COMMUNICATIONS

Albert W. Weinrich*

The paper describes the features of the NASA Space Station Communications System developed by Johnson Space Center here in Houston (Space Segment) and Goddard Space Flight Center (Ground Segment).

The Space Station Communications System supports the station and the constellation of satellites conceptually illustrated in Figure 1. These include communications with the shuttle and with astronauts during EVA activity, and with the OMV. Concepts for the future Space Station operations include functions such as relaying information from the Space Telescope shown in the bottom of the figure, and from environmental and communications satellites. Position location for the station is determined by processing signals received from the Global Positioning Satellite (GPS) shown to the left.

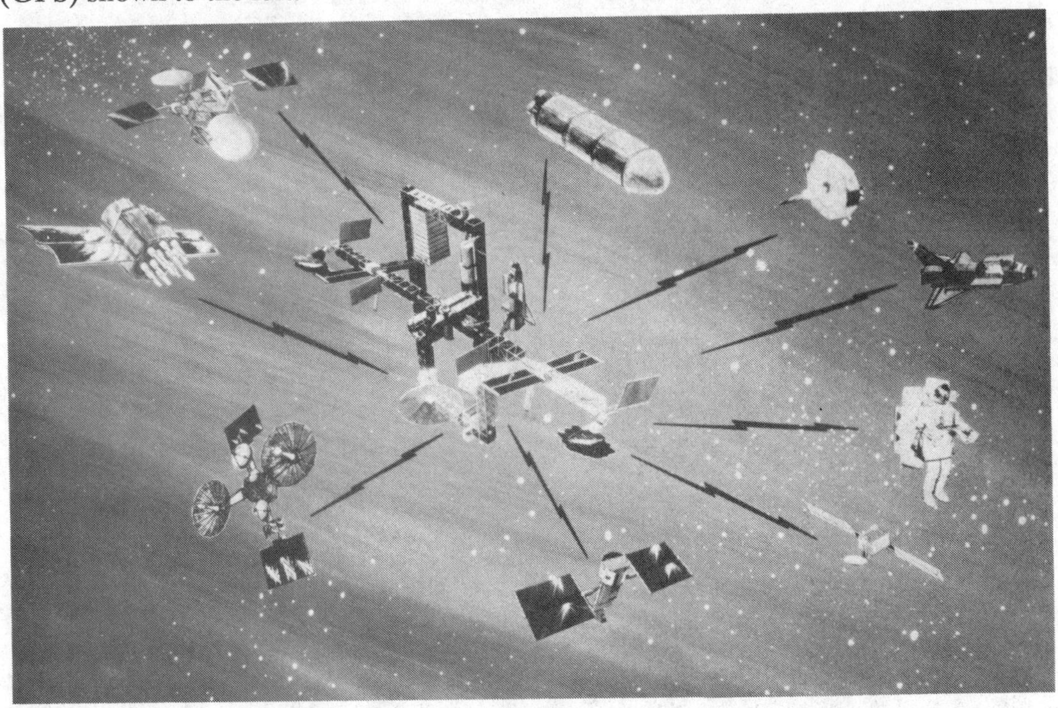

Figure 1 Space Station Communications System

* Dr Weinrich is associated with GE/Aerospace, RCA Government Communication Systems Division, Front & Cooper St., Bldg. 2-4, Camden, New Jersey 08102.

The Space Station Communications System is an end-to-end network supporting Space Station operations and user experiments on the station as shown in Figure 2. The space-to-space communications provides voice, video, and data communications between the spacecraft illustrated in the previous constellation picture; and through the Space Station Communications and Tracking System (C&T) for dissemination to, and use by, operators on the station; or for transmission through the Tracking and Data Relay Satellite System (TDRSS) to the ground for dissemination to users and control centers. Data from user experiments on the station pass through the data management system to the Communications and Tracking System for transmission to and from the ground. Tracking is performed by interpretation of GPS data available from all Space Station elements. Voice, video and data transmissions through the Tracking and Data Relay Satellite are received at the tracking station at White Sands and processed in the Data Interface Facility (DIF) at that location. User data is transmitted through either a domestic communication satellite or terrestrial lines to a Data Handling Facility (DHF) at Goddard Space Flight Center (GSFC). Here the data is demultiplexed and processed for distribution to Space Station users and operations centers. The communications between the users and Space Stations through these facilities accommodates user transmissions without special adaptation for operation in the Space Station. That is, the Space Station Communications System has internal protocols designed such that the communication network is transparent to users. Ground control and monitoring of the Space Station operation is accomplished through the Space Station Support Center at JSC.

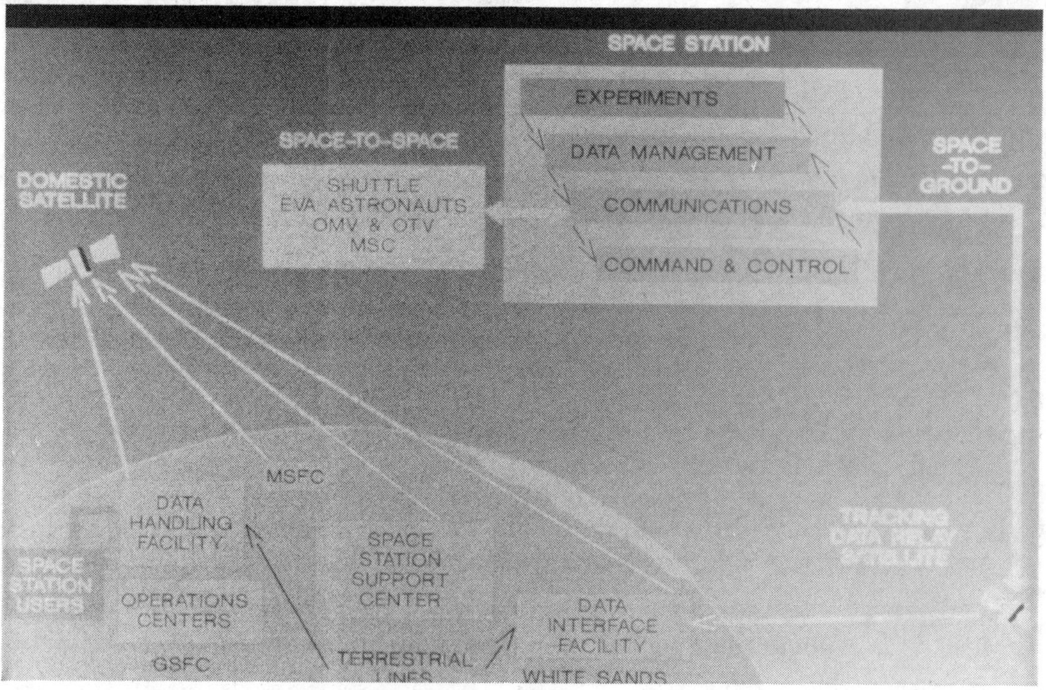

Figure 2 Transmission Systems - Space Station System

The space-to-space communications system shown in Figure 3 provides for all communications between the station and other orbiting bodies through which it operates. The EVA radio shown in the picture is an example of such equipment. The radio communicates through the proximity antennas or high gain antennas on the station at Ku-band. It is capable of 4 to 6 video channels operating in a digital format at 22 megabits per channel. Voice is also transmitted digitally with 64 Kb/s PCM. There are 12 narrowband data channels which transmit 500 Kb/s per second data streams on which the voice and data are multiplexed. The capability exists for secure communication with the data streams encrypted either by a commercial encoder or by NSA devices. Standard protocols are accommodated by the layered protocols and transmission formats pass through the system unaltered at user terminations. Convolutional coding compatible with TDRSS is available for applications requiring very low bit error rates. All equipment is solid-state. The architecture is of a modular design to permit growth in capacity by adding modules.

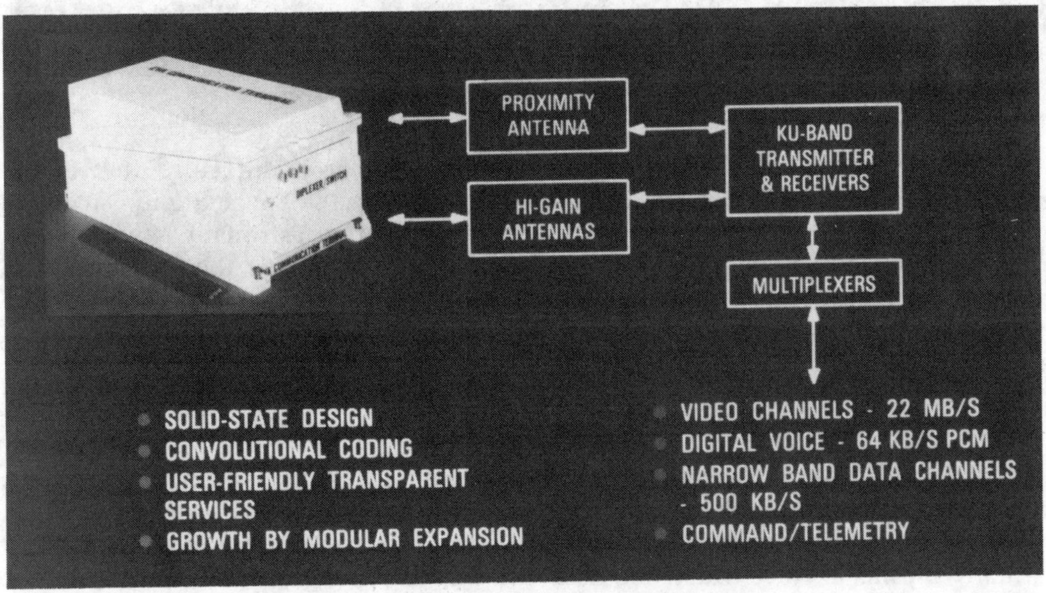

Figure 3 Space-to-Space Communications

The Video Subsystem shown in Figure 4 will have about 22 cameras at IOC. Additional cameras are planned for both the Service Facility (2 cameras and a monitor) and the Mobile Service Center (up to 6 cameras). The cameras are solid-state and will be of the new design for the shuttle as are the TV monitors. The transmission format is 525 line identical to commercial television. There is the capability for high definition TV (1024 line) TV as an optional feature. The video system has a high dynamic range approaching 60 dB with a bandwidth that provides quality equivalent to that from a broadcast studio.

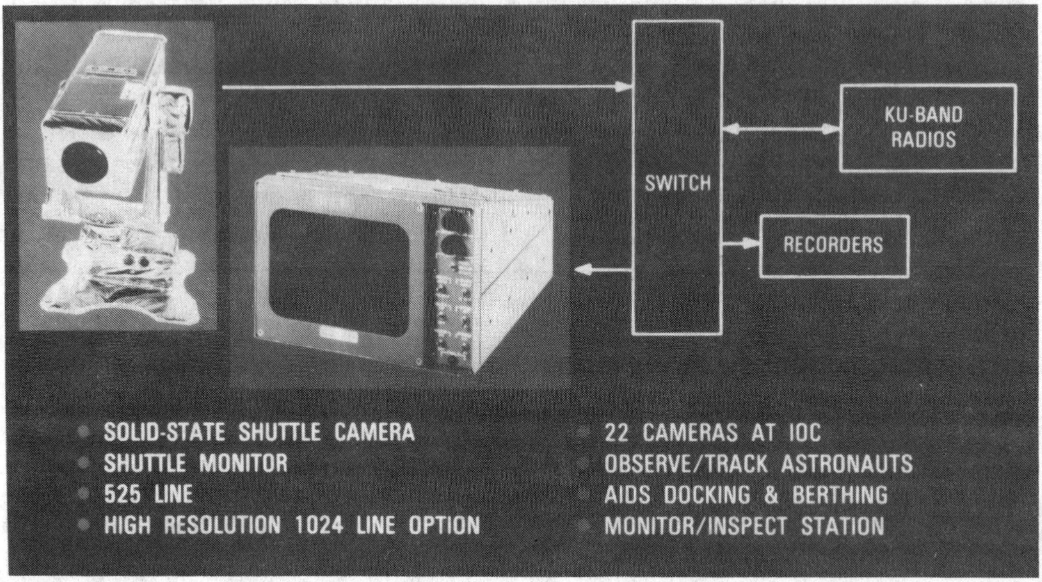

Figure 4 Video Communications

The video system employs switching in a modular architecture that allows growth to more than 100 cameras distributed throughout the Space Station structure and modules. The cameras will be used to observe and track astronaut activity as well as for monitoring and inspecting progress on experiments and the structure. The cameras are considered a valuable aid to docking and berthing of the shuttle. The video system offers enormous flexibility for the addition of cameras, wide variations in video network configurations through the modular switch and an opportunity for extensive video signal processing using the switch for access to special equipment such as that envisioned for tracking astronauts and laser ranging.

The space-to-ground communications through TDRSS shown in Figure 5 is a high data rate link from Space Station to the ground using a 300 Mb/s modulator that has successfully undergone field test at JSC and Goddard. The system can transmit virtually any mix of voice, video and data such that total data rate does not exceed the 300 Mb/s. For example, this high data rate link can accommodate four video channels at 22 Mb/s (88 Mb/s) with more than 200 Mb/s for data and digitized voice. The baseline system offers full TDRSS downlink capability with the user-friendly transparent services shown on earlier slides. It accommodates packet and circuit switched data from the user without constraint or alteration of the users formats.

The focus of the Space Station Communication System is to provide transparent services to the users with minimal or negligible constraints on protocol and format. All protocol processing internal to the Space Station communication system is handled within the system and poses no constraints on the user; it is analogous to the signaling and control systems within the commercial telephone networks. The system is designed for flexibility to accommodate a wide variation and mix of traffic. The links are high capacity digital transmissions; the space-to-ground link has a capacity of 300 Mb/s with any mix of voice, video and data. The architecture of the system is modular such that growth in capacity can occur by modular expansion subject only to the limitations of available frequency spectrum and power. All equipment designs selected for the Space Station Communication System baseline are derived from flight-proven equipment with full redundancy of components in all systems to attain the maximum possible reliability. In addition, the Control and Monitor subsystem uses expert system technology for automated fault diagnosis and correction without operator intervention. This system, as with the transmission systems, is expandable as the Space Station grows.

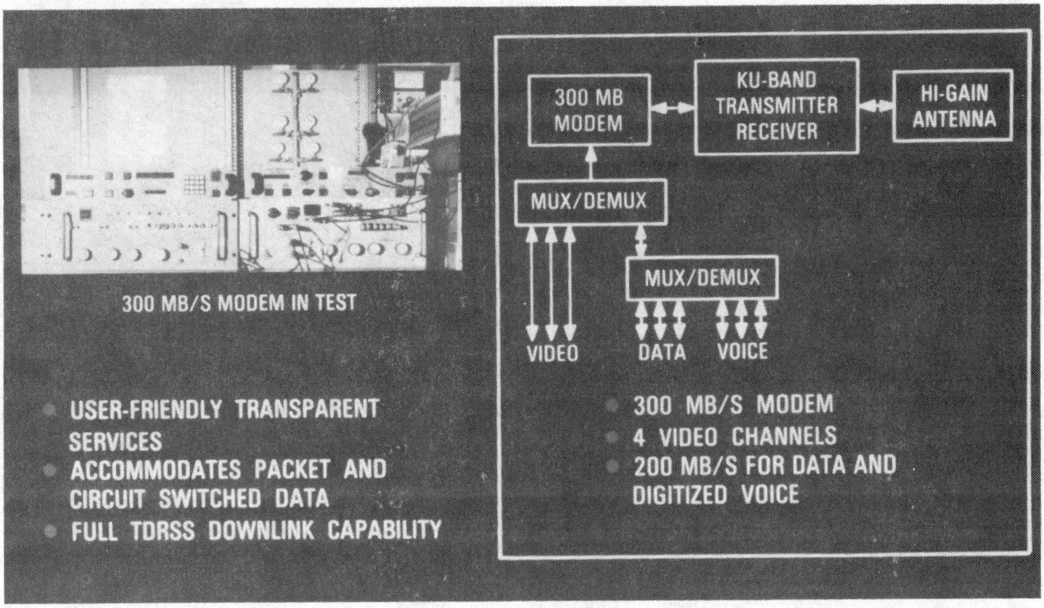

Figure 5 Space-to-Ground Communications

The Space Station Communications System, like the Space Station itself, is designed for flexible, reliable communications for users of this major NASA facility in orbit.

Charles T. Force: One thing that occurred to me as I was listening to Dr. Weinrich is a bit of trivia. There's a gentleman named Proxmire, who lives up where I come from and he posed a question to us about what we're going to do with data up to the year 2000. One of the contractors we had working on that came in the other day with a presentation, and they were talking about petabits of data. I wasn't sure what a petabit was. It turns out, we passed terrabits, which I suspect you have probably heard of, by three orders of magnitude. A petabit is 10^{15} bits of data. You hear numbers about how many times a tape would go around the world to hold this. I don't know what the answer is. But anyway, you might file petabits away as useful information for the future, because it looks as if we're headed that way.

AAS 87-654

TRACKING AND DATA SYSTEMS ASPECTS OF TELEROBOTICS

Arthur G. Stephenson*

I'd like to talk about tracking and data from the point of view of the Orbital Maneuvering Vehicle which is a subject that's near and dear to me. Before I get into the tracking and data, and particularly the communications aspects, which are a real challenge, I want to tell you a little about the vehicle itself and it's mission. The OMV came into existence to extend the Shuttle Orbiter reach capability. It's about 15 feet in diameter, and is about five feet in length. The goal was to make it as narrow as possible to leave room for other payloads in the bay and for frequent manifesting. It's designed to fly 100 missions, similar to the Shuttle. It has about 9000 pounds of propellant on board and 1000 pounds of hydrazine which will enable it to go to about 1400 nautical miles above the Orbiter or, for a very heavy payload, to about 300 nautical miles above the Orbiter. It will be operational, as Charles said, in 1993 and basically its mission is to deploy satellites from the Orbiter, to retrieve satellites, and de-orbit debris or used satellites that are no longer needed. And it's also going to have attached to it the Flight Telerobotic Servicing (FTS) system, which will allow it to remotely service satellites, not having to bring them back to the Orbiter.

In the Space Station era, it will be used for bringing things in and out of the Space Station, just as it does for the Orbiter. It will also be used to transfer logistics modules between the Space Station and Shuttle.

The OMV is being developed by TRW under the direction of NASA's Marshall Space Flight Center. Let me describe a typical retrieval mission. OMV goes to orbit on the Orbiter. The communications on orbit are through TDRS to the White Sands Ground Station and then either by land or DOMSAT back to the Mission Control Center at JSC. It will be flown by a pilot and a co-pilot, with operations similar to those you see in the cockpit of an aircraft. Both pilots will have control of the vehicle. Once the OMV is deployed from the Shuttle, it will transfer autonomously to the orbit of the satellite that it's retrieving. This is accomplished through the use of GPS, and on-board guidance and navigation, based on the known ephemeris of the target.Once OMV is about four miles from the payload it uses radar to acquire and approach the payload. All this is accomplished without any involvement from the ground. Piloting operations take over at about 100 ft from the payload.

* Systems Manager for the Orbital Maneuvering Vehicle, TRW, Room 2357, Bldg. R-11, One Space Park, Redondo Beach, California 90278.

Now, what I'd like to do is really concentrate on the last part of that mission, which is the docking phase where the communications become critical. Since the pilot is in total control of the vehicle and has the ability to do bad things as well as good, we're very concerned about the links that support this, both the NasCom as well as the TDRS links. Because of the links involved, we're talking about up to three seconds of delay. So you can imagine the slow response when the pilot moves his control lever, and about three seconds later, he see's the effect of his command. He's got hand controllers that give him complete six-degree-of-freedom control of the vehicle. The docking operation will be very slow and the last part of it will be less than a tenth of a foot per second. When we get the FTS on the OMV, I will predict that the ground controlled operations of moving arms around and doing servicing functions will also be very slow, very narrow-band operations because of the three-second delay.

Let me talk now about the communications aspects of the OMV. We communicate through TDRS, as I mentioned. We also communicate through GSTDN as a backup. For most communications we will be on the high-gain antenna, either through single access or multiple access with the TDRS. And for pilot operations we'll definitely be on the high-gain antenna. The low-gain omni antenna is primarily used in a backup mode, and a way of communicating to the vehicle when we're in the powered down state; when we're left on orbit for long periods. The transmission return link is critical in terms of providing the pilot with information that he needs to pilot the vehicle. We have both right and left circular polarization that enables communication through TDRS at the same time we're docking to a payload which is also communicating through TDRS on a different polarization.

Let me talk now about some real issues that I see with communications for the vehicle. First of all, the question of S-band versus K-band carrier frequency. We have a baseline system of S-band which requires a video compression unit on board (to reduce bandwidth) and reconstruction on the ground. We find that to be very satisfactory for the narrow band-widths that we're working with where we don't have a lot of dynamics. The operational constraints that come with S-band, particularly the radio frequency interference from Russian radars, is of concern and we've added Reed Solomon coding, which helps a lot to reduce burst errors. We also use error-containment techniques, which allow us, if we lose a segment of a picture, to recognize that and repeat the last frame, so that the pilots really don't see much change. The real advantage of the S-band system is the low cost relative to K-band. On the other hand, K-band allows us to get away from any compression issues and also away from the RFI system problems. The real disadvantage is the cost, which is significantly more with K-band, so we are baselining S-band at this point.

Another concern is, that we need to be able to dock to spinning or tumbling satellites. In the case of spinning satellites, we are talking about up to a half rpm, or in tumbling, about a sixth of an rpm. But having that as a requirement, we do have a peculiar situation where, if TDRS were directly above the vehicle, the antenna coverage for the high-gain antennas would come in and out as you go around. So we've been working on that and proving to ourselves that we can actually switch antennas and not lose receiver lock, so that the pilot doesn't notice that we're actually switching from antenna to antenna.

The piloting aspects of this vehicle are of critical concern and for the last four years we have been simulating it on computer graphics which fully simulates the OMV characteristics, as well as the target vehicle that we're docking to.

The last thing I want to mention is the collision avoidance issue. Certainly when we're flying into the Space Station, there's concern about colliding with the Space Station. We have to be able to tolerate two critical failures on board. And we also have to be able to tolerate the loss of a link. So if the pilot is flying in and if the down-link is lost, the pilot detects loss of his display and commands a back away maneuver. If the up-link is lost, the vehicle recognizes loss of signal and does the same backaway maneuver on its own. So there are a number of challenging issues here, not only the delays but also collision avoidance. I think, in the long run, the system will work very well, whether we stay at S-band or go to K-band, and it will be a key player in the time frame of the Space Station. Thank you.

QUESTIONS AND ANSWERS - TRACKING AND DATA SYSTEMS

Charles T. Force: Thank you very much, Art. Do we have any questions?

Question: A two-part question: What service requirements will drive space communications and tracking developments in the next 15 years? What technologies must be developed to provide these services?

Charles T. Force: I consider the high data rate obviously is one of the things that come through in this session. It comes through in almost any form we've discussed. I mentioned, petabits; the data rates are getting where they're almost unbelievable and what we do to accommodate them. I think that's one of the things. The need for reliability of transmitting very critical data, be it up-link or down-link, and trying to contrast with the need for rates for SAR, for instance. What techniques must be developed to provide the services? I think recording is one of the primary issues. The state of the art in recording technology is advancing very rapidly. But still in terms of the amount of data that we're faced with to archive, analyze, and process it gets to be beyond what the state of the art provides as we can see it right now. Here's one for Art.

Question: What is the largest module that the OMV can bring to the Station for berthing and docking? What is the role of OMV in the Station in the terminal phase?

Arthur G. Stephenson: The size of payload that we're talking about is about a 50,000 pound logistics module As far as the de-orbit maneuvers, we have a requirement to deorbit a 75000 pound external tank. I think the second half of the question had to do with the role of OMV in the Space Station. The Space Station will have a control center for OMV and we'll be controlling the vehicle from these as it is flown to the Space Station. So at that point you get around the three-second delay issue.

Question: Wouldn't infrared vision technology significantly aid or reduce the image-processing requirements for Space Station robots?

Why Important and of Interest: Simplifying the image processing problem should ease the requirements of the communication system if remote processing is needed and IR technology is well understood.

Recommendation: Some combination of IR and visual would probably make the optimal cost/performance mix.

Dr. Kumar Krishen: It would, to some extent reduce them; it depends on the application. You know robotic vision is such a broad scene. In certain applications you don't need high resolution, when you are trying to locate something, for instance, in space. However, on the other hand, if you're locating a nut or bolt on that scene, a real precise point, then you need high resolution. IR technology would certainly help to give you high resolution and therefore make it more possible to do robotics in space.

Charles T. Force: Thank you. The question suggests that some combination of IR and visual would probably make an optimum cost/performance mix, which sounds quite logical.

Question: High-gain antenna switching during the docking with spinning satellite, is this automatic or commanded?

Arthur G. Stephenson: It has to be accomplished totally on board because of the delay situation. The way it works, basically you can tolerate up to a 50 millisecond interrupt in the signal and remain locked. Beyond that you lose spread spectrum code synchronization and it takes about a minute to reacquire.

Charles T. Force: That concludes this session. I would like to thank you all for joining us. And my special thanks to the members of the panel.

STRUCTURES AND COMPOSITE MATERIALS

PANEL EIGHT

Chairman:	Anthony Amos Program Manager, Aerospace Sciences U.S. Air Force Office of Scientific Research Wright-Patterson AFB
Host Chairman:	Donald C. Wade Chief Structures and Mechanics Division NASA Johnson Space Center

The following paper number was not assigned:

AAS 87-659

The following paper was not available for publication:

AAS 87-660 "Role of Composite Materials in Future Space Systems," by Dr. Darrel Tenney, Chief, Materials Division, NASA Langley Research Center

AAS 87-658

STRUCTURES AND COMPOSITE MATERIALS - INTRODUCTION

Anthony Amos,* Donald C. Wade,† and Henry N. Bowes‡

Henry N. Bowes: Good morning, I'm Hank Bowes. I want to welcome you to this, the 34th National Conference of the American Astronautical Society. We have a very good program planned for you this morning, and I have a special request. Those of you who have been here before, perhaps this will be a little redundant, but I suspect there are quite a few new folks here. You should all have received, what we call, an Audience Participation Form. Does anybody have a blank look on his face when I say that? You all have a copy of the Audience Participation Form?

We looked at what this conference might be able to do, to make it meaningful. You and I have all attended many conferences over the years, and we listen to a lot of very good presentations and when we get all through, we think there's going to be a proceedings volume, and some of us look at it and some don't. We wanted something different for this conference. The space program is at a crossroads and we felt that this group was uniquely qualified to make its knowledge of the space program available to policymakers. Now, when you say "policy", some of us shrink away and say, "Well, that's the other guy's job". I used to think that myself, and I had the opportunity to participate in some other professional societies' activities in Washington, and know, that you and I can in fact make a difference about what goes on in that 'wonder world' called Washington, and 'the Hill'.

Therefore, we, the American Astronautical Society, national officers, and the host committee have taken the position that we will create a recommendation about our concerns and issues relating to the space program, its policies, and the technology, recognizing that they are, in fact, entwined. The policy drives the technology and having the technology will drive policy. And you and I can make that difference. We have asked, in this Audience Participation Form, that you express your personal concern, put it down, indicate why you think it needs attention and if you have a recommendation, add that to the list. These will be used to help the panelist speak to the

---

* Dr. Amos is Program Manager, Aerospace Sciences, U.S. Air Force Office of Scientific Research, AFWAL/FIB, Wright-Patterson AFB, Ohio 45433-6553.

† Chief, Structures and Mechanics Division, NASA Johnson Space Center, Houston, Texas 77058.

‡ General Chairman, 34th AAS Annual Conference. He is Vice President and Program Manager, Lockheed Engineering and Management Services Company, Inc., 2400 NASA Road One, Houston, Texas 77058.

issue. And as the originator of that, when the topic comes up and if you want to enter into a dialogue with the panel, I urge you to do so, because it is this feedback that, in my view, will make the difference. The Society has agreed to create a summation of this panel and all the others, coupled with what will happen this afternoon at the final session recap called, "Where from Here?" And it will become the recommendations that we will present to the governing agencies of the United States and, if appropriate, other international organizations. So, it is your challenge to put that down on a piece of paper, two or three pages, with the assistance of the chairman and the others. So I'll turn this over to Don Wade, Division Chief of Structures for Mechanics for the Johnson Space Center.

Donald C. Wade: Good morning. Welcome to the panel on Structures and Composite Materials. This morning it's my honor to introduce the chairman of this panel, Dr. Anthony K. Amos, who is the Program Manager for Aerospace Sciences at the Air Force Office of Scientific Research in Washington. He is responsible for the planning and implementation of all Air Force basic research in structural mechanics and he is also the science and technology agent for the Space Defense Initiative and the innovative science and technology offices in the area of structures and dynamics. Tony is a graduate of McGill University, MIT, and Princeton University with specialties in structures and applied mechanics. He has had prior professional experience as senior engineer at both RCA Astro Electronics Division and Boeing Vertol. He has also been in technical management at Boeing Vertol and NASA Headquarters before going over to the Air Force. He has a number of publications in the dynamics and aeroelasticity of aerospace systems. It is my pleasure to introduce Tony Amos.

Dr. Anthony Amos: Thank you, Don, and good morning, ladies and gentlemen. In the last couple of days of deliberations a number of issues have been raised concerning our national space program. They ranged from policy concerns, Government policy concerns, through mission system and subsystem options to disciplinary technological procedures in such areas as propulsion, life sciences, tracking and data systems.

This morning, we continue with the disciplinary issues focusing on the technical areas of materials and structures. One recurring theme during the last two days has been that of cost. We have heard about low-cost objectives in the Space Station planning, in space communication, in space transportation development. Among the many factors that impact the cost issue we cited the issue of weight, specifically with respect to transportation systems as well as the spacecraft and payload systems. Reliability was also cited as a cost-impacting issue. We need to add to that the issue of durability for the materials and structures that compose most of those systems. All of these issues have very serious ramifications for the materials and structures disciplines lightweight requirements currently driving the development of advanced materials. Conventional metallic materials are being replaced in ever-increasing amounts by advanced composite materials. This morning we are going to hear from one of the nation's experts in these advanced materials who will be talking about some of the issues that will make application in space either possible or impossible. Included are the open lattice type structural configuration that characterizes current designs for the Space Station, some of the other systems, such as a large aperture antenna backup structure designed specifically to provide coverage of a large expanse of space at very low weight per area or per volume as the case may be, other drivers in the materials and structures development and configuration, structural configuration development, in the area of high-temperature capability as was brought out in the Aerospace Plane development. The stability of structural systems involved in optical systems, the reflecting surfaces of telescopes, radar systems, laser systems, and so on, are causing very stringent technical challenges to the structures and dynamics as well as materials communities and we're going to hear some of that discussed today.

On the issue of lightweight structures the lattice type structure that I developed to cover large areas as well as large volumes of space, it has become evident that the only way to get those into space is to build them so that they can be folded and launched and then deployed or assembled in space. The requirement for space assembly or space fabrication was raised a couple of days ago in connection with the role of robotic and teleoperator devices in the space program. It is evident to most of us in this room that we cannot rely exclusively on EVA for doing this kind of operation in space.

To give a little background to this issue we've decided to show a short film from the assembly experiment that was conducted on the Shuttle a couple of years ago, just to remind us what is involved in EVA operations regarding space construction. That gives us some idea of how we're going to build the Space Station and other structures in space.

One implication of this type of space construction is that for the systems we put up there, we're not going to have the opportunity, as we do with some of the smaller spacecraft we will be launching, to build them on the ground, to test them thoroughly on the ground before we commit to launch. There are very serious implications for this. It means that we're going to have to rely more on analytical validation of the designs. We're going to have to resort to test data from previous flights as well as technology experiments conducted together -- the sort of information that will give us the confidence to launch newer and bigger systems.

With that as a background I would now like to introduce our first panelist who will be talking to you about the Role of Composite Materials in Future Space Systems. He is Dr. Darrel Tenney. Darrel is Chief of the Materials Division at NASA Langley Research Center in Hampton Virginia. The Materials Division conducts basic and applied research on advanced materials for aerospace applications including spacecraft, Space Station, space transportation vehicles, transport aircraft, rotorcraft and hypersonic vehicles. Dr. Tenney has been primarily involved with the development of composite materials technology for space applications, with an emphasis on long-term materials durability in the space environment. He has served as Assistant Chief of the Materials Division for three years, and prior to that he managed the overall operations of the Applied Materials Branch within the Division.

He has co-authored over 40 reports in the areas of space materials durability, metal matrix composites, high-temperature materials and environmental effects on composites. Prior to joining NASA, Dr. Tenney was Assistant Professor of Materials Engineering at Virginia Tech. He holds a Ph.D. degree in materials engineering from Virginia Tech.

Our second speaker will be Dr. Keto Sossaar, who will address some of the challenges in the design and qualifications of large space systems. Keto is a structural engineer by profession. He's currently the Director of Cambridge Research, which is a small firm specializing in structural dynamics and control interaction. His professional experience was with MIT Instrumentation Lab which subsequently became the Draper Lab. He was Division Leader in Structures and Dynamics. His academic training parallels mine in some ways. He and I started our undergraduate work at McGill University. We both shared in the mischiefs as members of the Phi Epsilon Alpha Engineering Society. And today he has decided to pull another big mischief. Most of us engineers cannot say a few words to an audience without viewgraphs. He has chosen to talk to you on this very technical subject without a single viewgraph.

Our third speaker will cover the panel topic of Large Scale Computational Issues in the Dynamics of Large Space Structures. Our panelist is Professor Edward Haug of the University of Iowa. He is Professor and Director of the Center for Simulation and Design Optimization. The Center is co-sponsored by the National Science Foundation, the Army Tank Automotive Command and NASA Goddard Space Flight Center, with participation by 30 federal and corporate laboratories. Work is in advanced simulation capabilities in dynamics of articulated structures in machines. His academic background consists of a Ph.D. in applied mechanics from Kansas State; he has ten years' experience in armament R&D and eleven years as Professor of mechanical engineering at the University of Iowa and he has authored nine books and some 165 papers in design optimization of dynamics.

Our final panelist will be Lt. Col. William P. Witt, III. Bill is the Program Manager for Acquisition, Pointing, Tracking and Fire Control Technologies for Directed Energy Weapons for the Strategic Defense Initiative Organization. He received his bachelor's in engineering sciences from the U.S. Air Force Academy, an M.S. in astronautical engineering, and Ph.D. in aerospace engineering from the Air Force Institute of Technology. Bill is going to be addressing you on the topic of Ground and Flight Experimentation in future Large Space Systems Development.

AAS 87-661

STRUCTURAL DESIGN AND VALIDATION CHALLENGES OF LARGE SPACE SYSTEMS

Keto Soosaar*

I think it's appropriate before I begin my main discussion to state why I am not going to use any visual aids. A couple of weeks ago I encountered a former colleague in an airplane. We had worked together in the early and mid-70s on a number of programs after which he left the aerospace industry completely. We were catching up on our personal and professional lives and he suddenly stunned me with a question. He asked, "How is your large satellite structures controls on orbit experiment going? And it was a stunning question because we had believed in 1976 that, in fact, we would have an orbiting experiment by '82-'83. All of our issues were resolved and my friend had simply, naively assumed that because it had existed on a viewgraph, it was eventually going to come to pass. So when I came home and started to prepare for this presentation, I realized that the technical issues really had not changed. The design and validation cycle is still as incomplete today on many of these important issues and ten years have passed since we first identified them. So rather than mislead everybody with pretty charts and hypothetical schedules, I would like to evoke my prior presentations by reference and proceed to the problems that I think we are truly facing. I hope, too, in the process, to suggest some solutions.

Design and validation are, in effect, the *ying* and the *yang* of engineering. They are complementary, yet opposing, parts of the same process. In design we postulate a solution to a stated operational requirement. In validation we observe this solution in that operational environment, judge its adequacy, and if necessary, suggest modifications to the goals and the constraints of the design process. A vital element therefore, of this process is rapid feedback of validation on design and also of design on validation. In some of our high-technology industries this occurs rather rapidly as products move from research, development, production and into operational phases. The engineer receives his feedback very quickly and this helps him hone the product very well for his intended market niche; also the engineer develops in the process.

This, unfortunately, is not the case in our community today. In the 1960s we designed, built, launched and operated within a span of a few years so that this important validation cycle would conclude while the engineer who had designed it was still in his original position. Those were, of course, much simpler satellites.

* Dr. Soosaar is President, Cambridge Research, 1033 Massachusetts Avenue, Cambridge, Massachusetts 02138.

I would like to mention, anecdotally, that my first job after finishing graduate school, in the late '60s involved structural modeling techniques and active optics concepts for the Space Telescope. Twenty years later, I'm still waiting for an on-orbit validation. Fortunately, there have been for me other systems with less of a lead time. I fear, however, that in the last ten years or so, we have created a community of aerospace engineers who have developed repeated concepts, with none of them ever achieving on-orbit validation, not to mention operational status. We really don't know how good large space structures are, and whether the design approaches that we are using are the right way to go. Simultaneously, through this last ten-year period, the prior generation of experienced engineers is beginning to evaporate through retirement.

Now, clearly, why has the feedback not occurred? We've had few opportunities due to Shuttle operational difficulties initially, apart from the recent catastrophe. But we have also developed, I think, satellite concepts that were bold, imaginative, ambitious, risky, and ultimately, expensive. As a result, the country may not be able to afford many hundred-meter structures experiments to validate a single structural issue of concern. Yet, we seem to have pursued this 'bigger is better' goal.

Certainly a larger optical system like the LDR resolves a smaller object and intercepts more photons. But this largeness leads, necessarily, to a lower average mass, because of our launch constraints. We get a lower stiffness structure, we wind up with reduced natural frequencies and we begin having many of the subsystems, that previously were independent, become coupled together through the elastic behavior of the structure.

In earlier times, the main goal of the structural engineer of the satellite was to connect the different pieces and assure launch survival without going plastic. The vibrational modes were well separated from on-board disturbance frequencies, so that structural dynamics, in an operational sense, was rarely a problem. Right now we have multiple structural modes in the control bandwidth. We can't notch-filter them out anymore; we have to feed back on elastic motion to quell it. All of these phenomena begin to have an important effect also on the performance metrics of our system, such as pointing, figure control and antenna pattern. Not only are the problems complex, but our understanding of the design process is still incomplete and we may not be able to begin validation except on orbit.

Given, though, that we regain our access to orbit, and we still assume some of our ambitious missions, what must we do to assure the integrity of the design and validation process? I'd like, very briefly, to touch on a couple of areas; these include educational preparation of the engineer, the availability of suitable analysis and design tools, appropriate organizational structures for these technologies to reside in and our national long-term commitment to the process. I think these are important issues here.

The current crop of structural engineers is extremely well prepared in analysis methodology, and intensely computer literate. There appears, however, not to have been inserted into the education process the "the sanity checks", that earlier engineers worried about because they had less experience with the computer and more experience with hand calculations. Similarly, there seems to have been, in general, very little education of the design process. Some of our younger engineers are extremely adept at moving models and data bases from a PC to a VAX to a CRAY in a matter of minutes, but have been known to go into extended catatonic states when asked to design a truss. There certainly are less design courses left in engineering schools. This is not surprising, since most of the engineering faculty today have not had an opportunity to get industrial experience. The engineering student often does not encounter real engineering until the first job. And many of them react to that by signing up for MBA courses. Certainly, our community cannot survive very long without engineers who have had some introduction to the design process. One suggestion that has been made, that I'd like to repeat here, is to turn the master's thesis, which traditionally has been a subdoctorate type of investigation, into an extended project that exercises the primary areas of interest of the student. This would be supervised jointly by the faculty and by industrial engineers and would result in a background similar to that of the 'diploma engineer', a grade used in some European countries.

Given that the student has been given the introduction to the design process, is he getting the multidisciplinary education necessary to deal with control structure interaction processes, for instance? Ten years ago, when structure control interaction first became an "in" area, we often saw structures and controls groups eyeing each other like equal-sized piranhas in a small tank with fear, suspicion, and greed. But, fortunately, the research market-place has regulated this, and in many universities now, it is possible to obtain multidisciplinary educations. As a user, however, of young engineers, in my own firm I would wish that all astronautics students could be exposed to a course where space missions, such as communications, radiometry, imaging, meteorology, radar, would be covered to the degree where these are shown to interact with structures, dynamics and controls requirements. Often this process has to become part of on-job training, but I think it would be very well served to have this as a university background.

Given now, that we have established a design curriculum, the next phase is a preliminary validation of the design concepts through analysis. The question that we face here concerns the appropriateness of the analysis tools that we have at hand. As a first order, we do have appropriate tools, but one is tremendously haunted by scale and by appropriate phenomenology. For dealing with our control structure interaction problems we will need to extract with reliability, hundreds of elastic modes or observe our structures responding to modeling error spill-over. In other cases, the same thing may be caused by our making linearity assumptions on joint dominated structures. Our spacecraft dynamics tools have experienced, also, substantial recent growth, but whether they are adequate in scale, scope and speed, remains still to be seen. My colleague, Professor Haug, will expound on this in more detail in a subsequent paper.

In the controls area for instance, it's not clear that we know properly how to integrate several independently designed control systems that are coupled through the same elastic structure. Along with technology-specific analysis and design tools, we must also consider multidisciplinary codes that integrate specifically separate phenomena like heat transfer, structural dynamics, controls, and ultimately, optics and radar into single packages that eventually provide direct understanding of how a structural performance metric affects wavefront, far field pattern and satellite pointing. I think, the IAC of NASA has been an excellent start on this and the ISM of DOD is building upon that.

Good analysis tools can go a long way toward validating design concepts but only if they themselves have been validated. Therefore, the modeling of lab and space tests is a particularly crucial part of analysis tool validation,and has to be planned for, as part of the experiment. My colleague, Lt. Col. Witt, will deal with the ground and flight experiments in his paper.

Another issue is organization. Organizationally we are still segregated into structures and controls groups. Universities, as we've seen, have been most successful in blurring the lines; industry has not made formal strides, but it successfully handles that through project organizations. There is still a horror story of a large aerospace firm which long anticipated the need for structure control interaction, and, several years in advance, invested many, many dollars in internal research. When the time came to bid, the structures, the controls, and Advanced Project offices became locked into a power struggle that was not resolved until after the proposals were due. No proposal was submitted, and key people left. Like industry, Government too, has kept structures and controls separate. There's less opportunity there, to create project organizations, particularly where smaller research efforts are concerned, and that does result in technical Balkanization and often duplication.

Finally, I have genuine concern that the design validation cycle is beginning to fall apart. The major cause, as I've mentioned earlier, is the uncertainty and stretch-out of space projects, unfortunately, currently attributable, I think, to an indifferent Administration and Congress. We have made technical progress, however, and that is a tremendous tribute to the perseverance and ingenuity of the governmental technology program managers.

The risks, however, exist. As I mentioned earlier, the projects are stretching far beyond the professional lifetimes of the key advocates and the technical managers. The important design validation cycle will not occur on an individual basis anymore, but on a team basis. And there is always that risk of finger pointing rather than learning from less-than-successful ventures. In the absence of a quick feedback mechanism on design validation, we need to seriously begin thinking about developing technology and flight data bases in depth.

We were attempting, in our company, to collect and correlate data on satellite deployment anomalies. We knew of a number that had existed on account of personal experiences with satellite programs. When attempting to uncover the data we were unable to find it or when it existed, it was so superficial as to be useless for planning purposes. Maybe the data exists, but we certainly could not find a way to get to it. We have to begin collecting. We also have to be able to access and generate data of opportunity to instrument new satellites. Yet this is most difficult. In experiences with the Space Telescope over the past ten years, where on three separate occasions there were attempts to get additional instrumentation put on it to look for unknown interactions of structure and control; all three times these were turned down because it would delay the program. Therefore, we have to institutionalize such a data collection process. There has to be an official well-known collector of such data, perhaps an Office of Chief Engineer or Chief Scientist, with sufficient clout and resources to insert himself into a program management process and say, "We have got to have this data. Here's some money to measure it with, but you are not going to be able to tell us to go away."

In summary, in attempting to give a paper without any visual aids, I don't know how successfully, in reviewing the structural design validation process, the following issues have emerged. We've observed that the design process is less and less a part of the young engineer's training. But with good will, that is fixable. We're finding that multidisciplinary people are emerging, as well as multidisciplinary tools. The organizational structures are adapting to the needs about as quickly as could be expected. But the validation feedback process, in my opinion, is in trouble. We've had few space flights but we find that the dedicated experiments we would like to configure are far too expensive. We have difficulty in accessing flights of opportunity and we must, therefore, develop fundamental policies of systems level data collection. The technical problems, whether of design or validation are of course still with us. But we can make progress there only if we also attack some of the nontechnical concerns.

AAS 87-662

LARGE SCALE COMPUTATIONAL ISSUES IN THE DYNAMICS OF LARGE SPACE STRUCTURES

Edward J. Haug*

My discussion focuses on some computational issues in dynamics of space structures that are emerging, but are not yet resolved, as indicated in Fig. 1. A fundamental issue regards the distinction between structures with and without articulation. The conventional view of a structure is a building, in which any displacement of one component relative to another leads to positive strain energy. In this sense, apart from control surface motion, an airplane is a flying building.

- Flexible Manipulators
- Nonlinear Deployment Dynamics
- Order-N Multiflexible Body Dynamics
- Parallel Processor High Speed Computation

Fig. 1 Some Computational Issues in Dynamics of Space Structures

Emerging important applications that involve flexible manipulator dynamics, nonlinear structural deployment dynamics, and dynamics of broad classes of machines require transmission of large motion at articulated joints. Such structures are called articulated structures, or flexible multibody systems. They exhibit very different behavior than conventional structures, hence they require special treatment.

One topic to be highlighted in this presentation is emerging algorithms for multiflexible body dynamics that involve a number of computations that is proportional to the number of joint relative coordinates in the formulation. These new algorithms are shown to be well suited for parallel processing high speed computation, leading to enormous potential for dynamic simulation using emerging parallel computing hardware.

* Professor and Director, Center for Computer Aided Design, College of Engineering, The University of Iowa, Iowa City, Iowa 52242.

The schematic in Fig. 2 shows a space station lattice structure, with habitation and other modules attached, and a new habitation module being manipulated by the remote manipulator system (RMS) from the hold of the shuttle.

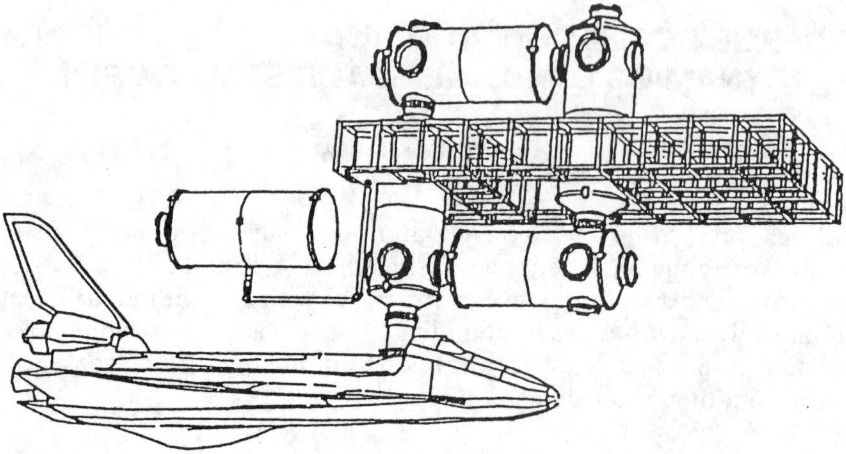

Fig. 2 Space Structure With Module From Shuttle

As shown in Fig. 3, the RMS transports the large habitation module through large displacements and rotations, in preparation for placement and attachment to the very flexible space structure. Such large translation and rotation applications are unique and potentially very important in control of space construction applications. The dynamic simulation technology for such applications, however, lags that of conventional finite element structural analysis.

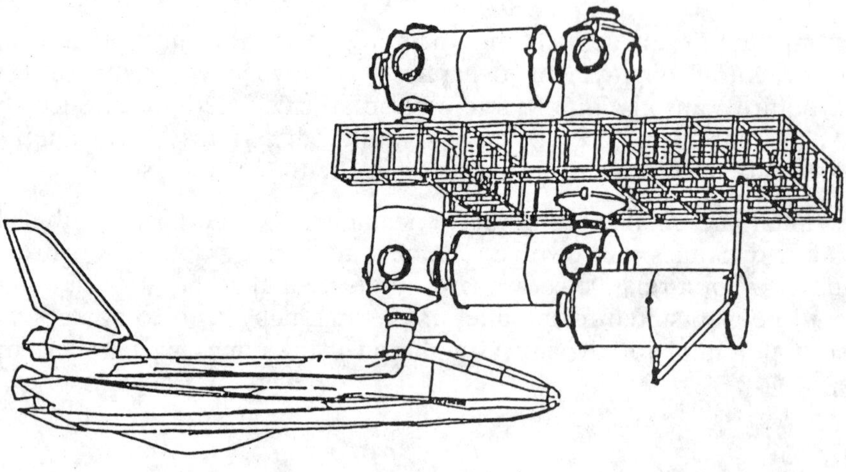

Fig. 3 Module Translating Along Structural Guide

Another example of a unique space structure is shown in Fig. 4. It is a mast that extends from the hold of the space shuttle to support a remote experiment. In the deployed configuration, it behaves as a conventional elastic structure, which can be simulated using existing finite element tools.

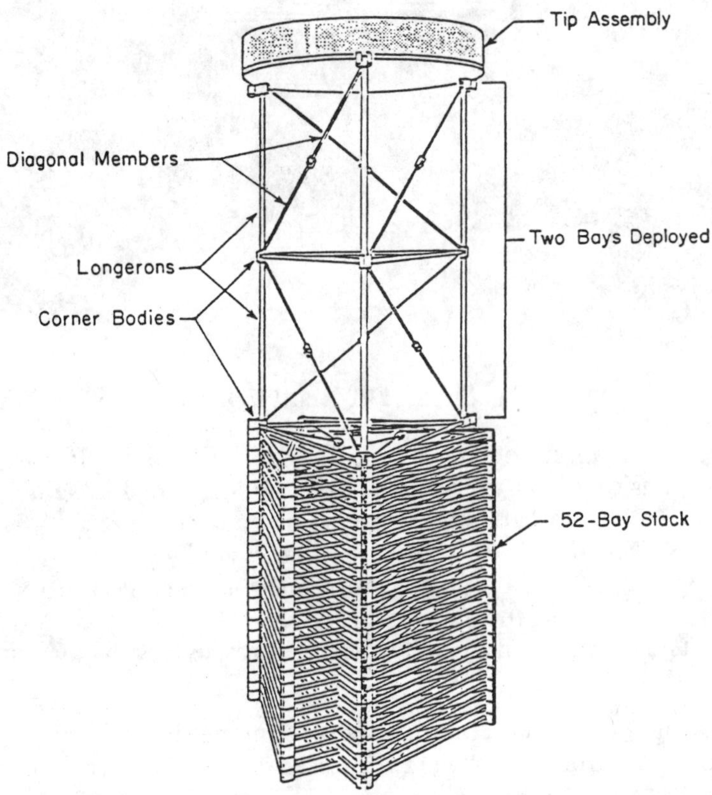

Fig. 4 MAST Partially Deployed

Extensive space structures must be stored in a compact form for launching into space and either deployed or assembled in space. The mast is shown in Fig. 4 in a partially deployed state. Very large rotational motion of components of the structure must occur dynamically, in order to deploy the structure to its intended configuration. One feature of this application that contrasts with the previously discussed manipulator is that it is statically indeterminate. In the language of multibody dynamics, it has closed kinematic loops. Deploying structures such as this are extraordinarily complicated and require a new generation of tools for simulation and design.

To indicate that articulated structures are not unique to space applications, consider the Army's new High Mobility Multipurpose Wheel Vehicle shown in Fig. 5 that is intended for offroad use.

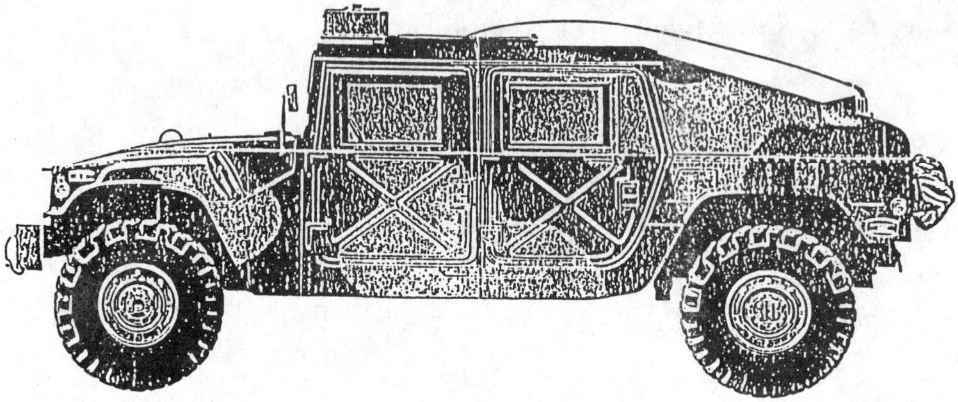

Fig. 5 An Off-Road Vehicle

The schematic diagram of the vehicle shown in Fig. 6 displays upper and lower control arm suspension subsystems that couple spindle-wheel assemblies and permit the suspension to adapt to irregular terrain. The rods connect the wheel assemblies to the steering rack (body 2) in the front of the vehicle and to the frame of the vehicle (body 3) at the rear. Large suspension displacements are required, to carry out the off-road functions of the vehicle. Flexibility of the frame and selected suspension components are important in determining stresses due to dynamic behavior of the vehicle.

To briefly highlight some emerging developments in simulation of such articulated structures, the graph theoretic [1] representation of the vehicle is illustrated in Fig. 7. Circles represent bodies that make up the system and lines that connect the bodies represent kinematic joints in the suspension and steering assemblies. This diagram shows that there are eight closed kinematic loops in this vehicle system. Using cut-joint concepts of graph theory [1], a spanning tree can be constructed, as shown in Fig. 8.

Based on such spanning tree graphs, recursive dynamics algorithms have been developed [2,3,4] that permits simultaneous computations on each of the ten rays that emanate from the chassis of the vehicle (body 1). These algorithms are based on new articulation joint formulations that are not used in structural applications and are ideally suited for parallel processor implementation.

The diagram shown in Fig. 9 is a computational task graph that illustrates the potential for parallel processor implementation of articulated structural dynamics. In major analysis phases that are encountered moving down the graph, up to ten concurrent tasks (the square boxes) can be found in position analysis, velocity analysis, backward reduction to form the equations of motion, and finally in acceleration recovery.

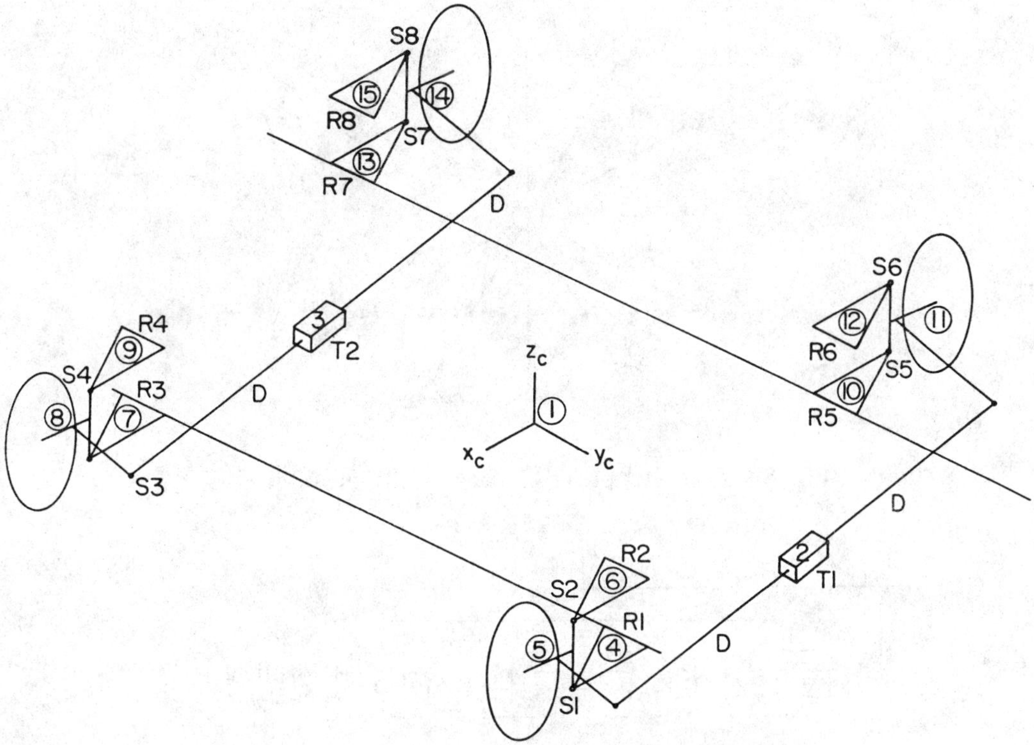

Fig. 6 Schematic Representation of Vehicle

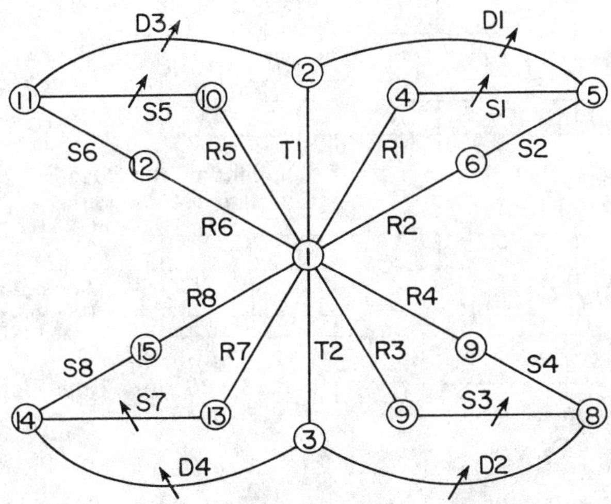

Fig. 7 Graphical Representation of Vehicle

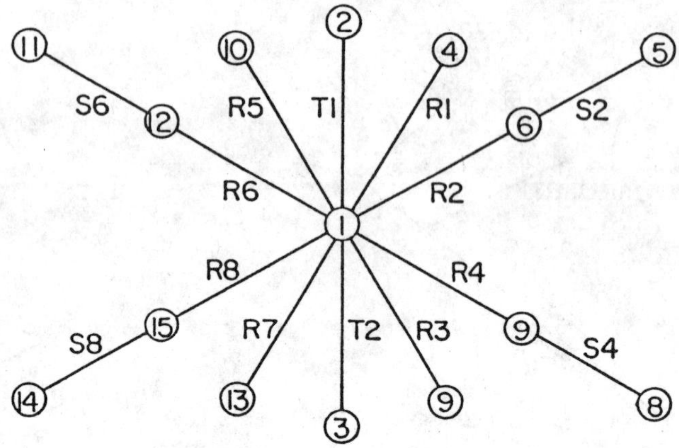

Fig. 8 Tree Structure Corresponding to Vehicle

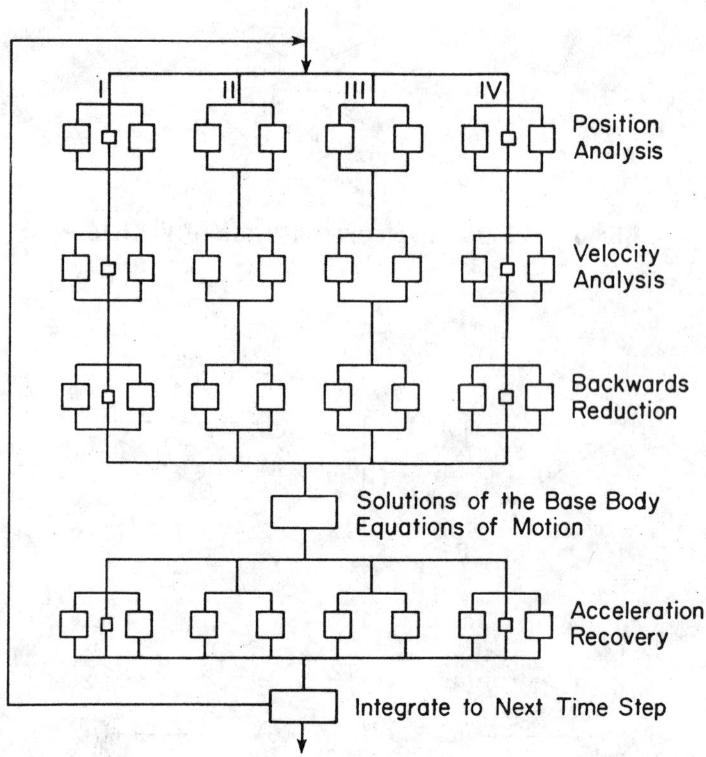

Fig. 9 Parallel Task Graph of Recursive Algorithm for Vehicle

An algorithm based on this graph has been implemented on an Alliant FX/8 mini supercomputer that has eight independent processors. It yields a factor of 36 speed-up over previous multibody simulation algorithms and even greater speed-ups, compared to nonlinear finite element codes. With the emergence of lower cost, high powered parallel processing architectures, we are just beginning to scratch the surface in exploiting the potential that exists for dynamic simulation of important new classes of articulated structures.

In summary, as indicated in Fig. 10, dynamics of articulated structures should be viewed as differing from conventional finite element structural analysis. Significant challenges and opportunities exist in creating simulation tools to support emerging applications of deployable space structures, manipulators, and large classes of mechanical systems. Realization of this opportunity depends on exploiting methods from both aerospace and mechanical engineering, to take advantage of unique developments in each field.

- Significant Multiflexible Body Simulation Challenges/Opportunities
- Synthesis of Emerging Methods From Aerospace and Mechanical Engineering Key
- Exploitation of Emerging Computer Architectures Crucial
- Broad Impact on US Engineering Technology Is Possible

Fig. 10 Summary Observations

As indicated with the vehicle example, tremendous potential exists in exploiting emerging parallel computing architectures that are ideally suited for dynamics of articulated structures. Achievement of this potential can have a substantial impact on a broad range of engineering applications, both in aerospace and mechanical engineering.

REFERENCES

1. Wittenburg, J., *Dynamics of Systems of Rigid Bodies*, B.G. Teubner, Stuttgart, 1977.
2. Base, D.S. and Haug, E.J., "A Recursive Formulation for Constrained Mechanical System Dynamics, Part I: Open Loop Systems, *Mechanics of Structures and Machines,* Vol. 15(3), 1987.
3. Base, D.S. and Haug, E.J., "A Recursive Formulation for Constrained Mechanical System Dynamics, Part II: Closed Loop Systems, *Mechanics of Structures and Machines,* Vol. 15(4), 1987.
4. Kim, S.S. and Haug, E.J., "A Recursive Formulation for Flexible Dynamics, Part I: Open Loop Systems", *Computer Methods in Applied Mechanics and Engineering*, submitted July, 1987.

AAS 87-663

GROUND AND FLIGHT EXPERIMENTATION IN FUTURE LARGE SPACE SYSTEMS DEVELOPMENT

William P. Witt, III*

Let me say a few words to begin with, really my own opinions, and make some observations based on some hard-earned lessons. So they don't really reflect any SDIO position. I'm going to talk about what I see happening in space to solve some of the problems that both Keto and and Ed have talked about; that is, validating some of our theory and some of our models. For structures, we're concerned with validating our concepts, different kinds of both, passive and active, isolation technologies for vibration suppression, different design approaches and structural behavior, and in that falls, validating models. For materials in space and ground experiments we are interested in what are the effects of different kinds of environments, primarily at this point in time, the atomic oxygen behavior on materials.

I want to talk about the thermal behavior. That's what's driving us, in my opinion, toward a lot of these high-performance composites for their specific strength and stiffnesses. That's really driving some of the choices and how we validate that in experiments. One question I'm constantly asked -- and you have to recognize in SDIO there are a large number of physicists who don't necessarily have an engineering viewpoint on how we solve problems -- Why do we even have to do space experiments, especially, why do you have to test structures or materials in space? We do work in a different environment. If we need to go into the space environment where the phenomenology, either background or environmental effects, like atomic oxygen, are present, we need to determine how those environments will interfere with the performance of a system. Sometimes the only way we can answer all those questions is testing in space. Experiments are too hard to simulate on the ground or they're too expensive. We could build a 60-story drop tower and vacuum chambers and do vacuum zero-G tests, but I suspect that's a little more expensive than doing space experiments, expensive as they are. Sometimes engagement geometries are important; that is, you want to look at things over very long ranges that we can't see or achieve on the ground. Sometimes with the aspect angle, it's really tough to look back on the Earth like a Landsat sort of satellite if you are on the Earth. So you've got to go up in space to do that. And then there are some aspects of size that affect going to space.

---

* Lieutenant Colonel, United States Air Force, Program Manager, Directed Energy Office, Strategic Defense Initiative Office, Washington, D.C. 20546.

These are just some rough numbers I've worked out for testing a big space structure. We take the classical approach to testing, and that is, we don't have an active suspension system but we want to get the fundamental frequency of the suspension system, far below the fundamental frequency of the spacecraft. Now in the past, on smaller satellites, we've worked with fundamental frequencies of 30 to 100 Hertz, as shown on the far end of the curve, where we dealt with very small displacements of these suspension systems. For larger structures, we move toward the knee of the curve, where if we try to get the suspension frequency down an order of magnitude from the fundamental, we're starting to get one, two, and three meter displacements on passive suspension. So what does that mean? We can also plot, for piano wires or rubber, and we are back in the one-to-three meter regime. What can I test on the ground? If I go with piano wire, it's kind of tough to do anything, or rubber and if we look at limits of existing facilities, there's just not much we can do with the spacecraft we're coming up with. So we're stuck on what happens on the ground; we can use ground test to validate models, and that may give us some confidence, but we'll never have enough confidence to go to our program manager and guarantee to him that the things are going to work. There are some reasons of size when you go to space.

If we look at the different kinds of experiments, we can do systems specific experiments; that is, I can build a Space Telescope and do my experiments and validation on that exact structure. I can do type specific experiments; that is, I can build something like a hoop column antenna and try to modify it enough to characterize a very flexible radar antenna or communications antenna. Sometimes I just do simulations which will apply to practically any system.

Now we look at how those experiments relate to some important things like cost. How well do they apply to other systems? What confidence do we have in results? And where do I have to do them? Systems specific experiments are going to be the highest cost. They are going to have the least amount of applicability. Typically, they will only apply to the specific system I'm trying to test. But if I'm trying to guarantee that my system will work, I'm going to have the highest confidence in those results. If it's a large space structure because of size, I'm probably going to have do it in space. But I may do it on the ground, depending on what it looks like.

Now this is my assessment of where we stand today. The type specific simulations just fits right in the middle of the range. Those can be hardware in the loop. There is one we're doing at SDI, probably to be done on the ground -- the simulations, things like those Dr. Haug is doing, and the integrated structural model we're working on in DOD. They're the lowest cost so they're usually the easiest to get started because it doesn't raise anybody's opposition.

As we do some system specific experiments and relate them, we gain confidence as we get more data. And we may, in the next 20 years or so, have done enough system specific experiments, related them back to some general experiments and simulation so that we have high confidence. Then we get all the advantages and low cost, and broad applicability. And we can do the experiments on the ground. But that's not where we stand today, as Keto pointed out. We haven't done these things, so we don't have the confidence. How do we get that confidence?

Let me speak to those of you who are involved in structures and materials and also any people who are in management. First of all, we have to quantify the benefits of large structures, and I'll talk a little bit about what those are or what I think they are. You need to identify likely space programs and then, this is my word, attack. You know structural engineers are very conservative, primarily because of safety reasons. If our structure fails, something catastrophic is going to happen. We take that design philosophy and fold it over sometimes in our approach to solving technology programs. We just need to be less conservative. If we see an opportunity, we need to attack. When you do that, you need to develop your argument and do the correct kind of analysis for whatever the primary mission is. You're not going to endanger it; maybe you can enhance it. Now there's a little hidden philosophy there.

It's my opinion that we're never going to see an experiment in space purely for structures and materials. I just don't see it happening. We just can't justify it. So we're going to have to add on to something that's already there. But if we do that, we need to get in on the concept development phase. We can't wait until some things are ready to go to the launch pad and say, "By the way, I'd like to put some accelerometers on this spacecraft". It just isn't going to work. So whenever you have ideas or see opportunities, you need to put in your two cents' worth and don't be conservative.

What are the benefits? Well, we've looked at several systems. Using mechanical fixes to vibration suppression problems is much less expensive than electromagnetic or electrooptic solutions. Now, that may fly in the face of reason because those are typically very small systems. But if you want very high precision and try to do it all with electrooptics, it gets down to a dynamic range problem. Then you can get up to a third to a half of your cost in something that is only one percent of your weight, and that's design, development, fabrication cost. So some of these high-technology structures and materials ideas really have a big payoff. But we need to point it out. We need to do cost-benefit analysis early with the ideas, and all we've done in the community up to this point is to point out that this is neat stuff.

The mechanical approaches to using structures and materials are typically more robust in these electrooptic or electromagnetic approaches. For example, if you're doing compensated imaging and you're trying to use a deformable mirror, one actuator failing on the mirror can throw your wave front out enough so that you lose significant amounts of information. So those are not very robust, especially if you're trying to push the dynamic range of those kinds of apparatus. We present alternative approaches to solving problems. I mean, we look at what happens in surveillance, which has been pushed to very small optics, very small cooled optics. It turns out that we can do some of the surveillance missions if we go to very large optics; that is, if we can have very big large precision structures, and we can get away from cryo-cooling. It's a complete change in design philosophy from what has existed for 25 or 30 years now. But it is an alternative, and it's an alternative that may be both cheaper and better. For the mechanical approaches, I don't mean adaptable control, but structures/materials approaches are perhaps easier to adapt. When we started designing electrooptic components, we picked wavelengths, we picked sizes and those, all of a sudden, started driving weight and everything else on the spacecraft.

There are a lot of opportunities coming up. I'm just going to quickly cover some in NASA, SDIO, and the Air Force. This is no guessing game as to who has the more current version of Space Station. Anyway, the Space Station is here; we're going to see something in space. Given that the stock market doesn't collapse and we don't have another depression, we'll get the Station in orbit. But here's an opportunity to do large structures and materials work. LDR is a little more iffy. I have to talk to my NASA friends about that, about another opportunity to add on and enhance a mission. Sam Venneri allowed me to sit in one of the recent meetings, and they listed structures experiments going on at NASA Centers. Virtually every NASA Center has some kind of large space structures experiment or materials experiments directly applicable.

We have taken some of the technology structure from Space Station and put together a rather large antenna which sits out in the Air Force Weapons Lab. We're doing system identification research on this right now. And so that is a ground test facility that's available. We're putting it in a big hangar. We're going to make this active now, put in control passive damping structure isolation, and we're going to make this out of graphite epoxy. This will be a ground test facility again. It will be out at Kirtland Air Force Base within the next two years. We'll open it up for different kinds of investigators to work on these kinds of optical structure problems. There will be a big test bed out there.

There is a hybrid sort of experiment in progress at Martin Marietta; we call it the rapid retarding precision pointing simulator. It's on a piece of sandstone that's 20 miles long and about three miles wide. It's one of the quietest places in the country. Unfortunately, it looks like somebody's trying to build a gravel pit on it. But it is a big simulation facility where we'll be able to do very large angle slewing and retargeting type of maneuvers....

We're doing some of the things Keto talked about out of the A Cost Program. We're doing an experiment called the Joint Optic Structures Experiment. We're actually going to implement active structural control on a graphite epoxy nonanalytic structure; we're getting away from beams and plates. It's a pretty good size. It is made out of graphite epoxy with end bar fittings. It is a deployable kind of structure. This experiment is being done for SDIO by TRW with *Itek* as a sub on it. We're going to do it both in air and in a vacuum. We are putting disturbances in; this is not just a quiescent experiment, and we'll use proof mass dampers, accelerometers, and we're also doing some line-of-sight measurements. We have the ability to do both, just the local kind of control and then look at the distributive control and see how that solves the problem. There are optics, so we get some of the interaction that others have talked about. It includes control structures also. This is an on-going experiment, probably to be accomplished within the next couple of years.

We are now working on a neutral particle beam experiment called the Integrated Space Experiment; it goes in the Shuttle....

We are interested in how laser beams propagate up through the atmosphere, so we're doing an experiment called the Laser Atmospheric Compensation Experiment....

We are interested in big flat mirrors in space -- monocles. To do that, we're doing some experiments. One is called the Relay Mirror Experiment. We got on the concept development phase of the Relay Mirror Experiment, and were able to instrument the spacecraft. If you get in early, you can do it. We'll be taking data from that to tell us how those things operate in space. The instrument we put on it measures down in the radians of angular motion. It has about a 0.5 to a kilohertz bandwidth.

We are talking about another experiment called the High Power Relay Satellite. We'll put up a satellite, again with a big boom, except now we're just finishing the concepts, and although it's not required for the experiment, we can do an active control experiment on it on a big boom. It's a free flyer and again, what we've done is just tack on to another experiment without endangering that primary mission. We have three monocle segments, trackers and some other things, and then we'll have a big boom, again so that we can change links on and we'll be able to do a structural control experiment. Probably NASA will be the principal investigator.

The Air Force, Lynn Rogers at Wright-Patterson, did a program using available equipment. And he was able to show how pass and damping treatments would improve the structural design. That's an example of somebody taking advantage of what was already there.

My final messages to you are: Remember the strategy. You have to tell people how good you are if you want to be successful. And there are good things that can come from advanced structures and materials. You have to identify what the likely targets are and then really approach them and attack. You can't endanger the primary mission, so it means you are going to have to have more of an understanding in structures and materials and you have to get in early. What we want to do is to improve the confidence of the program managers. And advanced structures and materials are going to have to do more than ground testing. We're going to have to go into space but we will have to follow a strategy to get there. If you do only space tests, they are at least an order of magnitude more expensive than ground tests, and probably approaching two. We have to take advantage of what is already there. And remember, when we come to benefits, advanced structures and materials are both enabling and enhancing. They give us alternatives and help us do much better whatever the primary mission of the satellite calls for. Thank you.

QUESTIONS AND ANSWERS - STRUCTURES AND COMPOSITE MATERIALS

<u>**Dr. Anthony Amos:**</u> Thank you, Bill. Do we have any issues from the floor? This Audience Participation Question is addressed to Keto Sossaar.

<u>**Question:**</u> Your proposal to enhance the students' education curriculum sounds like a good idea. What is Cambridge Research doing to see that this idea becomes reality?

<u>**Recommendation:**</u> Direct a portion of your resources to broadening students' background along those lines described, if you don't already.

<u>**Dr. Keto Sossaar:**</u> Delighted to respond to that. We've always believed in our organization that basic research is a part of our operational work. I must admit this model came to me from having worked at the Draper Lab, where we had graduate students working along side engineers, getting their degrees while they were gaining practical experience on national systems. We've had an excellent relationship with MIT, Northeastern University and Tufts University. We have undergraduate and graduate students in our offices as co-op students, part time students and working toward their degrees. It's very easy to justify; it's a very economical way for the Government to get manpower to put on projects. The contract offices love it, because they are getting a tremendous high rate of return on an investment like that. And we get a chance, as a company, to see how that student really works in an engineering environment. As a result, we don't get any surprises when we hire people.

<u>**Question:**</u> What should the role of industry and Government be in helping universities update, improve, design courses, emphasize multidisciplinary programs and develop mission or reentry astronautics. Conversely, how should current universities' reward systems be modified to recognize talented design-oriented faculty members, often without Ph.D. and without extensive refereed publications. Need to emphasize both design and design experience and multidisciplinary training before students enter work force; e.g., the control structures interaction challenge, control structures thermal physics interaction.

<u>**Recommendation**</u>: Design-oriented faculty together with ABET, Government and industry should help draft performance evaluation criteria separate from those used for research-oriented faculty. University administration should agree on the need for such dual criteria.

<u>**Dr. Keto Sossaar:**</u> I think, ultimately, the Government is the customer of all of our space programs. A very few of us really work on commercial projects. So therefore, it's in the Government's interest to try to get the best product, from the educational

process that eventually feeds into developing its satellites and space systems. I think perhaps there has been insufficient attention paid by the Government in integrating the engineering process into the aerospace education process. We do tend to get people who are superb analysts but with very little understanding that the function of an engineer is to synthesize solutions, not to analyze someone else's concepts. I don't think that most aerospace engineers would care about the role that they actually had, whether they are called an adjunct professor or visiting associate professor of design, as long as they had an opportunity to deal with motivated students and share some of their experience with the class. I think that it will require external forces on the universities nowadays to get these things to happen, because these are not a natural form of evolution. On the other hand, these may happen, because one may be able to get the appropriate engineers to come and do this at a relatively low cost to the university. It's not an integral new faculty person you have to bring in, it is a part timer. That's a comment I have here. I'm afraid I don't know what ABET is. Was there any other issue that I didn't it catch here?

Dr. Anthony Amos: I think in the interest of time, we'll take just one more question. This is addressed to any or all members.

Question: What are some of the techniques for NDI, Non-Destructive Inspection, of composite materials? More specifically the non-metals. Dr. Tenney listed X-ray and diopenetrant. Are there other methods, techniques, perhaps ultrasound? Darrell, do you want to start with that?

Dr. Darrel Tenney: All right, I think there are a number of other techniques but probably the most common one that has been used is the C-Scan, using ultrasonics to look for delaminations. And in the case of flat panels that is one of the primary types of failure that we need to examine. The presence of delamination or excess voids that are formed in the composite lamination processing. So I think we've made some progress there. There is still a lot that needs to be done, though. One of the problems is really having a good set of standards where we know if we see a certain type of defect in the material how that translates into changes in the properties? And in the area of NDI of composites, I really believe we have a long way to go, particularly in the area of trying to understand lifetime issues. We've made a lot of progress in understanding what constitutes a successful processing. But in terms of how that translates into lifetime, there's still a lot of opportunity there for additional advancements. And I really believe we'll see a window opening for people to do research in that area. I'm very optimistic about the prospects of building sensors in the composites. I think that that potentially could have a big payoff. It's an area that deserves a lot of attention, one in which we haven't done a lot of work. We've got a long way to go.

Dr. Anthony Amos: There are a couple more issues that we're not going to be able to address from the podium here; however, all the speakers, all the panelists are going to be around for a while and I would recommend that those that submitted these issues try and grab the appropriate individual and discuss those issues after this meeting.

I'd like finally to thank our panelist for all the information and for all the issues they have raised for discussion here this morning and for their participation. And before we close, I would like to call on Don to say a few words.

Donald C. Wade: By way of thanks to the panelists we, the American Astronautical Society, are presenting these plaques as an appreciation of their efforts. Thank you very much.

AUTOMATION AND ROBOTICS:

SPACE STATION

PANEL NINE -- SESSION I

Chairman:	Rodger A. Cliff Lockheed Missiles & Space Company
Host Chairman:	Jon D. Erickson Chief Scientist Space and Life Sciences NASA Johnson Space Center

AAS 87-664 to -665

AUTOMATION AND ROBOTICS - INTRODUCTION

Jon D. Erickson* and Rodger A. Cliff†

Dr. Jon D. Erickson: I would like to welcome you to Panel Session 9 on Automation and Robotics. I'm the host chairman of the panel. I'd like to introduce the Chairman of the panel who will conduct the discussion this morning and I want to invite you all to participate. We anticipate a rather lively session.

Let me introduce Roger Cliff who is currently senior scientist for artificial intelligence and robotics with the Advanced Technology and Research Company. Previously he was with the Defense Advanced Research Projects Agency as chief scientist in their Engineering Applications Office. Roger will be conducting this morning's session.

Dr. Rodger A. Cliff: Good Morning. I, too, would like to welcome you to this session.

This is Panel 9 on Automation and Robotics. I'll go over the logistics of how we will operate this panel and then introduce our panelists.

We're going to break it up into two main topics. One is Automation and Robotics for the Space Station because that is today's issue and one that is vitally important right at this time.

Part 2 will be Automation and Robotics for the Future, I've titled that Moon, Mars, and Beyond. In each of these sessions, I have given several questions to the panelists and asked that they address them if possible, not just technology or policy, because, of course, that's what this whole discussion is all about, as to how technology and policy may impinge on each other. We need to worry about what level of A&R is appropriate for the IOC station, given the current state-of-the-art technology, but we also have to worry very much about the fact that one of the fastest growing technologies is automation and robotics and, of course, the underlying computer technology that makes a lot of it possible. So the question is, what do we do now? What hooks and scars need to be put on the IOC Station in order not to preclude the kind of growth we'd like to see over a 30-year lifetime. I've asked people to identify what they thought was the most critical issue in a policy. What should we do right now?

---

* Dr. Erickson is Chief Scientist, Space and Life Sciences, NASA Johnson Space Center, Houston, Texas 77058.

† Dr. Cliff is a Consultant, P.O. Box 15, College Park, Maryland 20740.

In the future session, I've asked panelists to think about such things as what kind of automation and robotics could we see, say, by the year 2025? And what new missions could be enabled by that? And again, what should we do now? What are the most critical issues? And we're going to talk about the relationship of automation and robotics to manned missions.

Now, about the panel format. This is a rather broad subject, so we have a rather broad panel here. We want to cover this from many viewpoints. We want this to be a very interactive sort of discussion, because there are many issues here that have yet to be resolved, since it's a new technology. So I'm encouraging the panelists to ask questions of each other. After each panelist speaks, we'll have a few minutes for the other panelists to make their comments. Then after each section, we will take a few questions from the audience and at the end of the whole session, we will have 15 or 20 minutes for a general discussion with many questions from the audience.

Now, let me put what we're going to do here in context. Sometimes it's hard to decide answers to some of these hard questions, unless you know where you're going. And that, of course, means that we really need a very well-defined goal of what we're up to and, as we've heard earlier in this meeting, there are some immediate problems in defining just what this nation's space goals are. Now, what they can be, of course, depends on what is possible to accomplish. That's how a lot of this interaction will take place. But we really need to know what we are actually trying to do. And then, of course, there's the policy that derives from those goals.

There's one very big policy issue that I'd like to bring up right now, and that is the problem of up-front dollars versus life cycle costs. Many times we say, look at what we can do with this marvelous technology to save life cycle costs. But in the difficult budget climates we encounter, it is exceedingly hard to deal with the up-front cost of developing a new technology. So we'd like to know if anyone can shed some light on this problem. Perhaps there is a way to break this barrier. Now, of course, from the policy we derive programs, and then from those programs, we technocrats can make real systems. In the absence of goals and policy, the political process has to focus on what there is to focus on, namely, the system, for the actual engineering can get politicized and that confuses things quite a bit. So if you can decide what you're going to do at the goal and policy level, create a shared vision among all those working on the project as to what should happen, some of the controversy can go away.

I'll allude to two of those that I hope the panelists will pick up on. It has been mentioned before at this meeting that there has been controversy between manned and unmanned space flight. Some people say we don't need to send people, that we'll just send our machines, and we can get all the data we want that way and isn't that enough?Well, I don't happen to think so. I foresee a future for people in space. What this stems from, of course, is a different view of what we are trying to accomplish. Are we trying to get scientific data and understand the universe or are we trying to go live there? We have to decide what we are trying to accomplish.

When we are having manned missions, there's another controversy that sometimes comes up and that is to what degree things should be automated. And this can take the guise of how reliable automation is or whether we really want to do things automated any way, and since humans are so wonderful, why do we need machines? If we have a good crisp idea of what we're trying to understand, I think that this becomes a nonissue. In other words, if you really know what you're trying to accomplish, the people that I know are involved in these programs are very confident and I think everybody will come up with the right answer.

Let us introduce the panelists now.

We have Ardell Nease who's manager of the Systems Project Office of the Space Station Systems Division of Rockwell International. He's been at Rockwell for some 19 years managing various space programs. He was recently appointed manager of the Systems Project Office.

We have Stan Sadin from the Office of Aeronautics and Space Technology at NASA Headquarters. He's had experience in industry and at NASA Headquarters, has served in many inter-governmental agency panels and committees promoting technology advancement and has been very interested during the entire time I've known him in long-range plans and in bringing really new and exciting technology into use.

Donna Pivirotto is from the Jet Propulsion Laboratory. She is currently manager of the Mars Rover Study of the Mars Rover Sample Return Project. She has formerly been manager of the Automation and Robotics Programs at JPL, the Space Station Office, Mission Design Sections, Systems Engineer of the Autonomous Spacecraft Project and a long list here that I won't go into. She's certainly well qualified to be in this discussion with us.

Takeo Kanade is from Carnegie-Mellon University. He's is currently professor of computer science and acting director of the Robotics Institute there. He's principal investigator for the CMU Computer Vision Project.

Skip Nunamaker is from the NASA Ames Research Center. He's presently chairman of the Advanced Technology Advisory Committee that he will explain to us, and director for space at Langley. Before that he was chief engineer at NASA Ames.

Stan Weiss is vice president and general manager of the Lockheed Research and Development Division in Palo Alto, past vice president and assistant general manager of Astronautics and Space Systems Divisions. And has served in a number of government and NASA positions.

Jackson Yang is a professor at the University of Maryland. He's director of the Robotics Lab there and of the Structural Dynamics Lab.

Harry McCain comes to us from the Goddard Space Flight Center where he's presently deputy project manager on the Flight Telerobotic Servicer Project of the Space Station Office there. Previous to that he was leader of the Integrated Robot Systems Group at the National Bureau of Standards.

Dick Volz is from the University of Michigan. He is director of the Robotics Research Lab there. Member of the Aerospace Safety Advisory Panel, member of the ADA Board and was a member of the Automation and Robotics Panel.

William Bates is chief of the Avionics Software Systems Safety Branch, Safety Division, at the NASA Johnson Space Center. He's had extensive experience both in mission operations at JSC and now in the safety activity.

Our first speaker will be Skip Nunamaker from Ames who will tell us about the ATAC.

AAS 87-666

AUTOMATION AND ROBOTICS FOR THE SPACE STATION AN ATAC PERSPECTIVE

Robert R. Nunamaker*

The Advanced Technology Advisory Committee (ATAC) was established by a Congressional mandate for fiscal year 1985. The creation of this committee clearly indicated the strong desire on the part of Congress that development of technology in the field of Automation and Robotics (A&R) related to the Space Station Project effort should "spin off" to the American economy. This presentation surveys the ATAC experience to date and describes some lessons learned.

The Advanced Technology Advisory Committee (ATAC) was formed as a result of strong interest on the part of the United States Congress that new technology be developed by the Space Station Program in the emerging field of Automation and Robotics (A&R), and that this new and evolving technology spin off to terrestrial applications in the American economy. Conference Report 98-867 of the House of Representatives, 98th Congress of the United States, was submitted as documentation of agreements by the two houses concerning funding for fiscal year 1985 (House of Representatives Bill S713). Amendment No. 39 establishes the Advanced Technology Program. The congressional desires were stated in the documented words of the mandate as follows:

> "The ATAC is mandated to identify specific Space Station systems which advance technologies not in use in current spacecraft. Additionally, it is the intention of Congress that automation and robotics implementation

* Director for Space, NASA Langley Research Center, Hampton, Virginia and Chairman, Advanced Technology Advsory Committee (ATAC).

will not only promote the efficiency of the Space Station, but by enhancing the technical and scientific base, will also lead to more productive terrestrial applications."

NASA Headquarters initially interacted with Congress and their staffers to implement the intent of the mandate. Figure 1 traces the process that took place prior to delivery of the first ATAC Report to Congress April 1, 1985. A NASA advisory committee assisted NASA Headquarters in setting the process in motion by working with the California Space Institute (CAL Space) and a newly formed NASA Space Station Automation Study Team. The primary purpose of these parallel activities was to establish for reference the present state of the art for the fastly emerging and very promising field of Automation and Robotics. To achieve this objective, CAL Space formulated the Automation and Robotics Panel (APR) which called upon the A&R expertise of academia and industry to document their perception of the A&R state of the art in 1985. As indicated in Fig. 1, several

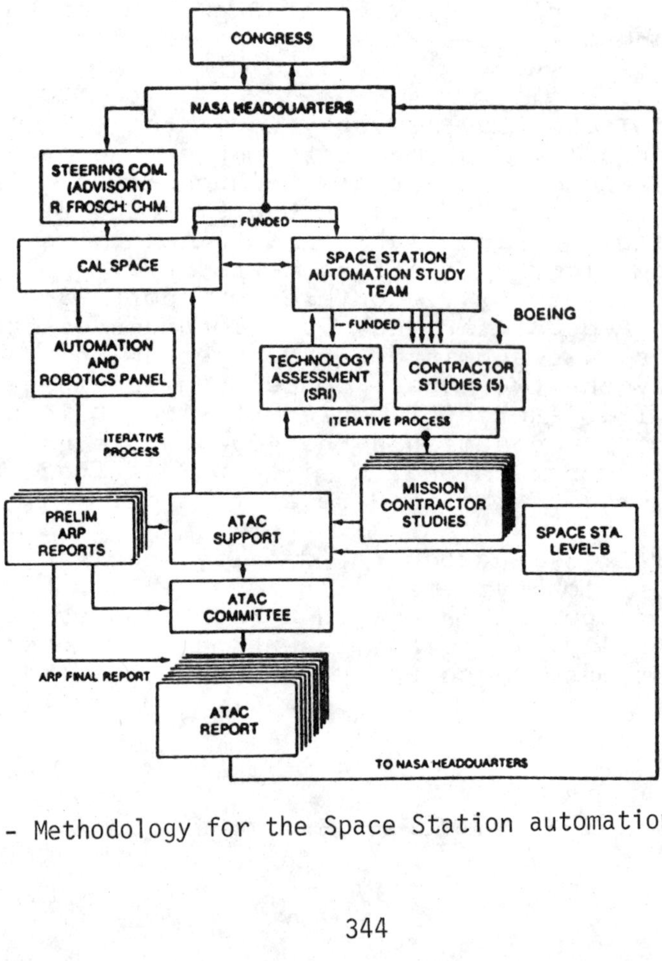

Fig. 1 - Methodology for the Space Station automation study

contracted studies with industry were initiated in parallel to establish the base reference for ensuring ATAC assessments. Stanford Research Institute (SRI) was funded to perform a technology assessment. Boeing's interest in this activity led them to perform and input the results of their own study into the process of supporting the ATAC objectives. All NASA Centers provided strong support to the resultant iterative process. The NASA Johnson Space Center (JSC) Space Station Level B organization managed by Mr. Neil Hutchinson provided valuable day-by-day technical and administrative support to the entire process which by this time had drawn a great amount of interest with many experts and organizations involved.

The ATAC membership was drawn from each NASA Center, Jet Propulsion Laboratory and NASA Headquarters under the chairmanship of Mr. Aaron Cohen, Director for Research and Engineering, NASA--JSC. With the state of the art for the field of A&R being documented by the Nation's recognized experts, the ATAC held their initial meetings to discuss how they would conduct their mandated mission. The ATAC mission statement is to:

> "Independently review the conduct of the Space Station Program and assess the integration of A&R technology. Based on assessments, develop recommendations, review the recommendations with NASA management, and discuss their implementation with consideration for safety, reliability, and cost effectiveness. Report assessments and recommendations twice annually to Congress."

It became exceedingly evident that while ATAC's mission was now clearly stated, the achievement of this mission could not be easily accomplished if the maximum benefit to the Nation was to be the result. The bottom line was that a group of NASA managers, who became the ATAC membership, were given the mission to assess how well a NASA organization (the Space Station Program) was planning and implementing the integration of Automation and Robotics technology into their design requirements and specifications. As we will see, the ATAC has been able to "walk this fine line" successfully because of the procedures ATAC has followed from the start and because of NASA's interest in implementing the intent of the Congressional mandate. Open and regular communications between those assessing, i.e., ATAC, and those performing, i.e., Space Station Program, proved to be the key in achieving success even though all ATAC

Reports certainly were not positive. However, with the effort put into exercising open communications the ATAC was able to do it's job and NASA could clarify it's own position about what it would (or would not) do relative to each ATAC recommendation. No surprises and a clear understanding of each ATAC recommendation and assessment has to date led to success. This, plus the strong support of the process by not only the Space Station Associate Administrator, Mr. Andrew Stofan, and Program Director, Mr. Thomas Moser, but the NASA Administrator, Dr. James Fletcher helped ATAC achieve it's goals. With this process and management support maximum benefit will result and the intent of the Congressional mandate will be met.

The present ATAC membership is shown in Fig. 2. ATAC provides an assessment report to Congress every 6 months to chart progress of the Space Station activities and ATAC's assessments relative to integration of A&R. The nature of ATAC has changed with changes in the program phase of the Space Station Project. Initially ATAC reported assessments of what would be needed and based on the ARP Report and contractor studies what would be available. Following completion of the Space Station System Definition Studies and as the hardware phase (C/D) began, ATAC's assessment still addressed what they felt was needed, but now instead of discussing what was available, the reports assessed the suitability of what was being proposed.

ROBERT R. NUNAMAKER, CHAIRMAN, DIRECTOR FOR SPACE, LANGLEY RESEARCH CENTER (LARC)

JOHN H. BOECKEL, DIRECTOR OF ENGINEERING, GODDARD SPACE FLIGHT CENTER (GSFC)

WILLIAM C. BRADFORD, DIRECTOR OF THE INFORMATION AND ELECTRONIC SYSTEMS LABORATORY, MARSHALL SPACE FLIGHT CENTER (MSFC)

JON D. ERICKSON, ASSISTANT CHIEF FOR AUTOMATION AND ROBOTICS, LYNDON B. JOHNSON SPACE CENTER (JSC)

J. STUART FORDYCE, DIRECTOR OF AEROSPACE TECHNOLOGY, LEWIS RESEARCH CENTER (LERC)

LEE B. HOLCOMB, DIRECTOR OF INFORMATION SCIENCES AND HUMAN FACTORS DIVISION, NASA HEADQUARTERS

HENRY LUM, CHIEF OF INFORMATION SCIENCES OFFICE, AMES RESEARCH CENTER (ARC)

WALTER T. MURPHY,. DEPUTY DIRECTOR OF ENGINEERING DEVELOPMENT, KENNEDY SPACE CENTER (KSC)

DONNA L. PIVIROTTO, MANAGER OF AUTOMATION AND ROBOTICS OFFICE, JET PROPULSION LABORATORY (JPL)

GIULIO VARSI, AUTOMATION AND ROBOTICS MANAGER, STRATEGIC PROGRAMS AND PLANS DIVISION, OFFICE OF SPACE STATION, NASA HEADQUARTERS

Fig. 2 - NASA Advanced Technology Advisory Committee

The ATAC recommendations shown in Fig. 3 were documented in ATAC Repost No. 1 and traced through Report No. 5 (September 1987) assessing progress.

1. A&R SHOULD BE A SIGNIFICANT ELEMENT OF THE SPACE STATION PROGRAM
2. THE INITIAL SPACE STATION SHOULD BE DESIGNED TO ACCOMODATE EVOLUTION AND GROWTH IN A&R
3. THE INITIAL SPACE STATION SHOULD UTILIZE SIGNIFICANT ELEMENTS OF A&R TECHNOLOGY
4. CRITERIA FOR THE INCORPORATION OF A&R TECHNOLOGY SHOULD BE DEVELOPED AND PROMULGATED
5. VERIFICATION OF THE PERFORMANCE OF AUTOMATED EQUIPMENT SHOULD BE STRESSED, INCLUDING TERRESTRIAL AND SPACE DEMONSTRATIONS TO VALIDATE TECHNOLOGY FOR SPACE STATION USE
6. MAXIMUM USE SHOULD BE MADE OF TECHNOLOGY DEVELOPED FOR INDUSTRY AND GOVERNMENT
7. AUTOMATION SHOULD BE USED TO ENHANCE NASA'S MANAGEMENT CAPABILITY
8. NASA SHOULD PROVIDE THE MEASURES AND ASSESSMENTS TO VERIFY THE INCLUSION OF A&R IN THE SPACE STATION

Fig. 3 - ATAC Recommendations

The implementation of the ATAC process to date has brought forth an understanding of several key factors that must be addressed when recommending integration of A&R requirements. On the other hand, these same factors must be considered before the Space Station supports or does not support (with specific reasons) the ATAC recommendations. These key factors are: (1) A&R is a new field promising great benefits; however, great missions of the past have not been built on promises; (2) commitment and cost are up front, benefits are downstream for technical performance and cost; (3) recommendations will not be implemented unless confidence can be developed and this can only be done by demonstration and testing; and (4) A&R is essential for the evolutionary station. The Space Station program managers know for certain they have a very constrained development budget and therefore it would be natural to dedicate all moneys and energies to make the initial Space Station complex operational and not pay for the "niceties" that primarily provide benefit downstream following the initial operating configuration (IOC). However, A&R is essential for the evolutionary aspects and objectives of the Space Station, so moneys, energies, and schedule time must be dedicated to testing and demonstrations that "prove their value" before Space Station managers integrate A&R designs into the Space Station system to evolve and improve performance.

Therefore, in conclusion, the ATAC has developed a modus operandi that has provided it with the capability to successfully achieve the 1985 Congressional mandate. ATAC recently completed Progress Report No. 5 as the Space Station Program moves into its C/D (hardware development/ operational) phase. ATAC will maintain a certain flexibility to change with the changing characteristics of the Space Station program phases to ensure successful achievement of the intent of the mandate. However, the ATAC experience to date has established that while there is intent, the "proverbial shelf" that all good program managers use to acquire as many designs for their spacecraft as possible, is sparse with A&R systems. Therefore, there is a real challenge to the A&R community to design, develop, and demonstrate A&R systems/subsystems that can be considered for integration into the selected contractors proposed designs. Needless to say the timing is severe. ATAC will continue to push for A&R integration and they will continue to receive NASA upper management support because of the important benefits A&R systems will provide to the evolutionary Space Station, but the A&R community must work expeditiously to move innovative ideas from the laboratory to the system designer's "shelf."

AAS 87-667

SPACE STATION INITIAL OPERATIONAL CONCEPT (IOC) OPERATIONS AND SAFETY VIEW - AUTOMATION AND ROBOTICS FOR SPACE STATION

William V. Bates, Jr.*

In order to minimize long-term operations costs, improve crew productivity, and to provide increased protection against hazardous operations, many of the functions performed by ground support elements or the crew in the NSTS program will be supported by automation and robotics in the Space Station program. A major challenge to the accomplishment of this task will be the integration of the individual elements and functions from the various contributing NASA Centers and vendors into a phased, evolutionary transfer of those complex functions now performed manually and on the ground to a cohesive, automated on-board system. As most of the NSTS Operations tasks are people-intensive and involve a very low level of automation, application of new technologies to the combined problems of minimal experience with automated functions and the increased complexity of systems in the Space Station design will demand a phased approach with extensive growth capability. A further challenge is afforded by the overlying requirement to maintain a safe or minimal hazard environment for the crew and overall Space Station configuration. Significant advances will be required in the present state-of-the-art of automated, distributed, real-time systems management and in the hazard analysis and risk assessment of structured software, and eventually AI-oriented, integrated systems. Inherent in both of these areas will be a requirement to totally integrate multiple functions that are now performed in standalone fashion with extensive human involvement. A dependence on performance models and new or extensive testing methodologies is envisioned for the considerable future.

I'd like to offer my thanks to Dr. Cliff and especially to the Society for allowing me to participate in this panel today. My subject gets a little bit narrower in scope and addresses the configuration at IOC, and you will notice it covers both safety and operations. I am in the safety world now, but I found the two were inseparable on trying to address them, so you'll see the two linked here.

Because of the time limitation, I'm going to have to ask you to cooperate with me and be good active listeners and readers, because I will not have time to address every one of the bullets on the chart and also my comments are going to be 95% directed toward automation, because that's what my background is and there are several others here on the panel that are going to address robotics.

---

* NASA Johnson Space Center, Houston, Texas 77058.

The first question we were addressing was what level of automation and robotics should the IOC station have? I tried to list things here not in any order of priority but in somewhat logical steps (Figure 1).

The first thing we need is to provide an integrated environment with sufficient performance; that is, variety of horsepower, because what you're trying to do here is take what we now have in a nonautomated people-and-paper intensive environment in the Mission Control Center and working with the crew and moving that to an on-board environment.

The second point concerns an eight-person crew to perform a high percentage of the tasks. After hearing Dr. Helmreich's comments yesterday about the people element, I'm not so sure whether I ought to worry more about the type of the crew as opposed to the percentage of the time they will be working. We need to provide at least a minimum of first-level capability of the system that is defined as the Operations Management System. That is defined in a document JSC20792 and is essentially a definition of what goes on with an integrated team in the Mission Control Center, from flight director on down. You have to be able to do automated fault detection, and prioritization, enunciate that, do auto switch over, do auto safing, and those sorts of functions.

The next most important facet is to provide an automated or a directed fault analysis capability. And the real key here is, you need not 95%, but actually 100% of the configuration. You can't offer a reconfiguration recommendation if you don't know what the configuration is. That has been a major problem for us on Orbiter because of the way it was designed. If you don't design it for safety up front, you'll probably never get it in, just because of the high cost.

Another thing that you need to have is that definable paradigm or an analysis capability, because you have to recognize that systems are different and will require different methologies to analyze the problems. We have some basic capability now in analyzing multi-failures for the Orbiter. We need to move that to a function that is on board. The rules for hazards and safety? That's going to be a constraint boundary and we need to make sure that, as we change things while we're going along, we do not violate the rules.

The next point is, we need to have a configuration on board to where we could have at least a relative level of support on the ground that was equivalent to what we had for the Shuttle Operations Flight Test (Figure 2). And then that would taper off as we matured and moved more functions on board. We need to be able to provide a design that allows the systems to gracefully degrade or fail, and to have autosafing where it's required for time-critical failures. You need to be able to have query through the functions on the "hows", and the "whys", and the "what ifs". That will be mandatory because you're going to be doing your final testing as you put this in use on orbit.

You need to provide the crew or the ground with capabilities to interrupt automated functions but with the clear understanding that such interruptions should not violate the hazard safety rule set.

The other points here and in Figure 3 are about higher level language (including "NASA English"), traffic control, on-board planning/scheduling and reduction of EVA requirements.

What is the most important thing to do right now? I think the most important thing we need to do is to define automation as a system level discipline (Figures 4 and 5). It needs that identity. Wherever it's a secondary element, it will always get left behind. You need to have the focus on this. It requires an Agency-integrated approach (Figure 6). You need to have the same focus on this that we had on going to the Moon. If you don't get that level of focus on this, then we're not going to be building in the type of system and all that the ATAC has directed and Congress thinks that we're building.

The second point is that we need to start making use of the various systems and facilities that we have around to mature the technology so that we do have this shelf of tools around for the engineering communities to use. And we have some of that now. The best example I could give you is in the Mission Control Center and the upgrade that we're trying to do there. If we can't make distributed systems and communications work in that environment, then I'm not very confident that we'll be able to make them work in an on-board environment.

- WHAT LEVEL OF A&R SHOULD IOC STATION HAVE?
    - THE A&R LEVEL SHOULD BE SUFFICIENT TO PROVIDE AN INTEGRATED ENVIRONMENT WITH SUFFICIENT PERFORMANCE, FLEXIBILITY AND COMMUNICATIONS TO ACCOMMODATE A VARIABLE MIXTURE OF STRUCTURED, RULE-BASE, FRAME-BASE, DATA BASE, ETC., SYSTEMS TO ACCOMPLISH THE FOLLOWING:

- ALLOW AN EIGHT-PERSON CREW TO PERFORM A HIGH PERCENTAGE (>50 PERCENT) OF USEABLE (NON-M&O) WORK (MUCH HIGHER LEVEL THAN ORBITER DESIGN).

- PROVIDE A MINIMUM OF THE FIRST LEVEL OF FUNCTIONALITY OF THE OPERATIONS MANAGEMENT SYSTEM (OMS) - (AUTOMATED FAULT DETECTION, PRIORITIZATION AND ANNUNCIATION AND AUTO-SYSTEM SWITCH-OVER/SAFING.

- PROVIDE AN ONBOARD AUTOMATED-OR DIRECTED-FAULT ANALYSIS CAPABILITY WITH COMPLETE (>95 PERCENT) SYSTEM CONFIGURATION AND STATUS INFORMATION. (INCLUDE DEFINABLE PARADIGM CAPABILITY WITH TRANSPARENT ACCESS TO ALL REQUIRED DATA BASES; ONBOARD AND GROUND.)

- PROVIDE THE CAPABILITY TO COMPLETELY ANALYZE THE IMPACTS AND EFFECTS OF MULTIPLE, (CROSS-SYSTEM) UNRELATED ANOMALIES/FAILURES AND TO PROVIDE RECOMMENDED, PRIORITIZED RECONFIGURATIONS (OR AUTO-IMPLEMENT IF POSSIBLE).

- PROVIDE A HAZARDS/SAFETY CONSTRAINTS RULE SET FOR TRANSACTION MANAGEMENT.

Figure 1

- ALLOW A LEVEL OF GROUND SUPPORT (SSSC) AT THE RELATIVE LEVEL OF SHUTTLE OFT.

- PROVIDE THE CAPABILITY TO OPERATE SAFELY WITHOUT THE FUNCTION OF INTEGRATED SYSTEMS MANAGEMENT SOFTWARE.

- PROVIDE SYSTEMS AND SOFTWARE THAT ARE PREDICTABLE AND "GRACEFULLY" DEGRADE OR FAIL. PROVIDE AUTO-SAFING FOR TIME-CRITICAL FAILURES.

- PROVIDE COMPLETE AUDIT TRAIL OF ALL TRANSACTIONS AND TRACEABILITY AND TWO-WAY VALIDATION FUNCTIONS FOR ALL SOFTWARE. (QUERY AS TO "HOW?", "WHY?", AND "WHAT IF?")

- PROVIDE A CREW MANUAL OR GROUND VIRTUAL OVERRIDE CAPABILITY (WITHOUT VIOLATION OF THE HAZARD/SAFETY RULE SET) TO ANY AUTOMATED FUNCTION-DOWN TO THE INDIVIDUAL SYSTEM LEVEL.

- PROVIDE A MINIMUM OF TWO METHODS OF ACCESS (ABOVE LOCAL SYSTEM DISPLAY AND CONTROL) TO ALL LIFE CRITICAL OR CORE FUNCTIONS (SOFT/HARD SWITCH AND MULTI-LEVEL/ACCESS PORT).

- PROVIDE A PERSONAL OPERATING INTERFACE MODE IN "NASA ENGLISH" FOR A VARIETY OF USERS, AND THE CAPABILITY TO EASILY SWITCH FROM ONE MODE TO ANOTHER.

- PROVIDE SPECIFIC AUTHORIZATION CODES/RESTRICTIONS FOR SAFETY AND SECURITY OPERATIONS.

Figure 2

- PROVIDE AUTO-TRENDING OF TRACE CONTAMINANTS, AS WELL AS EXTERNAL CONTAMINATION.

- PROVIDE SUFFICIENT LEVEL OF INTEGRATION AND AUTOMATION IN THE "TRAFFIC CONTROL" AREA TO ASSURE SAFE AND CONTROLLED, ONBOARD DIRECTED, RENDEZVOUS AND PROXIMITY OPERATIONS.

- PROVIDE THE CAPABILITY FOR ONBOARD, SHORT-TERM CORE SCHEDULING AND PLANNING FUNCTIONS WITH CONSTRAINT VIOLATION DETECTION AND NOTIFICATION.

- EMPLOY ROBOTICS TO THE MAXIMUM EXTENT POSSIBLE TO GREATLY REDUCE EVA REQUIREMENTS ESPECIALLY FOR HAZARDOUS OPERATIONS SUCH AS FUEL TRANSFER, LONG EXPOSURE, EXCESSIVE WORK, ETC.

Figure 3

- **WHAT HOOKS AND SCARS SHOULD BE IN IOC STATION?**

  - ANY SYSTEMS WHICH INITIALLY REQUIRE DEGREES OF MANUAL OPERATION SHOULD BE MODULARLY DESIGNED TO ALLOW MINIMAL HARDWARE/SOFTWARE GROWTH FOR INCREASING LEVELS OF AUTOMATION ("DESIGN TO AUTOMATION").

  - PROVIDE COMPLETE MODULAR DESIGN - PARTITION OF FUNCTION (OPERATIONS - SYSTEMS PERFORMANCE MODELS - TRAINING - ETC.). COMPLETE IV&V AND CONFIGURATION MANAGEMENT TRACEABILITY PER MODULE.

  - CAPABILITY TO GROW TO FORECASTED TECHNOLOGIES AS PART OF OPERATIONS MANAGEMENT SYSTEM (OMS) FUNCTIONAL TRANSITION FROM GROUND TO ONBOARD.

  - INDUSTRY (OR MIL (OR) NASA) STANDARDS ON INTERFACES AND PROTOCOLS SHOULD BE MAINTAINED. ALSO PROVIDE FOR NETWORK/COMMUNICATIONS GROWTH. (DO NOT CUSTOMIZE COTS)

  - EXTENSIVE BITE OR MODEL COMPARE TEST CAPABILITY.

  - TAKE ADVANTAGE OF NEW TECHNOLOGY DEVELOPMENTS IN HARDENED SYSTEMS AND ALLOW FOR FURTHER GROWTH IN THIS AREA OVER STATION LIFETIME. (FAULT TOLERANCE)

Figure 4

- **WHAT IS THE MOST CRITICAL ISSUE IN A&R FOR IOC STATION?**

  - TIME TO LAUNCH AND PRESENT STATE-OF-THE-ART. UNIQUE, SCHEDULE-DRIVEN GROUND AND FLIGHT DEVELOPMENT SHOULD BE AVOIDED TO THE MAXIMUM EXTENT POSSIBLE.

  - DEVELOPMENT AND PROOF-OF-TECHNOLOGY EARLY ENOUGH TO DEFINE DESIGN (HOOKS AND SCARS).

  - PROVIDE A SUFFICIENT LEVEL OF AUTOMATION TO MAKE THE SYSTEMS OPERATIONALLY MANAGEABLE WITH A HIGH PERCENTAGE OF ONBOARD CONTROL.

  - PROVIDE SAFE SYSTEMS AND A SOFTWARE ENVIRONMENT THAT IS HIGHLY FLEXIBLE AND EVOLUTIONARY.

  - EMPHASIS ACROSS THE AGENCY ON INTEGRATION STARTING WITH FACILITIES - (MCC --> SSSC, MSIF, SS SAIL, SSTF AND SSSC COMMONALITY AND CONFIGURATION MANAGEMENT).

Figure 5

- **WHAT IS THE MOST IMPORTANT THING TO DO RIGHT NOW?**

  - DEFINE AN <u>AGENCY LEVEL</u>, INTEGRATED, SYSTEMATIC APPROACH TO DEVELOPMENT OF THE REQUIRED TECHNOLOGIES.

  - PROVIDE A FOCUS FOR MATURING OR PROVING THE TECHNOLOGY IN AS REALISTIC AN ENVIRONMENT AS POSSIBLE. (MCCU --> SSSC --> SS)

Figure 6

DISCUSSION

Dr. Rodger A. Cliff: I'd like to invite some comments from the panel. While you're thinking of comments, I would like to say that you just struck on two of my favorite items here. One is the idea that we need to be able to control things from an automated viewpoint and that we have to plan up front to do that. I think that's vitally important. I call this sensibility and controllability. Each piece that goes into the Station needs to have the capability of its state or condition being sent by remote equipment, automated equipment -- it could be on the Station or remote equipment on the ground. There is a tendency which is absolutely the right way to go, when you have a simple system, to say, I want a toggle switch and a wire from here to there because, in that way, I know what's going to happen. But the problem is, when you get to thousands and thousands of toggle switches and thousands and thousands of wires, that may no longer be the most reliable way to do things.

Robert Nunamaker: I agree very much with what Bill has said, relative to putting things in demo and on the shelf, as I indicated, but I think we have to make sure that it's really automation. I've seen too many things called automation in ground operations which really are not.

Dr. Rodger A. Cliff: Our next speaker will be Harry McCain.

AAS 87-668

THE FLIGHT TELEROBOTIC SERVICER PROJECT

Harry McCain*

Good morning. Very quickly, I'm going to try to familiarize you with the Flight Telerobotics Servicer Program that is being done at Goddard Space Flight Center. The FTS Program is the primary robotics element of the Space Station. Essentially, it came to be as a result of the realization that there were a lot of things that could be done robotically and should be done robotically or telerobotically on the Space Station. And that each of the work packages was involved in one way or another. The Agency concluded, and the Space Station Management concluded, that maybe we ought to put together one program that could try to address a major portion of these activities. Let me just talk a little bit about that program. The system that we're building is a flight system that is to be launched on the first element launch to support assembly of a Space Station. We want to reduce the required crew EVA, which is expensive. EVA is, in fact, a dangerous endeavor and there are hazardous things that need to be done, and hazardous materials that need to be handled, which is much safer to be done robotically. This also leaves the crew available to do more intellectual tasks. Providing maintenance and servicing for platforms is another task that will be done *in situ* on polar platforms, which is a very hostile environment. That will probably be done with the FTS or some version of the FTS mounted on the Orbital Maneuvering Vehicle.

As a fall-out from this process, we want to accelerate transfer from the research world to terrestrial industrial robotics technology.

The FTS project was kicked off in May of 1986, and we've gone through a Phase A study and we are embarking on Phase B studies now. This is essentially, the concept that came out of the Phase A study. We aren't talking about a specific machine, with specific capabilities, two arms, a third appendage for grappling at the work site and camera booms. The thing sticking off the back is a battery pack which allows the FTS to operate without being connected or tethered to the rest of the Station. The FTS will employ state-of-the-art technology. I want to emphasize state-of-the-art right now because we are a flight program, and in order to get something ready to fly in the mid-90s, we have to deal with existing state-of-the-art technology. We feel that technology does exist, that there are no missing links required in order to build a very viable system. But the system must be evolvable for future growth, so we have to worry about that. And again, the development process should support technology

* Deputy Projects Manager, Flight Telerobotic Servicer Project, NASA Goddard Space Flight Center, Greenbelt, Maryland 20771.

transfer. One of the major things that we've done is to adopt an architecture which we hope will become standard throughout the industry, which provides generic hierarchical structure and functional definition for the FTS system. It is a key element in our program. What we're drawing from are two different worlds that we're going to try to integrate: the teleoperation world or the unstructured world that you see in the nuclear facilities and the underwater capabilities and then combine that with the capabilities that you find in the manufacturing environment, which is a very structured environment and very autonomous. We want to merge those two worlds into telerobotics.

Robotics really is a conglomeration of a number of technologies. We're bringing these technologies together in NASA and providing systems engineering expertise, software engineering and systems integration. Initially, we are recipients of the technology that exists in industry. We start to have some value added as we integrate these things and, maybe on out into the '90s we really become a source of technology to industry. We feel a lot of that has to do with the standardization. For initial tasks we're looking at things like truss assembly, utility tray assembly, inspection and ORU change-out.

I just would like to reiterate the theme that has already come up; it has to do with design for automation and robotics. Robotics is really an element of automation, which is the general topic. But if you don't start from the grass roots with the design concept, that includes automation, and robotics, it's not going to work. Certain things need to be done, certain physical characteristics need to be in the work pieces in order for the task to be robot friendly. Fortunately, these configurations tend also to be EVA friendly, since an EVA astronaut is really not as dexterous with his suit and his pressurized gloves, as a human sitting in a shirt sleeve environment. So these things do tend to work together but it's very important that the system be designed right from the beginning to be automated. Thank you.

DISCUSSION

Dr. Rodger A. Cliff: Thank you, Harry.

Question: Can you say something about whether you're going to have any in-flight experiments or demonstrations prior to flight element one?

Harry McCain: Yes, as a matter of fact, we have a mission that we're calling the development test flight that was recently manifested for mid-'91 and we will be testing a number of the subsystems involved with the eventual Space Station system.

Question: How are you getting the designers of the basic Space Station to include in their concepts the design of your Flight Telerobotic Servicer?

Harry McCain That's a good question. We are really just getting started on trying to work with the people at Work Package Two, because Work Package Two is responsible for putting the initial Station on orbit. We have a number of different organizations trying to develop techniques, manuals and guidelines in order to accommodate robotic capabilities. But right now, unfortunately, it is somewhat of an *ad hoc* process that we'd really like to refine.

Question: I think when you say state-of-the-art that people are going to get misled by that. You can't go out and buy off the shelf the hardware that's space qualified. So you're not talking about state-of-the-art space items, you're talking about state-of-the-art control technology, I think, and software technology. Is that right?

Harry McCain: That's right. I guess, what I'm trying to say is that these things exist on the bench, but they're not qualified, and there's a very long lead time to take something from a ground base technology into something that can be flown and be reliable on orbit. We're sort of taking a snapshot in time and saying that what exists now on the bench and in the research laboratories is what we're going to have to evolve into a flight system. So whenever you're in a flight program, you're always really behind the state-of-the-art almost from day one. But we feel that we can put together a very viable system and with the architecture that I mentioned, we feel that it can evolve fairly easily.

Question: What would you do specifically to make the Space Station robot friendly?

Harry McCain: Well, I guess I'm not an expert on that subject. I don't think anybody is yet. But some very practical things would be to have pre-defined grip points, to have power and data outlets all over the Space Station, and to mark the work pieces, so that you don't require a very sophisticated vision system to recognize things. Things of that nature are what I would recommend.

Dr. Rodger A. Cliff: Thank you, all. Our next speaker will be Jackson Yang, who will be talking about some of the technology issues.

AAS 87-669

RESEARCH AND DEVELOPMENT VIEW - AUTOMATION AND ROBOTICS FOR SPACE STATION

Jackson Yang*

First of all, I would like to announce that I am very happy to be back in Houston. As a mechanical engineer, I was heavily involved with the offshore platform technology development that dates back about ten years or so. Then things were a little bit more prosperous here in Houston when I attended yearly the Offshore Technology Conferences. Now I hope that space and robotics will bring to Houston the prosperity that existed during the offshore oil technology boom period.

There was a communications breakdown with Rodger, in that I was supposed to receive a package earlier to inform me of the format and the critical issues we are supposed to talk about. Well, that package never got to me, so I am just going to make a few general comments about the Research and Development.

Before I start I want to share with you what I feel is a major problem facing the academicians, one which I think has hindered academicians from contributing to the extent we should to the space effort in automation and robotics. Hopefully this discussion will give you a little food for thought so that you can continue or begin evolving solutions.

What I see as a major problem is that we really should educate the academicians on what are some of the unique problems that operation in a space environment poses. To say that the space environment is hostile to man is a gross understatement. Manned Extra-Vehicular Activities (EVA) are not only hazardous but pose large constraints on performance and capability. A man inside a pressure suit is confronted with fighting the cumbersome suit, which is exhausting, and an all "thumbs like" glove (poor dexterity). There is a definite need for *enhanced* automation and robotics in space to enhance safety and capability.

As you know, automation and robotics are very interdisciplinary involving the aerospace people, mechanical engineers and so on. The aerospace faculties might be very aware of problems, like zero-g vacuum environment and the thermal and the radiation problems and constraints, but some other academicians, such as mechanical engineers and computer scientists, are not. For instance, I personally have a hard time with space acronyms that are thrown at you during a task. Is there some way

* Dr Yang is Associated with Advanced Technology & Research, Inc., 14900 Sweitzer Lane, Suite 104, Laurel, Maryland 20707.

whereby academicians could become more familiar with space terminology and space requirements?

I can see a similar situation where I'm very heavily involved right now. There is a recent push by Congress and National Science Foundation in the area of construction automation. I participated in a number of workshops and served as head of several panels to review proposals, where we try to help the civil engineers to automate the construction industry. If the civil engineer does not really understand the problems at construction sites, for instance, he/she can drive to the site and get "educated". Now it gets a little bit more difficult with space problems -- no routine inspection trips....

Regarding the automation and robotics requirements through years 2020 I would like to discuss several areas with emphasis on space. Autonomous operation, I see as a must, especially for free-flyer and platform payloads. It is the major driver to efficient operation of the Space Station. Now dexterity improvement is also a big driver in large structures. Antennas, reflectors, collectors, requiring high accuracy surface control, emplacement or repair of instruments are required and also support elements, maintenance and repair are important. The various areas in autonomous operations are very standard. These are the components on which research and development work is needed....

Sensors need increased sensitivity, 3-d vision, and fusion of visual, tactile and proximity sensors. This will require instant reaction response, micromotion, extremely high reliability, and efficiency. The parts for labor interpretation control and generation are the software, and require improvements in dexterity and mobility for a space robot. The ... A&R and KB, the Knowledge Base, the CAD Data Base, are essential for autonomous operation. I think that Professor Kanade will talk more about the A&R. There are a number of things we should look into. Some of these are developments like the Stanford JPL hand and the Utah MIT Jacobson's four-finger hand.

The automated ground base system is the other area of space system activity that will offer real potential in enhancing the NASA mission.

Automated manufacturing launch and servicing at the launch site activities must be an integral part of the program.

In summary, free-flying autonomous robot technology will be required to maximize the value of Space Station Operations. And, the two major technology drivers, dexterity, and A&R, will require quantum jump advances. The advances can only be achieved by involving the best talents available in the Government, industry, working together with the academic community. Thank you very much.

DISCUSSION

Dr. Rodger A. Cliff: Thank you very much, Jackson. I would like to put a question to Stan Sadin. Do you have any suggestions about the initial problem he brought up?

Stan R. Sadin: Yes. I was going to volunteer. Regarding what NASA can do with respect to the academic community and their involvement in the space program and learning all the acronyms and whatever else goes with it, we've been pursuing that in great depth, and in fact the University of Maryland is an active participant in a program that we have at NASA, which is a university designed program, in operation right now at 30 universities around the country, and we wish we could involve them more in the process of advanced mission design. And we are processing, I would say, about a thousand students a year at the senior and graduate level, conducting system design programs, much the way we do in Government and in industry. The students form teams and they do advanced system studies. They are on the phone with their sponsoring centers and they learn the acronyms and and get all the help we can manage to give them. So we're headed in that direction. Right on the heels of that, and in the process of response to a solicitation right now, is a call for establishing, at the universities interdisciplinary system-oriented research centers that will be funded to the tune of maybe a half a million dollars each initially, but eventually, we're talking about having some 25 to 30 universities funded to the tune of one to two million dollars each. And I would expect to see A&R play a very large part in that activity. So we are working on the problem.

Panelist (not identified): I think there's another area too, that's very important. As Stan indicated, one of the quickest ways to learn the technology and the acronyms is to have money cross hands. And another area that that's going to occur, other than what Stan indicated, is the Outreach Program. A little over a year ago OAST sent out a solicitation to the universities and industries to propose doing technology engineering experiments, not space science experiments, but getting onto the business of actually doing engineering experiments on the Station and on the Shuttle. We have received 231 proposals from industry and the universities just a couple of weeks ago and 41 of those were selected to be funded. And out of the 231, there were a little over 20 of them that had to do with automation and robotics. I don't remember the number out of the 41 that were actually funded. Not very many. But I think that's an area that really ought to be nurtured and people ought to start pushing to propose what they would do in this field.

Rodger A. Cliff: I think this is a very important point, so I would like to ask both Dick and Takeo, if they have any ideas on how to facilitate this process or what is needed, or a different viewpoint.

Prof. Takeo Kanade: I think this new program to establish the centers in universities is a very good idea. But in addition to that, when it comes to a specific problem of space research, I wonder if each university wants to maintain large equipment, as a sort of the infrastructure, that might be necessary if, for example, university people are interested in doing space research. I don't know whether universities want to bear the large cost of maintaining buildings and so on. Is there any idea to establish national or regional test beds for that kind purpose?

Panelist (not identified): I wasn't going to answer your question that way, but the answer to your explicit question is, yes. There will be test bed programs of all sorts, and I'll talk to some of that in my presentation later. But there are also individual programs that don't require the school to develop a complete research facility. Also there are individual grants that come from the Centers that go to universities. And we do have another program that will be oriented toward funding of innovative individuals as opposed to centers for research and that's in the planning right now.

Panelist (not identified): Stan got to some of the things that I was going to suggest. I agree very much that the programs that are coming out are excellent and will do a lot to address that problem. But I think, rather routine interaction of a large number of universities with all of the NASA Centers is an important thing to do. Getting students and faculty to go to NASA Centers for the summer time, personnel exchanges and then projects being carried out in the universities during the school year.

Dr. Rodger A. Cliff: We could probably spend another half an hour, at least, on this problem, but we will have to move on to our next speaker, who will be Ardell Nease, who will now switch to an industrial view of things.

AAS 87-670

SYSTEMS INTEGRATION VIEW - AUTOMATION AND ROBOTICS FOR SPACE STATION

Ardell Nease*

Thank you. I want to take a look at this whole A&R question from a little different point of view; primarily from the point of view of a system integrator and how some of the things that have been talked about earlier by some of the speakers can be incorporated into systems.

Specifically, I want to bring out four points.

Initially, what kind of A&R can we put on the IOC Station? I've got a couple things to say about how A&R can be incorporated into initial systems.

Then following right behind that, I will talk about some of the hooks and scars which will be required to make sure that we continue to incorporate A&R into the growth station.

I have a few thoughts on the critical issue of how we maintain confidence in A&R technology.

I will finish with some things we have to do right away.

Initially the Station will have robotics in two areas. The first area will be assembly and the second area will be maintenance (Figure 1), primarily external maintenance.

There are a lot of opportunities to pick up robotics in the assembly sequence. We're going to have three robots, three teleoperators, available to us very early in the assembly sequence. We're going to have a Shuttle Remote Manipulator System (RMS). We're going to have a Flight Telerobotic System (FTS), I understand, on the first element flight. We are going to have, very soon in the assembly sequence, the Space Station remote manipulator systems. Each one of these systems will perform a vital role in putting the Station together.

Once we start getting this Station up and running, we're going to have to maintain it. And the FTS is being built to perform many of those functions. The systems, on the other hand, are also being designed to incorporate the capability to interface with the FTS.

* Manager, Systems Project Office, Space Station/AC86, Rockwell International Corporation, 12214 Lakewood Blvd., Downey, California 90241.

Looking at the other half, the automation part, expert systems will both advise and control IOC Space Station Systems (Figure 2). There are two aspects to the kinds of things that expert systems can do. Expert systems can select schedules and resource options for the crew to make the final decision. In terms of systems that have a particularly long time constant, as far as the decision and the effect are concerned, we can do things like detecting, enunciating, isolating faults, at least to the ORU level. These are things that can be done in the systems from day one. We've got to be real mindful that we design hooks and scars into the system right from the beginning.

There are four basic hooks and scars (Figure 3) that the IOC Station will have. First of all ORU design standards for autonomous maintenance. This goes hand in hand with the Goddard FTS design development. We design our systems, our ORUs, to be compatible with being maintained by the FTS.

Second is a distributed expert system hosting environment on the DMS. The Data Management System needs to be capable of handling expert systems and their design. There's no problem in doing that; it's just a matter of being mindful of what the requirements are for experts for unique purposes or for expert system hosting.

Third, our Standard Data Processors need to be able to accommodate, not only expert systems, but advanced processors. And again, there's no real technological issue with doing that, it's just a matter of making the initial plan and the initial commitment to do it.

Then finally, we will have an easily upgradable, man-machine interface. A very important part of making, not only robotic, but telerobotic and expert systems or automated systems work, is the man-machine interface. We've got to be very careful when we put together the first man-machine interface requirements that we don't rule out some of the hooks and scars that will be required to operate the automated systems that will come along in the growth Station.

The most critical issue that I see in A&R right now, is getting the confidence in the A&R technology (Figure 4). That means getting the system designer mindful of what's going on in A&R research, and then, the other way around, getting the A&R researcher mindful of what's going on at the system design level. The system designer has design problems or design opportunities that the A&R researcher can respond to. He's got to be mindful of what those real world, real time, right-now system design problems are and he's got to provide solutions to those design problems. That means real solutions, not something from a research lab that may or may not be space-qualified or may not work in our environment. It's got to be a thought-through system engineering approach to solving the problem. So there's a loop, there's a loop that has to be closed and that's the designer talking to the researcher, and the researcher talking to the designer to build confidence in A&R technology.

Then finally, we've got to keep technology development on track (Figure 5). We've got to get the money into those areas where it makes sense to develop not only robotic but the expert system technology, too. We need to solve the real problems that we have in the real systems.

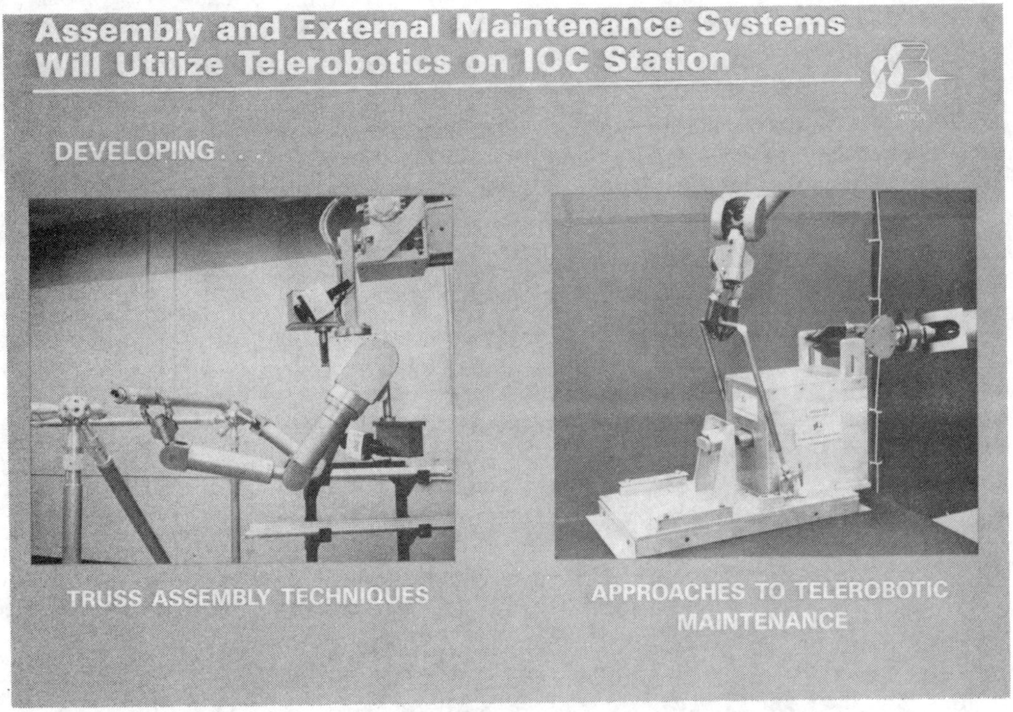

Figure 1

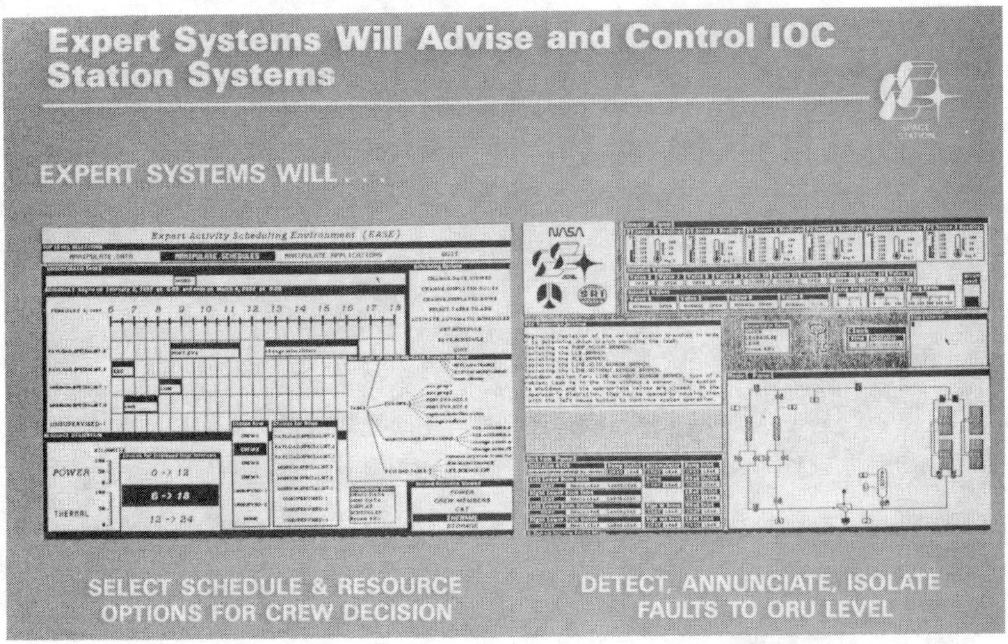

Figure 2

Figure 3

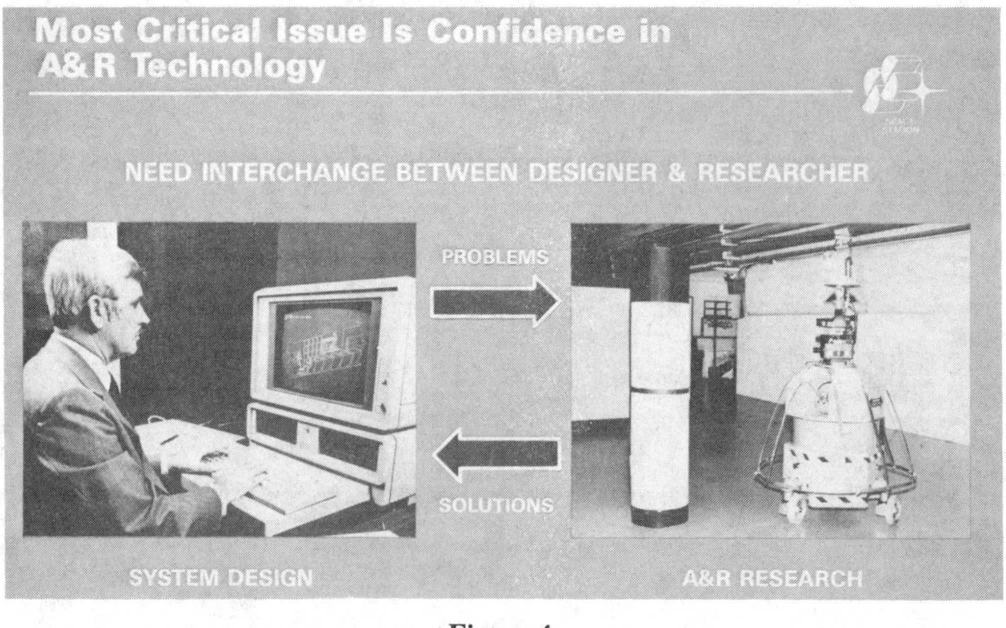

Figure 4

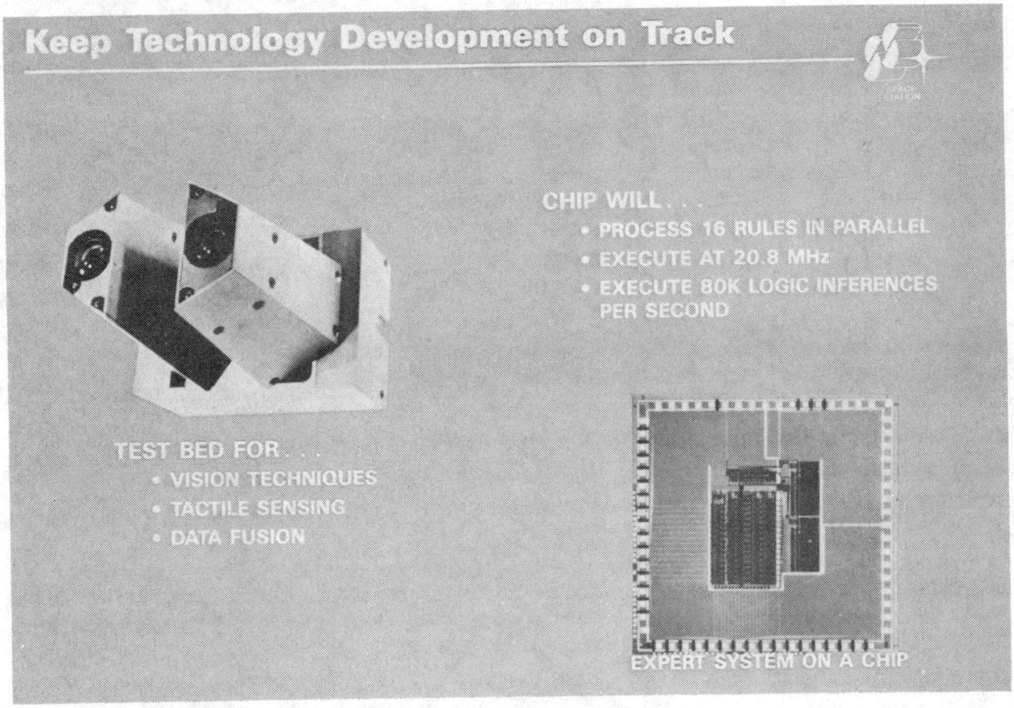

Figure 5

In summary, A&R will be integrated into the IOC Station. It will be there from the beginning. I've talked to you about the early A&R implementation. We've got to get the confidence into the A&R technology by the system designers and the system decision makers. We've got to keep our technology development on track. Thank you.

AAS 87-671

SPACE STATION GROWTH: UNIVERSITY VIEWS - AUTOMATION AND ROBOTICS FOR SPACE STATION

Richard A. Volz*

Thank you, Rodger. I'll probably say a few things that will also go beyond Space Station and look at other programs that might be coming up as well. I'm really just going to try to cover two things. One is just some application areas that one might look at and the second will be some of the technological issues that I think have to be looked at in terms of what we need to do now.

Up to now, the plans in the area of automation and robotics have been largely, at least in the robotic side, on construction, servicing and maintenance which are largely external to the Space Station. I think that in terms of other application areas for growth in A&R we need to look more to the inside of the Station. We are going to have a very small number of people, relatively, that can be in that Station at any time. We need to take a look at how we best utilize the humans that we do have there. Anything that isn't in the nature of a routine operation might be one of them. We're going to have to try to look at how we can automate anything routine. What do we want our humans to be doing in space, and how do we automate the rest of things, particularly within the Space Station?

We're going to need to look at different ways of coordinating with Earth. We'll certainly have much more in the way of expertise and intelligence on Earth than we're going to be able to put in six or eight people that we will have in the Space Station. How are we going to use the intelligence on Earth to back up and augment the intelligence of people that we have in the Space Station? We want to take a broader look at the problem. What we're trying to accomplish, I think, is intelligent action at a distance. We work in a Space Station and we want to achieve a number of intelligent things there. We have a system that involves some automation, some robotics, and also some people. In some of the other space missions we have been trying to achieve some intelligent action, on the Moon or on Mars, where there may or may not be people present. But, for the Space Station we have to consider how to use the people on board.

Looking more broadly at the dimensions of the problem, I think there are really three principal dimensions in here that we use in terms of trying to achieve intelligent action at a distance. See Figure 1.

* Professor Volz is Director of Robotics Research Laboratory, University of Michigan, 1101 Beal, Room 110, Ann Arbor, Michigan 48109.

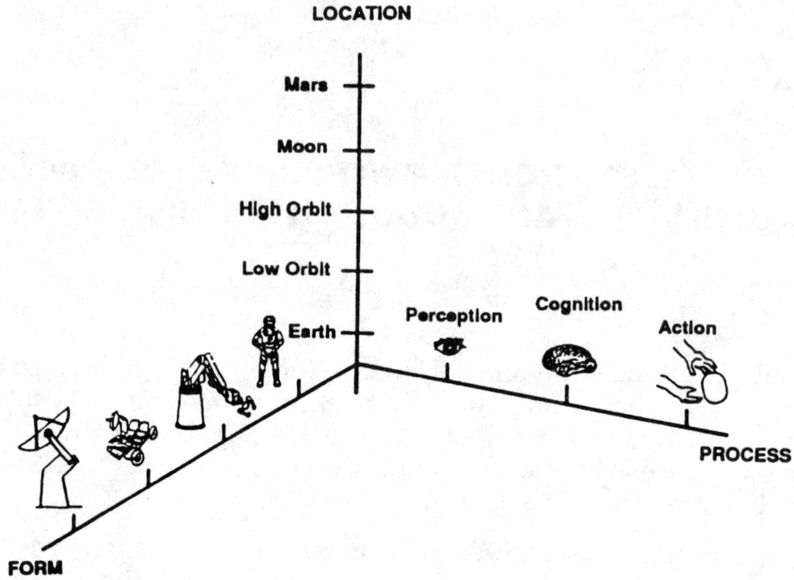

Figure 1 Dimensions to the problem space of intelligent action at a distance

There's a process dimension which has coordinates of perception, cognition or action.

There is a form dimension which is what kind of an entity is it that is doing that process.

And there's a location dimension indicating where the entity is located. You can think of a point in space representing some element of intelligence. And a system, really, is a collection of points at different locations, with different forms of doing the different processes that would take place. Things like automation tend to be rather limited just one plane, say, perpendicular to the form axis; then telemanipulation is, say, the human doing perception in cognition in one place and some device in another location doing some action.

There are obviously many, many more combinations that one can look at. By looking at the broader problem, sometimes you find that it actually solves even your restricted problems. For example, in trying to deal with issues of time delay in telemanipulation, we've found that we were able to almost double performance, even in situations where there were no time delays. To illustrate, here are a series of things that might be done.

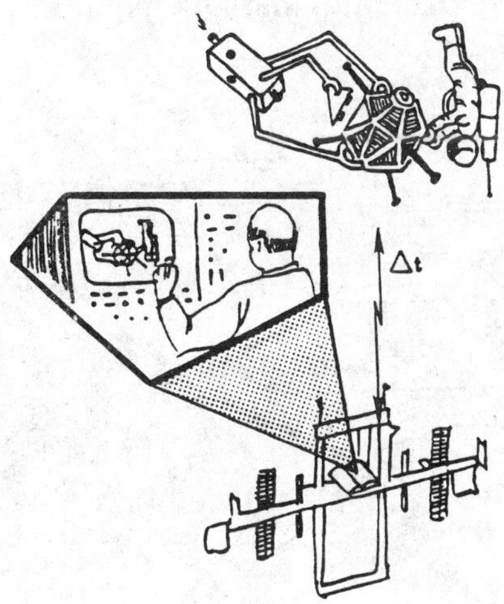

Figure 2

You might have someone in the Space Station controlling the Flight Telerobotics Services (FTS) or some device that's servicing a satellite. There may even be an astronaut present who's helping.

You may have, on the Moon or in the Space Station, rotating drums containing plants and you may have a human either in the Space Station or on Earth, trying to control not one but perhaps a number of robots or devices controlling them.

Figure 3

In terms of trying to do things requiring coordination with Earth, you can't have all the expertise on the Space Station. You may have someone on the Space Station controlling a robot, but in case of a diagnostic problem that can't be solved in space, someone on Earth who is hooked into that link might resolve the problem.

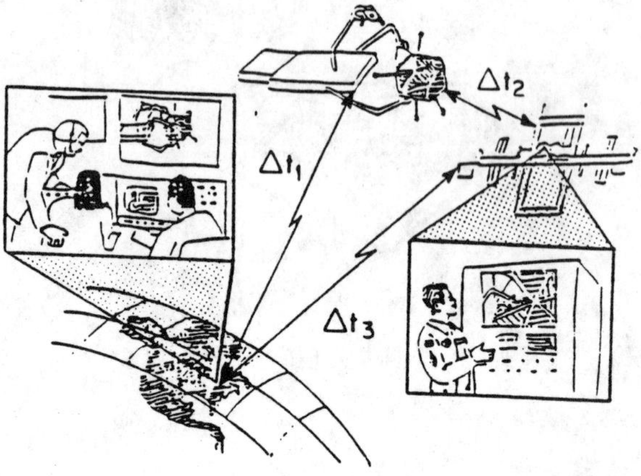

Figure 4

Here are some of the problems that we need to look at in dealing with this. There are questions of distance and time delay. There are some safety questions -- obviously, physical safety, if we have humans and robots at the same time. There are also some questions of software safety that we haven't really begun to address very carefully yet. The types of work that are done now in analyzing software and checking out software for flight are tested extensively. But what do you do when an expert system or some AI system, that changes itself, fails as a human does? How do we qualify software in those kinds of systems? Perhaps there are some new ideas in terms of AI and learning, rendezvous and interaction between humans and control agents, simulation and collaboration, not just between one human and machine, but perhaps between groups of actors that may include several humans and several devices. Thank you.

Dr. Rodger A. Cliff: Thank you very much. In the interest of having time for more questions from the floor at the end, we'll move right on to Donna Pivirotto.

AAS 87-672

NASA VIEW - AUTOMATION AND ROBOTICS FOR SPACE STATION

Donna L. Pivirotto*

It's handy that Dick has just given half of my pitch. What I'm going to talk about is the NASA view. What I'd like to say about the NASA view is that it's schizophrenic. There are a lot of people in the manned space world and a lot of laborers and factories, which means that automation is their enemy. It's going to take their jobs away. They are not going to get to do all the neat things they want to do. If we get too automated, we won't be able to sell these big manned programs and generally they are seen as competitors.

Actually, I think the true picture -- and the one we had to turn people around to -- is that people and machines are powerful collaborators. Some of the issues are technical, as Dick Volz was talking about, and a lot of them seem to be psychological. I come from the unmanned side of NASA -- one side of the schizophrenia and we, of course, we like automated systems very much, because it's very hard to put a person on a spacecraft going to Jupiter. When I first started interacting with the JSC, Marshall, and Kennedy community, there was a real cultural shock going on. When I asked, for instance, where the system executive control on the automated spacecraft is, they said it's in the flight controller, and it's Jean Krantz. They had no concept at the time of the fact that there could be a hierarchical control system for machines. Now what Harry McCain was talking about was the attempt to apply the hierarchical control system to see where human beings actually intersect in controlling the machine systems. Those kinds of issues are really key to being able to make effective use of machines in space applications.

Then there's the view of having direct hands on the controls. For instance, there is a caution and warning light that goes on if the robot happens to be in the Space Station. And the astronaut can flip the toggle switch. So there is a perception of, "We'll let the system identify the problems, but they don't get to activate anything. We can't trust the robots." The assumption is that people make fewer mistakes than machines, which is a very interesting assumption.

Since it costs a lot to put a person in a spacesuit and it's dangerous, you'd really like to have a person sitting inside this Space Station with remote control, or better yet, since it's quite expensive to put people into space, you'd rather have a person sit-

* Manager, Automation and Robotics Office, Jet Propulsion Laboratory, 4800 Oak Grove Drive, Pasadena, California 91109.

ting on the ground. So, instead of one hundred thousand dollars a day, it's a few hundred dollars a day. The problem is, that once you get physically away from the system you're trying to control, you have a problem with teleoperation. That is the kind of thing that, where you move a handle on the ground and there's a joint encoder in the handle and a similar joint encoder on the robot which moves the same distance, you have a time delay problem. For instance, if I have a two-second round trip light time communication delay, and I'm trying to catch a pin, I'm going to miss it every time. So the problem is, as the delay gets longer, like going to Mars, how do you build a system where you can exercise control over the robots in space at a distance?

There are two ways to do it. There is the teleoperation school of thought that says, "We're going to build up. We're going to teleoperate and we're going to add stuff to the teleoperator". The other school of thought says, "We're going to automate and we're only going to let the operator interfere with the system when necessary".

What we're trying to do right now within NASA is to reconcile the psychological view that says, "You have to teleoperate because I don't trust machines." and the robotist's view that says, "Hey, automation's the way to go". And so the architecture of autonomous systems and the way the teleoperation research community and the autonomous control research community work together are the key issue, I believe, in designing systems that are really going to function in space. Thank you.

AUTOMATION AND ROBOTICS:

THE MOON, MARS AND BEYOND

PANEL NINE -- SESSION II

Chairman:	Rodger A. Cliff Lockheed Missiles & Space Company
Host Chairman:	Jon D. Erickson Chief Scientist Space and Life Sciences NASA Johnson Space Center

AAS 87-674

THE PROMISE OF AUTOMATION AND ROBOTICS FOR THE MOON, MARS AND BEYOND

Rodger A. Cliff*

Thank you, Donna. We'll move right on now into the future opportunities that can be provided to us by automation and robotics....

In view of the goals stated in the report of the National Commission on Space and the Ride Report supporting a mission to Planet Earth, exploration of the Solar System, an outpost on the Moon, humans on Mars, some or all of those are certainly wonderful programs that support the goals we have seen before and it doesn't take a lot of imagination to see how any one of these will benefit very much from automation and robotics.

I will move on now and Stan Sadin will tell us about some of the programs that OAST is developing to help the technology come into being.

* Dr. Cliff is affiliated with Lockheed Missiles & Space Company, Organization 90-02, Building 253, 3251 Hanover Street, Palo Alto, California 94304.

AAS 87-675

OAST RESEARCH PROGRAM - AUTOMATION AND ROBOTICS FOR THE MOON, MARS, AND BEYOND

Stan R. Sadin*

Thank you, Rodger. I would like to address the subject of where the Agency is going and tell you that there has been a lot of discussion about the Commission Report and the Sally Ride response. There have been several other responses, including that of the AIAA. There have been all sorts of committees studying the problem of how the Nation can recover from the deep trough that we're in and move on to these visionary and exciting futures.

I'm going to talk to you about some things that are being done. It certainly is upsetting to me that people view the Ride Report as strictly a move toward studies, because there are some real things coming out, and investments that I believe the Government and NASA are about to make to the tune of somewhere between one and a half to two billion dollars, over a period of five years for technologies that will both help us recover our national technological posture and build toward the future. That's more than studies, and as they say, a billion here, and a billion there, and pretty soon you're talking real money.

We're going to discuss two programs, one of which is called the Civil Space Technology Initiative, and if you want the acronym, it's CSTI. It's the civil equivalent of another program that sounds somewhat similar, except this is oriented toward recovering our capability for getting to Earth orbit (SDI) and for operating in Earth orbit. There's a lot that A&R, Automation and Robotics, can do for us there.

Then it's followed by Pathfinder which takes off from low Earth orbit and does the kinds of things like the four missions that are mentioned in Sally Ride's Report, the four focused missions.

I'm now going to and show you what these initiatives are about. The first one, an FY88 start, which is very real to the system right now, is CSTI. It has four major thrusts. One of them is called Operations and for many years we struggled with the question, "What do operations mean?" We finally came to the conclusion that it is related to this discipline called A&R. Ten years ago, as Rodger remembers, when we first started studying A&R for NASA, we didn't have a line item budget covering civil programs. We're now going to be talking about line item budgets that are in the tens and the hundreds of millions for the Agency.

* Assistant Director for Space Program Development, Office of Aeronautics and Space Technology (OAST), NASA Headquarters, 600 Independence Ave., S.W., Washington, D.C. 20546.

Regarding, CSTI, there's a long list of tasks which will be done. There are two elements, one for robotics and one for autonomous systems that will deal with the on-orbit operations, with automated launch operations, with the growth Space Station, with making the Shuttle more effective, making Space Station more effective and supporting a lot of the things that you've heard about in these past sessions. Within the other tasks, there are built in a lot of autonomous operations. Many of our propulsion systems will have built-in diagnostics and real-time control systems of the type Rodger was saying he wished we'd develop.

I'd like now to move on to the Pathfinder Program and show you that there are four elements there. One has to do with transportation from low Earth orbit out to the Moon and out to Mars and elsewhere in the Solar System. It's a matter of operations for getting ready in Earth orbit and then moving forward to the Moon and Mars from Earth orbit. That means operations in space, exploration at the surface, and then the question of how to make humans capable of surviving some of these conditions both in-flight and making them productive when they get there .... not just having basket cases when they arrive.

I will now cite some of the examples of focused technology projects. And these individual projects within Pathfinder are much like the list of projects that I showed you under CSTI. Each is a major project which has a focus, with deliverables. Pathfinder goes on for five years and probably for many years after that, to continue development of these capabilities.

The first one is an Autonomous Lander, landing at Mars, in the presence of hazards, at interesting sites.

The next one is Surface Rovers, both unmanned and manned.

The next is Autonomous Collection, Analysis, and Preservation of Samples.

The next is Autonomous Rendezvous and Docking.

The next is On-Orbit Assembly of the massive systems that will take off for the missions to the Moon and Mars. A lot of this will be delivered by heavy-lift launch systems and will require major construction jobs on orbit, things we are only beginning to fathom, and that's beyond what's done in Space Station.

The next is a Fluid Depot for cryogenic fuels, both at Earth orbit and probably, eventually, at lunar orbit.

The next is Lunar Materials Processing and we will focus on a pilot plan with emphasis on oxygen production at the Moon.

Those are only seven of the 17 task elements; they are the ones that have direct applications to A&R. And A&R is built into almost every single one of the remaining tasks. Within the Ride Report is the statement that unless these technologies are pursued, you can forget about all the exciting missions that we talk about for the future. So Sally encourages you, and I encourage you to support our programs.

Dr. Rodger A. Cliff: I've asked Takeo Kanade to speak next and to deal with what technology we can expect and how to attain it and how the universities can help.

AAS 87-676

UNIVERSITY RESEARCH - AUTOMATION AND ROBOTICS FOR THE MOON, MARS, AND BEYOND

Takeo Kanade*

I'd like to make some short general comments for advancing automation and robotics for the future space program. I believe that autonomy is the key to future space missions, but since the previous panelists have discussed that point, I won't elaborate on that further.

Now, how to achieve the autonomy? Of course, there are many, many technological issues, including sensors, controls, and materials, and so on. But, we have to look at one higher level, that is, to think of the reliability at the level of intelligence. Historically, the reliability was in regard to mechanical, electrical, and computer software. However, this is now functioning well. What we are talking about is reliability as an autonomous system. So we have to think of a system which can actually have some concept of self-awareness; that is, the system must know what it's doing, how well it's doing, and when it should call for help, and how to repair itself, if necessary.

Whenever we talk of this kind of thing, it comes down to AI. Does the current AI help us? I have some doubt about it. The reason is that today's AI is too abstract. It's apart from the physical world; I think we have to bring this current AI down to physical reality. AI needs the physical world. Historically, robotics was thought to be a particular application area of AI. I think that's wrong. Actually AI is a particular small area of robotics. Unless we convince AI people of this, we may not achieve completely intelligent autonomous systems.

One specific thing that we have to be aware of is the huge computation gap between Earth and in space. Whenever we talk with the space people, they are talking about, at most, one MIPS on the space-qualified computer, while even today's AI program needs a hundred MIPS, at least, or a thousand MIPS. We have to overcome this particular specific technological problem.

What can the university or the basic research community of robotics contribute to the advancement of the true automation in space? The first thing is, we have to encourage and expand, what I call autonomous systems research. We have to realize that an autonomous system is not an extension of telerobotics. I think those two are completely different things. In fact, in some sense, telerobotics might be more dif-

* Professor Kanade is Acting Director, Robotics Institute, Carnegie-Mellon University, Pittsburgh, Pennsylvania 15213.

ficult than automation. The basic difference is that, automation is not the extension of telerobotics, because in telerobotics, the basic assumption is that the self-awareness is in the operator. According to my definition, self-awareness of autonomous systems must be in the system; there's a fundamental difference between those two approaches. So we have to abandon the view that automation is an extension of telerobotics. Now, there are numerous research programs on components, such as vision, pathfinding, sensor control, and so on. But actually, there are not many programs on autonomous system basic research going on.

In fact, when we met at the JPL Mars Rover workshop, we were astonished to find out that probably, as a community as a whole, we have total collective experience of running outdoor autonomous mobile robots, at most, 50 to 100 kilometers, while, when it comes to the Mars Rover which is planned, say, at the end of the century, for which we may want to freeze the technology in five to seven years, we'll probably need a thousand kilometers of test run. So we are way behind on this kind of schedule.

We really have to encourage the research, not the development, of autonomous systems. There are some activities going on, most of them actually sponsored by DARPA. As a matter of fact, one of the programs that we haves at Carnegie Mellon is called NAVLAV. It is a computer controlled van with sensors. We have quite a bit of computation on it by several on-board computers. Well, of course, you know, this alone will not solve the space problem. But I think it's a start and there are some other research problems, so I think we should really encourage the research in this direction.

Finally, in order to encourage that, it's important to loosen the research constraint. It's silly to ask people to do innovative, bold imaginative research while we specify that the only computation that you use is one MIPS on space-qualified material, and so on. Let people hang loose and do whatever they want to do and then see what innovation comes out. Thank you.

Dr. Rodger A. Cliff: Thank you very much, Takeo. We'll move on now to Stan Weiss. I promise I didn't put him at the end because he has a "W" at the beginning of his last name.

AAS 87-677

INDUSTRY RESEARCH - AUTOMATION AND ROBOTICS FOR THE MOON, MARS AND BEYOND

Stanley Weiss*

Being the last speaker of the last morning struck me as being an early invitation for me not to come, Rodger. But it's better than being the last person of the last afternoon, and I have had that experience on one occasion but it ended up with a very spirited discussion with a six-person audience. But they were devoted.

I have understood that I am to address what we in industry are looking for, when we view the types of future that have been discussed here earlier in the sessions and also from this podium.

First of all, we tend to look for a clear understanding of requirements. I don't know if any of you have them written down now, but I haven't been able to develop them, and I don't expect we will be able to, for anything approaching the period between, say, the end of this century and through the first quarter of the next century. I don't think we really should.

On the other hand, we do look for a clear understanding of long-range goals, because that's what we have to base our own assessment of requirements and our own assessment of investment on. And I guess that really applies to Government investment as well. We have heard from my colleagues here at least some speculation in that area, also from Sally Ride, from the National Commission. And we have noted the exciting nascent leadership of what Stan Sadin has described in the Pathfinder Program. But I think, within industry and within the entire community, we have to speculate rationally and without inhibitions that are inflicted by our own narrow histories and experiences and vision that generally are affected by boundaries determined by where we are in today's state of the art.

One of the problems we have, of course, is that we are in a period of incredible technological growth with a pace that outraces most of our imagination. And how we really orient ourselves toward the next century in 2025 is something that demands very open thinking and probably very young thinking. Many of us like to think we still have that immaturity, and I hope we do, in terms of looking toward these advanced technologies. I see us looking toward systems ranging in automation and robotics from manned interactive telerobotics to fully autonomous systems designed for manned and unmanned missions, as we've discussed. Many of these will be for

* Dr. Weiss is Vice President and General Manager, Research and Development Division, Lockheed Missiles & Space Company, 3251 Hanover Street, Palo Alto, California 94304.

specific use and as well as those, which will have generalized use and a variety of appendages and methodology that permit use in wide groups of missions and circumstances and therefore lead very strongly to terrestrial spinoff. Much of this will take place with ever-increasing expectations of autonomous capability. By the 21st Century, we should also have a different sense of confidence in automation and robotics, versus our current view, that man must be in the loop for reliability. If you take off a little bit from Donna Pivirotto's comment, there is a variant that says, the assumption today is not necessarily that people are less mistake-prone than machines, it's really that people can correct mistakes better than machines.

Looking further technically, I think we'll see, not only the seven degrees of machines that we talk about now in the robotic side, we'll talk end degree of machines. We'll talk end degree, because we won't be constrained, by anthropomorphic designs. We'll certainly have to deal with this reliability question by means of self-repairing. We'll have to deal with heavy and light duty systems, solar and stored-energy powered systems and appropriate technologies that we must be looking at today in order to have them available and introduceable into systems at that time.

A little while ago, I gave a talk on some of the technologies I felt that we had to look at, from an industrial standpoint, in our laboratories in order to get there. And I thought I'd mention some of them here, because they seem to extend over the range of technologies that have been discussed previously. They involve autonomous operation for extended periods, recognition and resolution of errors, clearly the vision systems that are necessary involving a wide variety of sensors, shared human and machine control, high level communications, decision making in uncertain environments, real time knowledge transfer and learning with true knowledge base planning and control, multi-armed interactive controls and nonanthropomorphic systems.

At any rate, we can look at these technologies, but if they are not developed in a fashion that there is conjunctive design of the automation and robotics systems involved with the systems that will interact with them or utilize them, we will not have progressed as far as we should in dealing with advanced technology of this ilk, nor as a matter of fact, in dealing with other advanced technologies that will make it possible some of these incredible missions that we must indeed do.

Parenthetically, as it is with most advanced programs today, we talk about spinoffs. And frankly, I think that the first spinoffs we'll see for terrestrial operation will be those that derive from introducing the methodologies and the capabilities that will have to be operative in the Space Station efforts of the latter part of the century.

Looking toward the future that we're talking about for the 21st Century, Mars, other planets, a lunar base, things that most of us believe very strongly in, we need a national commitment; we need continuity in those commitments, we need an expectation that our civilization, indeed, has a responsibility for these missions, which will capture the American spirit and will to "incentivize" those of us leading technology development with a definitive purpose and give us the basis for using unbounded imagination to get there. Thank you.

AAS 87-673

AUTOMATION AND ROBOTICS - DISCUSSION

Panel Moderator: Rodger A. Cliff
Panelists: Robert Nunamaker
William V. Bates, Jr.
Harry McCain
Jackson Yang
Ardell Nease
Richard A. Volz
Donna L. Pivirotto
Stan R. Sadin
Takeo Kanade
Stanley Weiss

Dr. Rodger A. Cliff: Now I'd like to invite any of the panelists to ask questions about the IOC portion of automation, while we take a few written questions from the floor.

Panelist (not identified): I'm very concerned when I hear a lot of animated talk about robotics. And I don't see the focus as to where it is going beyond what we looked at from the ATAC point, two, three, and four years ago. There are sporadic things being done on that, but I don't really see that it affects what I do to robotics. I wonder if Ardell or anybody else would like to comment on that?

Ardell Nease: I think we identified some specific areas where expert systems can be used very effectively to solve some very real problems. Particularly in the FDIR, Failure Detection, Isolation and Reconfiguration, I think in that area where the configuration of the system is going to have to be controlled, and in the areas where the crew needs to have many of its tasks offloaded, such that an expert system can act in an advisory manner. There are some techniques that satisfy many of the standard structured software verification requirements that can be used to begin putting expert systems into the distributed environment of the Space Station Data Management System. I think we can do it if we start with a very structured approach to expert systems and then build from that, build on what we have learned. Let's just not get all hung up on AI technology being something weird and strange and different. It's just a new approach, it's like moving from one form of mathematics to another or from one language to another. It's just a new, better way to do it. And if we approach it from that standpoint and take the problems one step at a time, we can incorporate it.

Donna L. Pivirotto: One of the things I'm concerned about is when we talk about on-board automation, everybody immediately says expert systems and AI. As a matter of fact, my friend, William Bates, looked at the Shuttle and, I think, determined that some 80% of the functions, on the Shuttle could be done with automation. One of the things that bothers me about the Station design is that I have never heard anybody talk about what we need to do to make the Station automated. For example, I haven't heard that anybody has stated that you need to have redundant systems in place hooked up to a computer network, because whether you switch them by AI logic or by other means, they have to be in place and they have to be hooked up to a computer to make them work. And I think those kinds of things, as far as I can see, are important and I feel insecure about whether those aspects of the Station design are really happening.

<u>Rodger A. Cliff:</u> I have a question here, for Skip Nunamaker.

<u>Question:</u> What impact has the ATAC report had on the current budget negotiations on the Hill?

<u>Robert Nunamaker:</u> On the current budget it had very little impact, other than the fact that when it gets down into the committees, they look at what NASA is proposing to do, and then compare with the report and see if NASA is really implementing these issues that have come up in the report. We have had occasions, for instance, where Dick who is a very powerful staff member for the Bollin Committee, will take our report and will see something in there that concerns him, and he will stop funding in a certain area. So we've had not only to keep our communications open with the Space Station Office, but with some of the Congressional people who literally go by the wording of the report and who change the budget. And we have to point out to them just what progress is being made on some of our issues. While it's probably one of the many things that are an input to the budget, it still is an item that Congress is watching. Basically, Congress thinks that it's going to be hard for us to do our job without more support, and I think they're waiting for NASA, to be a little more bullish in the budget arena.

<u>Rodger A. Cliff:</u> A question here for Bill Bates.

<u>Question:</u> How do you provide for technology update? Clearly, one of the major limits of the Shuttle is its 1960s computer architecture. There's an observation here: It is almost impossible to upgrade automation/computers to the state of the art in Government projects. This has cost immense capability and increases operating cost. And there is a recommendation: To streamline the certification process and develop an organizational group with sufficient clout to analyze and implement technology improvements.

<u>William V. Bates, Jr.:</u> Had I not talked so slowly and had I gotten to the second page, I had a couple of topics there about modularization. Really what it boils down to is that you need to know where you're going before you start so that you know what is in each building block. They need to be building blocks, so that, later on, if you need to pull out one of those on the bottom level and replace it with some new technology, you must have built-in capability to do that. My real concern is that, as mentioned a while ago about what has happened since the original ATAC -- and I've seen a lot of new technology come along -- I haven't seen it applied to benefit the

automation situation. I think, taking a very practical approach to the IOC to make sure that we use what we've learned the hard way about structured systems and structured software, you can try to design for automation as much as possible. Going to standard interfaces is another key part to that. Don't get caught in a trap. Because of budget or schedule, you take what's an off-the-shelf item, and end up customizing to a point where there's nothing in the free world that will interface with it. You run into those sorts of problems. But I think, if you know where you're going when you start, that's a key element, instead of evolving and always being a victim of budget and schedule.

Dr. Rodger A. Cliff: I certainly would like to second that one. Having a vision of where you want to go is essential to putting in the hooks and scars, and this is very much related to that in designing the process by which you update the computers and the software. We'll take one more question right now and save the others for a little later. We're getting some really good ones. This is for Harry McCain. It's from Lyle Jenkins.

Question: What is needed to qualify the Flight Telerobotic Servicer for use on the first element launch of the Space Station and to provide confidence in its operation? Use of robotic devices in conjunction with manned space flight has a limited experience base.

Harry McCain: Well, I think the way I'd respond to that is I'd just pick one of the subsystems, which I think is the most critical one, and that's the computing capability. What we're faced with right now, is really a lack of the kind of computing capability, flight-qualified computing capability, that we've gotten used to in robotics laboratories on the ground. The real big issue is getting more powerful flight-qualified computers. And the real driver there is in the radiation environment. For both the long-duration and the single event, we're very much struggling with that, and related to that are software verification and validation. That's a very, very hot topic these days. As we see the software become more and more complex and people get less and less confident about our ability to really check it out thoroughly before we fly. Those are the two major issues. I really don't see a problem in qualifying the mechanical manipulators and the other more physical devices that we'll be trying to fly. The real issue is the computers and the software.

Dr. Rodger A. Cliff: We still have time to answer a few more questions. We have a few left over and if you want to write up a few more, we'll give them our best shots. One for Harry McCain.

Question: What aspects of FTS design support the growth in robotic dexterity technology spoken of by Dr. Yang?

Harry McCain: When I talk about growth I come back to our architecture and say that because of the way we have designed the architecture, we hope to be able to support various elements. Something dexterous, like a hand, which in itself has a multiplicity of control capabilities associated with it, is then essentially subordinate in a hierarchical sense to the arm capabilities and so forth. So the architecture is really our key and the establishment of the interfaces, so that we can adopt the technology advances, such as the Utah-hand of Steve Jacobson -- a very complex system with very dexterous capabilities.

Dr. Rodger A. Cliff: I have a question here which is addressed to Bill Bates. It's, in some sense, a comment on one of the other questions and statements that went with it. It's fairly long so I'll paraphrase it.

Question: The assertion here, is that the present AP101 general purpose computers are being upgraded to Shuttle computers like those used in the B1B, and the significant capability exists to upgrade to the 175A architecture. The assertion is that the Shuttle data processing system does have the capability to increase its automated capability if it is needed and that the real question is need.

William V. Bates, Jr.: Well, I think the need is there. The only way to really make more efficient use of that would be to do some better instrumentation. Donna brought up a fact earlier that the Orbiter is about 75% deterministic. It has enough instrumentation to tell what's going on. There's about 20% or 25% that is probabilistic, and then there's about 5% to really manage it and to know what's going on. To back-fit that with the appropriate instrumentation and really make use of that is probably where the biggest problem and the biggest cost would be. You can make additional use of the computing capability to automate some functions, but I think the real driver would end up being the basic fact that it is not really designed as an autonomous vehicle.

Dr. Rodger A. Cliff: This question is addressed to anyone on the panel and I will expand it a little.

Question: What are the NASA future plans for supporting design knowledge capture and programs other than the Space Station? But I would like to include the Space Station there too. What plans are there for design knowledge capture? And to explain the importance, in order for AI systems to be able to efficiently do fault detection, isolation and recovery, they need to understand the system rather deeply, and this means machine-readable forms that encode the knowledge of the people that built the hardware. How is this being attacked in NASA?

Panelist (not identified): Let me attempt an answer to that. I think the current status is that NASA is behind industry in regard to the adoption of standards and facilities in an integrated environment sense. Air Force programs, NBS programs, industry volunteer efforts, etc., are moving in the direction of defining *de facto* industry standard for these types of activities. Both the Space Station program and the Shuttle program are at the point where they are about to jump into those areas, probably with both feet, but they are looking at their budgets at the same time to see just what they're going to get when they do these things. They are aware that, with the *de facto* industry standard, both the Shuttle program and the Space Station program will need to move in that general direction. They have not, as yet, committed to those activities with any extensive support.

Dr. Rodger A. Cliff: Thank you very much. Unfortunately, we'll have to cut off the questions here. I'd like you to show your appreciation to the panel for the interesting presentations and thoughts they have given you here and I really hope we can move forward and take advantage of the ideas that we have been talking about today.

SPACE STATION/LARGE STRUCTURES

PANEL TEN

Chairmen:	Dr. Edwin E. Speaker Space Station Program Office NASA Headquarters
	Dr. Martin Mickulas Branch Head, Structural Concepts NASA Langley Research Center
Host Chairman:	Dr. Robert Goetz Technical Assistant to Center Director NASA Johnson Space Center

The following paper number was not assigned:

AAS 87-680

The following papers were not available for publication:

AAS 87-685 "Human Performance in Space," by B.J. Bluth, Aerospace Technologist, Space Station Program Office, NASA Headquarters

AAS 87-686 "Deployable Truss Support Structures for Extremely Accurate Reflectors," by John Hedgepeth, Consultant, Astro Aerospace Corp., Carpenteria, California

SPACE STATION/LARGE STRUCTURES - INTRODUCTION

Edwin E. Speaker* and Martin Mickulas†

Two papers were presented at this session which continued the theme on structures and composite materials. The two papers reported recent results in advanced structures technology, on-orbit assembly techniques, and design analysis of deployable antennas. The papers were then supplemented by extemporaneous comments from a very distinguished discussion group consisting of Allen J. Louviere, Senior Vice President, Engineering & Operations, Space Industries, Inc.; S. Z. Rubenstein, President, Space Station System Division, Rockwell International Corp.; Robert F. Thompson, Vice President, General Manager, Space Station, McDonnell Douglas Corporation; and Charles P. Blankenship, Director of Structures, NASA Langley Research Center.

The first paper, "Extra-Vehicular Activity (EVA), Space Constructions, and Fundamental Issues", was presented by Walter L. Heard, Aerospace Technologist, Structural Concept Branch, NASA Langley Research Center. His co-authors were Judith J. Watson, Aerospace Technologist, Structural Concept Branch, NASA Langley Research Center; Lt. Col. Jerry L. Ross, USAF, Astronaut Mission Specialist, NASA Johnson Space Center; and Kathryn A. Havens, Aerospace Technologist, EVA Crew Systems, NASA Johnson Space Center. Their paper showed many excellent views of astronautics involved in on-orbit assembly of structures, handling large objects in space, and in training for the expected Space Station structural tasks. The need for a new neutral-buoyancy test facility was discussed as well as the need for continued technology development in conjunction with specialized Space Station crew training to allow in-orbit assembly of large structures quickly and safely.

The second paper, "Deployable Truss Support Structures for Extremely Accurate Reflectors" by Dr. John Hedgelpeth, Consultant, Astro-Aerospace Corporation, described a number of extremely ingenious design concepts for the creation of large reflector structures in space.

The comments from the discussion group were highly supportive of the advanced structural development and analysis work reported on in the two papers. All members of the discussion group agreed that successful long-term scientific and commercial developments in space would depend greatly on advances in structural design, assembly methods, and operational techniques. Proof-of-concept flights were cited as being absolutely essential to the advancement of structural technologies and to the qualification of structural designs for use in the Space Station.

---

* Dr. Speaker is associated with the Space Station Program Office, NASA Headquaters, Washington, D.C. 20546.

† Dr. Mickulas is Branch Head, Structural Concepts, NASA Langley Research Center, Hampton, Virginia 23665.

AAS 87-681 to -684

EVA SPACE CONSTRUCTION: EXPERIENCE AND FUNDAMENTAL ISSUES

Walter L. Heard, Jr.,* Judith J. Watson*, Jerry L. Ross†, Sherwood C. Spring†, and Kathryn A. Havens‡

On-orbit construction, including the role of the Space Shuttle, with EVA astronauts as space construction workers is currently under investigation for Space Station assembly. This paper addresses some issues that should be considered in this investigation. Principal results and astronaut appraisal of recent EVA space construction tests performed in 1-g, neutral buoyancy, and on-orbit are summarized. The use of EVA for Space Station construction, ongoing ground test programs, and some current EVA flight rules that may affect on-orbit construction activities are discussed. The paper concludes with astronauts' appraisal of and recommendations for EVA space construction.

INTRODUCTION

As NASA undertakes the mission of providing the nation with a permanently manned space station, on-orbit construction, including the role of the Space Shuttle, with astronaut extravehicular activity (EVA) as the primary construction method are being studied. Space Station is currently baselined to be an erectable structure. The benefits of erectable truss structures, i.e., those that are assembled piece-by-piece on-orbit by EVA astronauts (or robots), have been well documented,[1,2,3] and the development of high-fidelity quick-attachment joint hardware[4,5] enables EVA assembly techniques.

The purpose of this paper is to summarize the principal results and present astronaut appraisal of some recent EVA space construction studies performed or sponsored by NASA in 1-g, simulated 0-g (neutral buoyancy), and on-orbit. Some issues concerning the use of EVA for Space Station construction are discussed along with ground test programs currently underway to address these issues. Additionally, some current Shuttle EVA flight rules that impact EVA time available for space construction activities are presented. The paper concludes with the astronauts' appraisal of and recommendations for EVA space construction.

EVA ASSEMBLY EXPERIENCE

The EVA structural assembly tests summarized herein were performed by the NASA Langley Research Center (LaRC), Johnson Space Center (JSC), Marshall Space Flight Center (MSFC), and Massachusetts Institute of Technology (MIT). They involve ground tests performed in 1-g and simulated 0-g within the past 6 years, and a flight test performed in late 1985.

---

* Aerospace Technologist, NASA Langley Research Center, Hampton, Virginia 23665-5225.

† Astronaut Mission Specialist, NASA Johnson Space Center, Houston, Texas 77058.

‡ Project Development and Integration Office, NASA Johnson Space Center, Houston, Texas 77058.

Ground Tests

Mobile Work Station Concept.[6]

Description: This concept was developed to demonstrate the potential of mechanically aided astronaut assembly of space truss structures and to provide an experimental basis for estimating on-orbit assembly times. To form a 54-foot-long 3-bay tetrahedral truss beam, 38 identical graphite-epoxy tubular struts, each 18 feet long, were assembled by a pair of test subjects. Each strut tapered from a 4-inch diameter at midlength, to a 2-inch diameter at each end, and each weighed 3.5 pounds. In order to estimate assembly times for space operation of the Mobile Work Station concept, a series of tests were performed to isolate the effects of water drag and pressure suit encumbrance. Tests were performed both in 1-g with the test subjects in street clothes (Fig. 1), and in neutral buoyancy with the test subjects in either scuba or NASA A7L-B (Skylab) pressure suits (Fig. 2). The 1-g tests, performed at

Fig. 1 Mobile Work Station 1-g truss assembly test.

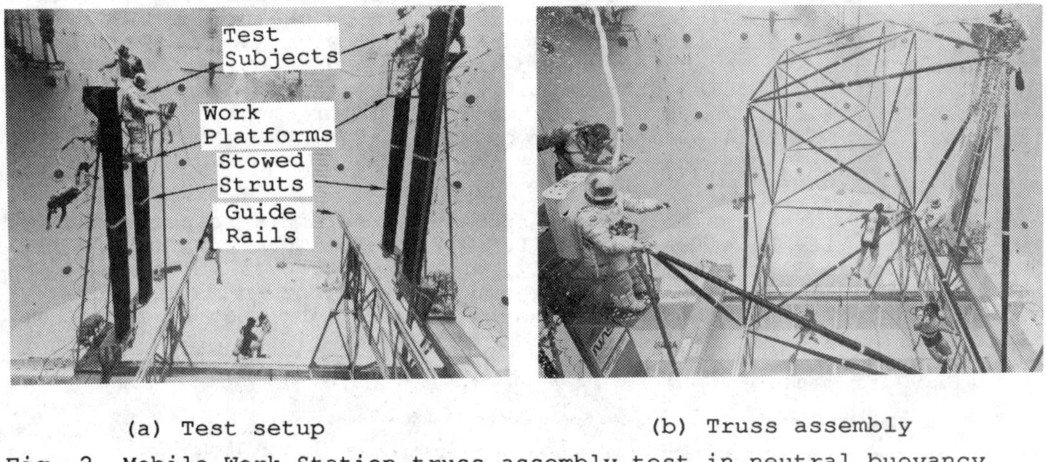

(a) Test setup (b) Truss assembly

Fig. 2 Mobile Work Station truss assembly test in neutral buoyancy

LaRC, enabled the manipulation of hardware without the effects of water drag, and the effects of gravity were inconsequential because the struts were lightweight. The restrictions of a pressure suit on the test subject, however, could not be investigated in 1-g. The neutral buoyancy tests were performed in the Neutral Buoyancy Simulator (NBS) at MSFC. The scuba tests permitted the manipulation of neutrally buoyant (weightless) hardware without pressure suit encumbrance, but included the effects of water drag. The scuba assemblies averaged 39 seconds per strut, an increase of about 15 seconds per strut over the result from the 1-g tests. This increase was attributed to water drag. Assembly in neutral buoyancy with the test subjects in pressure suits added another 14 seconds per strut to the scuba assembly time. This increase was attributed to pressure suit encumbrance.

The assembly procedures were based on an assembly line technique wherein a few simple repetitive tasks could be accomplished by hand, i.e., withdrawing struts and nodes from stowage and installing them on the assembly fixture. Throughout the tests the test subjects were secured in foot restraints attached to mobile work platforms and were, thus, always able to remain in place while exerting the necessary assembly forces. The mobile work platforms were remotely controlled by an operator familiar with the assembly procedures who positioned the test subjects at the various work stations so that the necessary strut-to-node attachments could be made. The mobile work platforms which transported the test subjects at a rate of approximately 1 ft/sec also transported the stowed struts and nodes. The truss was assembled one bay at a time on the assembly fixture. When a bay was completed it was mechanically translated along two guide rails out of the work area so that the next contiguous bay could be assembled.

<u>Results and observations</u>:

1. The assembly line technique in combination with the Mobile Work Station fixture enabled rapid assembly of the truss. All assembly tasks were accomplished in pressure suits with relative ease, and minimal fatigue was reported by the test subjects.

2. The average time for a complete truss assembly in 1-g with the test subjects in street clothes was 15 min, 12 sec (2.5 struts/min).

3. The average time for a complete truss assembly in neutral buoyancy with the test subjects in scuba was 24 min, 42 sec (1.5 struts/min).

4. The average time for a complete truss assembly in neutral buoyancy with the test subjects in A7-LB pressure suits was 33 min, 34 sec (1.1 struts/min).

5. By eliminating the effects of water drag, an average on-orbit assembly rate of 1.6 struts/min was projected for struts of comparable size.

<u>Astronaut participation and appraisal</u>: Two untimed, neutral buoyancy assemblies by NASA astronauts who were inexperienced with the Mobile Work Station were performed. The subjects were given only a short pretest briefing on procedures and operation of the quick-attachment joints. Astronauts Owen Garriott and Jerry Ross performed one assembly, and astronauts William Fisher and David Walker performed the other. Post-test comments were:

1. Tethering of the hardware may be a requirement during space operations and should be considered in subsequent tests. Use of the waist and wrist tether hooks currently used by EVA astronauts could cause significant hand fatigue.

2. The 2-inch diameter of the strut ends approaches the maximum size that can be easily handled with pressure suit gloves.

3. The required assembly tasks can be done in weightlessness, and the procedure is simple to learn. Test subject control of mobile foot restraint positioning may be desirable.

4. Joints should provide a visual means for verifying the locked condition.

5. The Mobile Work Station simplifies assembly tasks for EVA astronauts. The issue of resupply of truss struts and nodes to mobile work platforms for continuing assembly needs to be addressed.

Swing-Arm Beam Erector (SABER).[7]

Description: A second series of ground tests designed to study machine aided EVA assembly of space structure involved the SABER concept. This concept also used a mobile work station approach, although, only one astronaut was required to assemble the truss and much smaller struts were used. An assembly fixture, incorporating a truss-advance mechanism, supported and manipulated the truss while a telescoping swing-arm positioned the astronaut. A photograph of the setup for a 1-g assembly test is shown in Fig. 3. The struts were stowed in racks on the swing-arm and the nodes were stowed near the lower ends of the three guide rails. A multibay truss was assembled one bay at a time on the assembly fixture. When a bay was completed, the fixture advanced the truss upward one bay-length to clear the work area for assembly of the next contiguous bay. The truss advance mechanism and telescoping swing-arm were remotely controlled by a separate operator.

A series of 1-g and neutral buoyancy assembly tests were performed. Thirty 1-inch diameter, tubular, aluminum struts, 20 of which were 5.3 feet long and 10 of which were 6.2 feet long, were assembled with quick-attachment nodal joints into a 3-bay tetrahedral truss of triangular cross section. Fig. 4 shows photographs of the neutral buoyancy tests performed by a test subject in an A7-LB pressure suit.

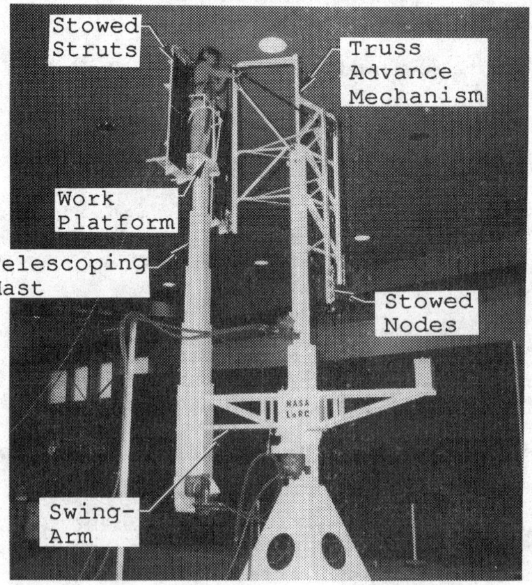

Fig. 3 Setup for SABER 1-g truss assembly tests.

(a) Removing strut from stowage (b) Completing 3-bay assembly

Fig. 4 SABER truss assembly test in neutral buoyancy.

Primary results and observations:

1. The SABER enabled rapid assembly of the truss by a single test subject.

2. All required assembly tasks were accomplished routinely by test subjects in A7-LB pressure suits.

3. The repetitive tasks of the assembly procedure were quickly learned by the test subjects

4. Five consecutive assemblies of a 3-bay truss were performed by one test subject in 1 hr, 44 min. Included in this time are four 10 minute periods after each 3-bay assembly during which the test subject was idle while the truss was disassembled and restowed by scuba divers. The total amount of structure assembled was equivalent to an 85-foot, 15-bay truss beam consisting of 150 struts and 48 nodal joints. The average rate of assembly was 1.4 struts/min, and no reduction in productivity due to test subject fatigue was evident from the data.

Astronaut participation and appraisal: Untimed 1-g assemblies were performed by each of two NASA astronauts who were inexperienced with the SABER concept and the assembly procedure. The astronauts were given a short pretest briefing on the assembly procedures and hardware. Astronaut Jerry Ross performed one assembly, and astronaut Sherwood Spring performed the other. Post test comments were:

1. The 1-inch diameter struts were easy to handle and should present no problem in pressure suit gloves.

2. The assembly procedure routine is simple to learn.

3. The handling of two different length struts during the assembly is acceptable provided the struts are clearly marked.

4. Tethering of hardware may be required on-orbit and should be addressed.

Ground/Flight Tests

The two experiments discussed in this section were performed both in neutral buoyancy ground tests and on-orbit. The experiments were launched aboard the Orbiter Atlantis on November 26, 1985. Although numerous neutral buoyancy development and training tests were performed prior to flight, data taken only during latter tests, after the astronauts were fully trained, were used for comparison with flight data.

Experimental Assembly of Structures in EVA (EASE).[8]

Description: EASE consisted of a series of structural assemblies designed to measure the performance of astronauts working in EVA. The assembly tasks were designed to test the astronauts' ability to adapt to working in the weightless environment; not to optimize assembly efficiency. In the baseline experiment, two astronauts repeatedly assembled an inverted tetrahedral framework consisting of six identical struts and four nodal joints. Each strut had a mass of 64 pounds and was 12 feet long. The EASE structure was attached by one nodal joint to a pallet in the Shuttle cargo bay.

Two astronauts worked together to perform the baseline assemblies. One usually worked from foot restraints at the base of the structure (low man). The other astronaut, for whom no body restraint devices were provided, worked at the top of the structure (high man). Thus, the necessary leverage needed by the high man for an assembly task could be gained only by holding on to the structure. In order to move to the various work positions, the high man was required to translate manually along the structure being assembled. Photographs of the EASE structure being assembled by the baseline method in neutral buoyancy ground tests and on-orbit are shown in Fig. 5.

(a) Neutral buoyancy

(b) On-orbit

Fig. 5 EASE baseline (unrestrained astronaut) EVA assembly tests.

In addition to the baseline experiment performed during the first EVA day, some assemblies were performed in an expanded experiment during the second EVA day. The Remote Manipulator System (RMS) and Manipulator Foot Restraint (MFR) were used in the expanded experiment to position the normally "free-floating" high man so that both astronauts were restrained while exerting assembly task forces. A photograph of an on-orbit assembly using the RMS/MFR is shown in Fig. 6.

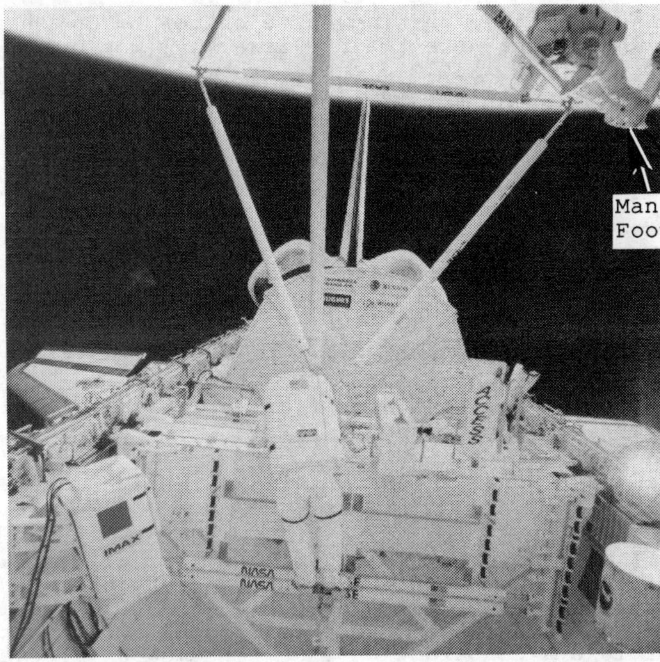

Fig. 6 EASE on-orbit RMS/MFR (restrained astronauts) EVA assembly test.

Primary results and observations:

1. Unrestrained EVA operations, including manipulation of large masses, are feasible though significantly more difficult and fatiguing than restrained.

2. On-orbit assemblies with one astronaut unrestrained were, on the average, 18% faster than comparable neutral buoyancy assemblies.

3. Although RMS/MFR assembly operations are much easier for the astronaut to perform, comparative analysis shows that these operations are only marginally faster than unrestrained operations.

Astronaut participation and appraisal.- Astronauts Jerry Ross and Sherwood Spring performed eight baseline assemblies on-orbit during the STS 61-B mission. Six EASE baseline assemblies performed in neutral buoyancy during an integrated simulation one month prior to flight were used for comparison with the flight data. One on-orbit assembly was also performed using the RMS/MFR to provide fixed body restraint for all assembly tasks. The following comments are taken from the postflight debriefing December 11, 1985, 8 days after the mission:

[baseline (unrestrained) assembly]

1. Ross: "We think that EASE is a practical way to do a limited task...we think that the approach of having both people in fixed foot restraints is a much more viable method of doing construction tasks. We found the free-floating person trying to maintain body position while torquing a 60-pound tube around and aligning and attaching to be extremely fatiguing in the forearms, hands, and wrists."

2. Spring: "I found the task of picking the beam up or being handed the beam and aligning it, free floating in orbit, to be extremely fatiguing...more so than the water tank by a fairly large factor."

3. Spring: "I think the learning curve on how to do these [strut manipulations during unrestrained assembly] was pretty steep. The first couple were a little bit easier because you were stronger, and then you got tired fairly rapidly and you started using your head a little bit more. I think I was learning pretty rapidly and at about the same time I was getting tired rapidly."

[RMS/MFR (restrained) assembly]

5. Spring: "...working off the MFR is easy. You get handed the beam, you can put two hands on it, or even one, but you're working out of a fixed foot restraint."

6. Ross: "The top guy could keep ahead of, or up with, the bottom guy during the MFR build. On many of the free-floating assemblies, the bottom guy tended to be ahead of the top guy, even though he was handling everything...time and effort both are multiplied considerably for unrestrained."

<u>Assembly Concept for Construction of Erectable Space Structure (ACCESS)</u>.[9,10]

<u>Description</u>: The objectives of ACCESS were to: (1) evaluate an assembly line technique for effective use of astronauts as space construction workers, (2) provide on-orbit data for correlation of assembly rates and techniques developed in neutral buoyancy simulations; (3) gain on-orbit EVA construction experience; and (4) evaluate assembly, handling, and maintenance of a large space structure in support of Space Station development. To meet these objectives ACCESS was divided into two parts, a baseline experiment, and an expanded experiment. As indicated in Fig. 7, the baseline experiment involved two astronauts working from fixed foot restraints, and the expanded experiment involved one astronaut working from a foot restraint (MFR) attached to the end of the RMS arm.

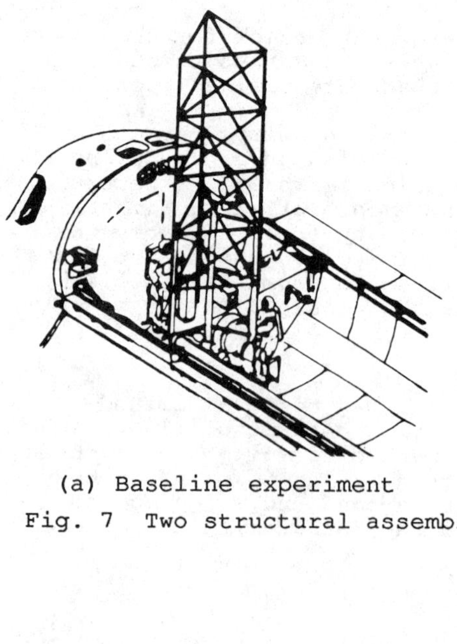

(a) Baseline experiment

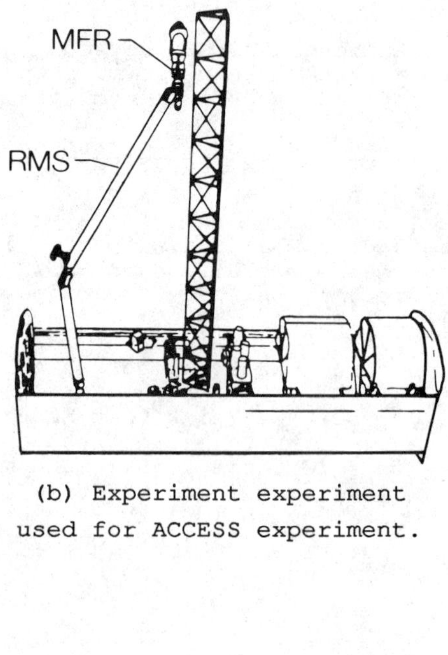

(b) Experiment experiment

Fig. 7 Two structural assembly methods used for ACCESS experiment.

For the baseline experiment (EVA-1), the astronauts assembled the entire 10-bay truss (93 struts) from fixed foot restraints. Fig. 8 shows the baseline experiment being performed in neutral buoyancy and on-orbit.

(a) Neutral buoyancy (b) On-orbit

Fig. 8 ACCESS baseline structural assembly tests.

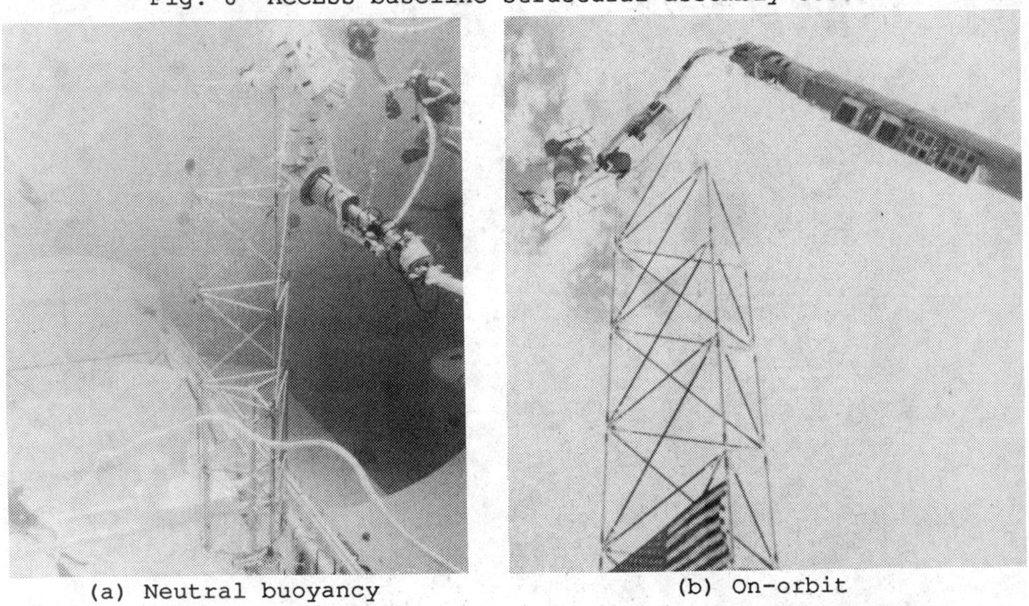

(a) Neutral buoyancy (b) On-orbit

Fig. 9 ACCESS assembly using Manipulator Foot Restraint.

For the expanded experiment (EVA-2), the two astronauts assembled nine bays of the truss using the baseline method. The rest of the assembly activities were performed from the MFR. The assembly tasks performed from the MFR are: (1) assembly, disassembly, and reassembly of the topmost (10th) bay (Fig. 9), (2) installation of simulated electrical cable (Fig. 10), (3) replacement of a strut and a node in an interior bay to simulate truss repair (Fig. 11), and (4) detachment of the truss from the assembly fixture, manual manipulation of the truss (Fig. 12), and reattachment to the assembly fixture.

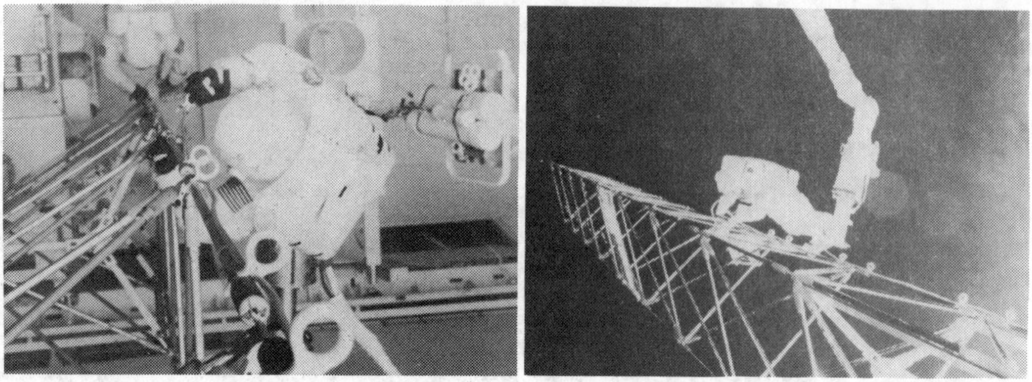

(a) Neutral buoyancy (b) On-orbit

Fig. 10 ACCESS installation of simulated electrical cable.

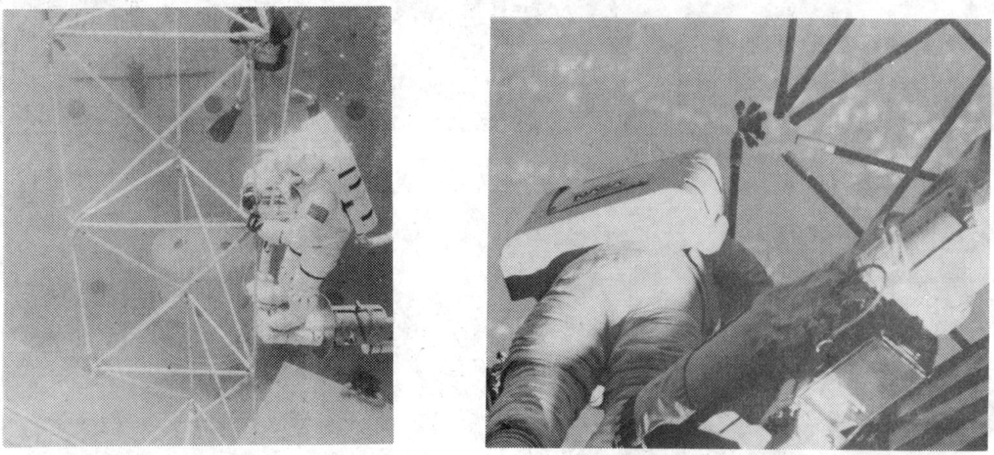

(a) Neutral buoyancy (b) On-orbit

Fig. 11 ACCESS removal of strut and node to simulate truss repair.

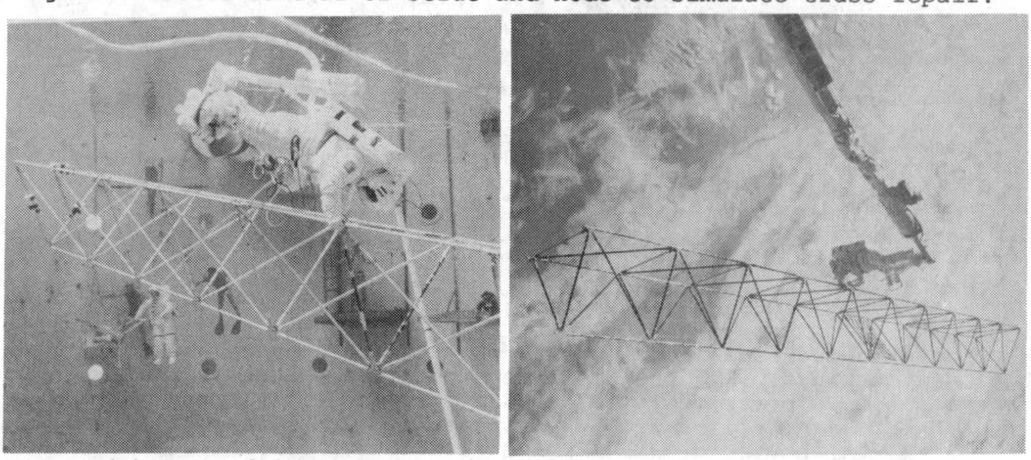

(a) Neutral buoyancy (b) On-orbit

Fig. 12 ACCESS experiment truss manipulation.

Primary results and observations:

1. All planned EVA space construction tasks were accomplished successfully.

2. Approximately 500 pieces of untethered hardware, ranging in mass from a few ounces to nearly 200 pounds, were handled on-orbit without mishap.

3. The average assembly time for the 45-foot truss was 21.7 min (4.3 struts/min) in neutral buoyancy, and 25.5 min (3.6 struts/min) on-orbit.

4. The work effort level to handle ACCESS size hardware was essentially the same as experienced in neutral buoyancy.

Astronaut participation and appraisal: Astronauts Sherwood Spring and Jerry Ross performed the neutral buoyancy tests as well as the STS 61-B mission flight test. The following comments were made at the post flight debriefing of the crew members:

[hardware]

1. Ross: "The sleeves [locking sleeves used to attach the strut to the nodal joint] didn't slide as well as they did in the water. They wouldn't [automatically] pop on, always."

2. Spring: "...when we first started working in the water at Marshall [MSFC] almost all the joints went together about the same way that they went together for us in space [the locking sleeves had to be closed manually]. It's just, after working that hardware for a couple of thousand iterations, we got it nice and loose where it would just [automatically] snap in [as it was designed to do]...but we ended up with new hardware again [on-orbit], and it was just like the first few times we worked at Marshall.

3. Ross: "Raising the bays [sliding the bays up the assembly fixture] went fairly routinely. They seemed to slide quite nicely. Bringing them down [during disassembly], there seemed to be a little bit of stiction that you had to overcome before they would start sliding down."

5. Ross: "The question was asked about the condition of the gloves after the first EVA. We could see very little wear...the abrasion level on the palms themselves was very, very minimal and for as much hand work as we did, I think that's a high compliment to that hardware."

[baseline assembly work effort]

6. Spring: "[During ACCESS did you ever feel overworked or particularly uncomfortable?] I think the answer is no, but ACCESS still requires a considerable amount of hand work, especially in the lower position ...nothing on ACCESS worked us any harder than [in] the water...I thought there was a good comparison to the water."

[tethering]

7. Ross: "We were not as cavalier when we were bringing struts out [of the canisters] as we were in the water tank; i.e., we were a little more cautious to make sure we didn't lose any struts...I guess the bottom line is it didn't seem to change the times that it took us to do things significantly."

8. Ross: "I think ACCESS times would probably at least have doubled, and the hand fatigue would have probably tripled, if we'd had to tether [with standard wrist tether hooks]."

[RMS/MFR assembly]

9. Ross: "...the MFR was used to demonstrate the capability to add on something to an end of a structure that's already in place, or the addition of attachment points for some hardware that you want to mount...it was not an efficient way to build bays of a truss such as that."

10. Spring: "ACCESS was designed to be worked out of a fixed foot [restraint] station. It worked beautifully there. We did it off the MFR to show that it could be done more as an add-on to something that's already existing. That's not the way to build ACCESS. The way to build ACCESS is the way it was designed."

[truss manipulation]

11. Spring: "Manipulating it was no problem at all. We were able to grab hold of it, rotate it, and translate it. At one point I got [both] a translation and a rotation in it. I think as long as we keep the rates down, we could do quite a bit more [than we were required to do with ACCESS]."

12. Ross: "I found ACCESS extremely easy to maneuver, both from the middle and from the ends. I could point it wherever I wanted to; I could start it and stop it whenever I wanted to. It felt very stiff--very rigid."

SPACE STATION MISSION

Results from the basic research in EVA construction, previously summarized herein, are being applied in the development of Space Station truss hardware and EVA construction techniques. In addition, the NASA LaRC has an ongoing, independent, in-house research program to study an assembly concept that could not only be used during initial Space Station build-up, but also for maintenance, repair, and growth activities after Space Station is operational. A general description of the current NASA proposed Space Station construction scenario is presented in this section along with the LaRC alternate construction scenario.

Baseline Construction Scenario

A phased approach to Space Station construction has recently been adopted by NASA to provide permanently manned capability in the mid 1990s. The Revised Baseline Configuration for Space Station from this phased approach is shown in Fig. 13. The basic structure is an erectable truss boom 110-meters long with a 5-meter square cross section. The current construction scenario is to use astronauts in EVA to assemble half the boom and associated subsystems during Flight 1 and the other half during Flight 2. The first pressurized module, a laboratory, would be brought up on Flight 7 to enable man-tended operations, although the Station would not be permanently manned until Flight 11. The full Revised Baseline Configuration would be completed during Flight 20. Flights 1 and 2 are critical because they not only require extensive EVAs, but the successful completion of virtually all of the scheduled construction tasks is essential to the mission. The interim configurations left on-orbit following Flights 1 and 2 must be autonomous with fully functional subsystems for reaction control, thermal control, power generation and distribution, etc.

Fig. 13 Space Station Revised Baseline Configuration.

The current NASA construction scenario proposes that the Orbiter be used as a construction base and two EVA crews of two astronauts each perform the assembly tasks. Only 24 EVA man-hours are currently baselined per flight. Up to 24 additional EVA man-hours per flight are reserved for growth and contingency activities. The structure would be assembled on an assembly fixture which, itself, is deployed or assembled and attached to the cargo bay and the truss would be moved on guide rails vertically out of the cargo bay assembly area as each bay is completed. Upon completion of the Flight 1 assembly activities, the assembly fixture would be separated from the Orbiter and left on-orbit, attached to the interim Space Station structure. On Flight 2, the Orbiter would dock with the assembly fixture/truss system to continue the vertical build out of the cargo bay. At the end of the Flight 2 assembly activities, the assembly fixture would, again, be separated from the Orbiter and left on-orbit attached to the Station structure to become the transporter for the Manned Servicing Center (MSC) to be brought up on a later flight.

Alternate Construction Scenario

This concept uses a transporter for moving building material and payloads and also serves as a positioning device for EVA astronauts and payloads. The conceptual design was first reported in reference 11 and hardware has been fabricated for 1-g and neutral buoyancy ground tests to further explore the feasibility for the construction activities planned for Flights 1 and 2. This concept is discussed briefly herein to point out some attractive features not mentioned in reference 11 that may be useful for Space Station construction.

Fig. 14 is a series of computer generated pictures showing the various stages of truss assembly using the transporter system. Fig. 14(a) shows the transporter, complete with strut/node stowage canisters, guide rails, and astronaut positioning arms folded for stowage and attached to specially designed support pallets that would be mounted in the Shuttle cargo bay. A deployment mechanism is shown in the right hand side of the picture and a 12-foot diameter rotary joint is shown at the left side. Before deployment, the rotary joint (and other appropriate payloads) would be attached to payload positioning arms on the backside of the transporter (probably by RMS operations). The transporter would be remotely deployed prior to EVA operations. The transporter is shown

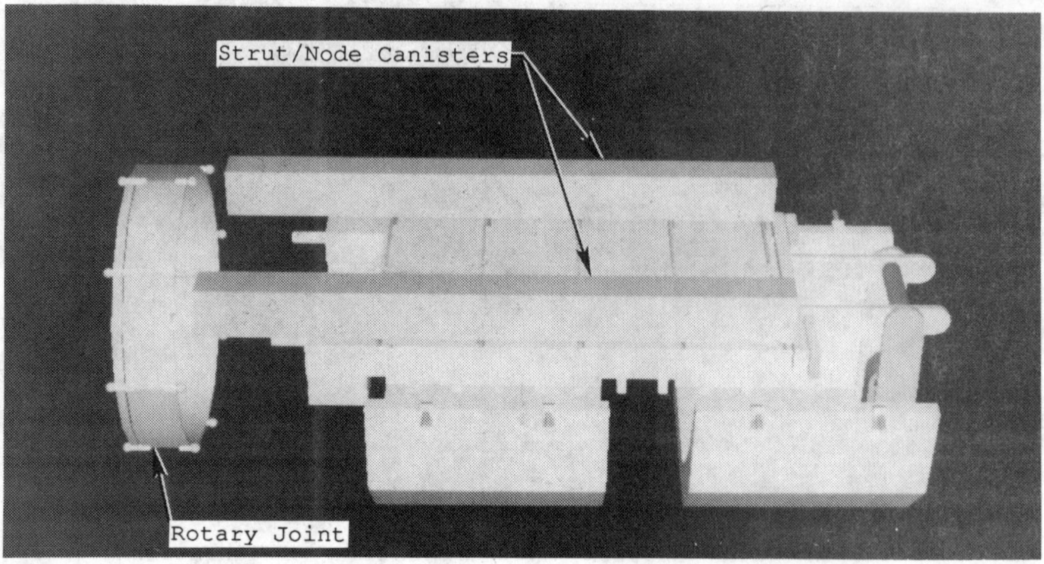

(a) Stowed configuration

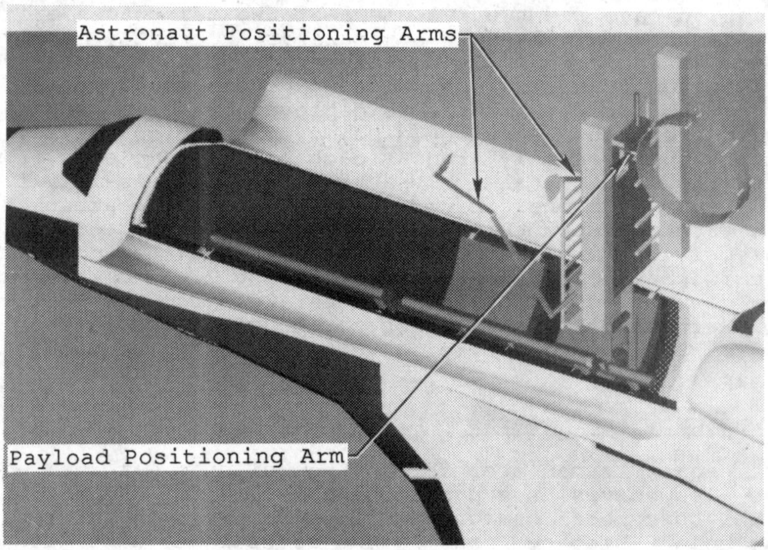

(b) Deployed configuration

Fig. 14 LaRC Transporter assembly concept for Space Station.

fully deployed in the Shuttle cargo bay in Fig. 14(b). Following transporter deployment, a transition truss is erected by the EVA crew. Then, self-deploying utility trays are installed using RMS operations and deployed one or two bay-lengths by critically damped springs at the hinge lines (Fig. 14(c)). The EVA crew would initiate and control the utility tray deployment by manipulating the appropriate latches or pins. The transporter "walks" along the truss (Figs. 14(d), 14(e), and 14(f)) by alternately extending and retracting a drawbar while the astronauts, attached to the positioning arms, assemble the structure around the utility trays. Fig. 14(g) shows the rotary joint being moved into position by the positioning arm. The EVA crew then installs the support struts (Figs. 14(h)).

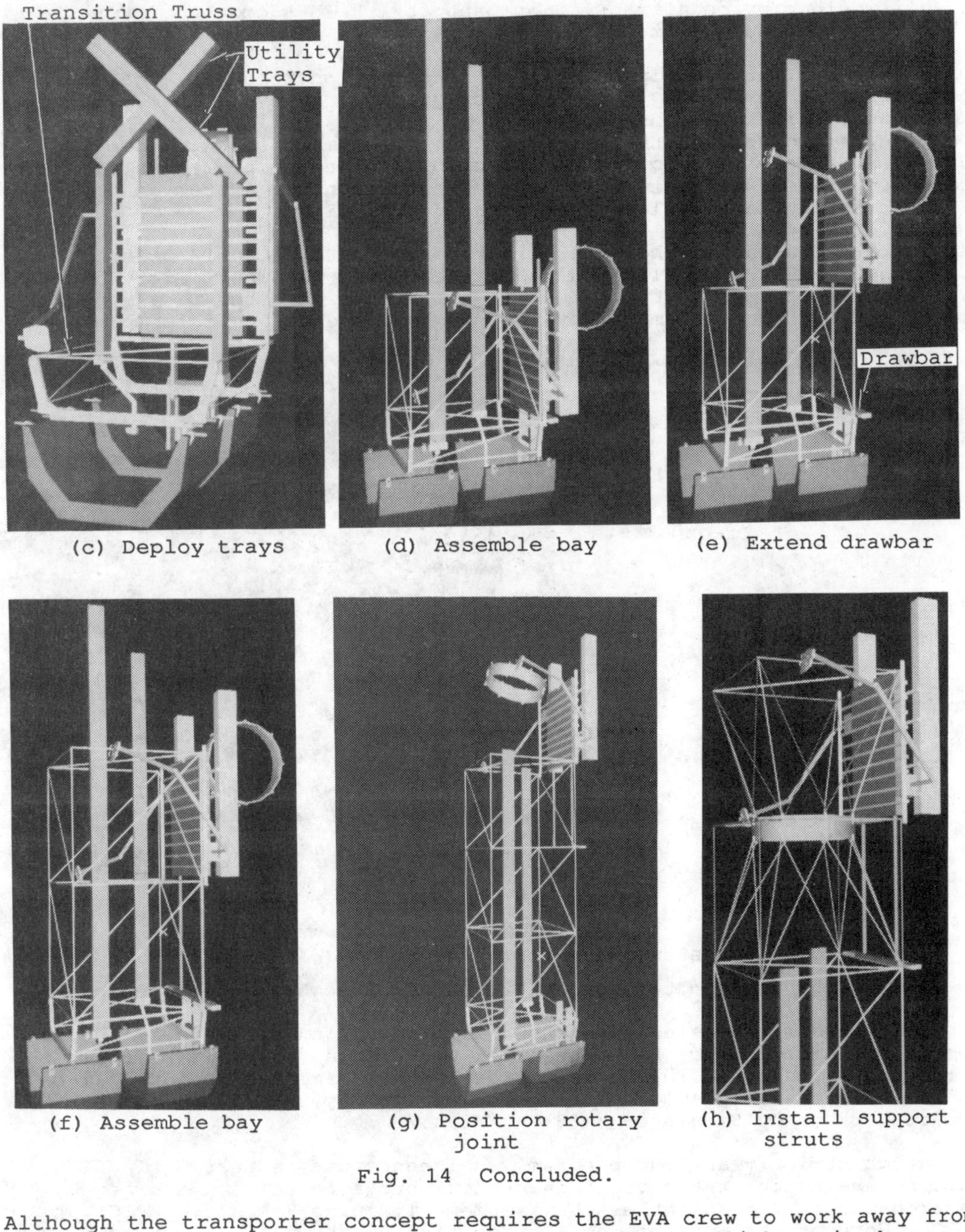

(c) Deploy trays (d) Assemble bay (e) Extend drawbar

(f) Assemble bay (g) Position rotary joint (h) Install support struts

Fig. 14 Concluded.

Although the transporter concept requires the EVA crew to work away from the cargo bay during the first two Station build-up flights, it does not impose requirements on the EVA crew that will not have to be addressed on later flights. This scenario, wherein the truss is connected to the Orbiter through a transition truss by standard trunnion mounts provides a stable interface between the truss and the Orbiter and may also facilitate jettison capability for emergency situations. The truss itself would provide a translation path for the EVA crew to return to the Orbiter in such situations.

Under the current NASA Space Station construction scenario, the entire boom is built vertically out of the cargo bay. However, it may be more advantageous in the transporter concept for the continuation of the build-up scheduled for the second flight to be made spanwise out of the cargo bay. This may require roll control of the Orbiter/truss system to be accomplished with the aid of the reaction control system on the Station structure installed during the first flight.

Hardware representing the transporter concept has recently been fabricated at LaRC for 1-g and neutral buoyancy testing. Fig. 15(a) is a sketch of the test setup. The transporter is attached to a support tower and the truss is moved by the drawbar as each bay is completed.

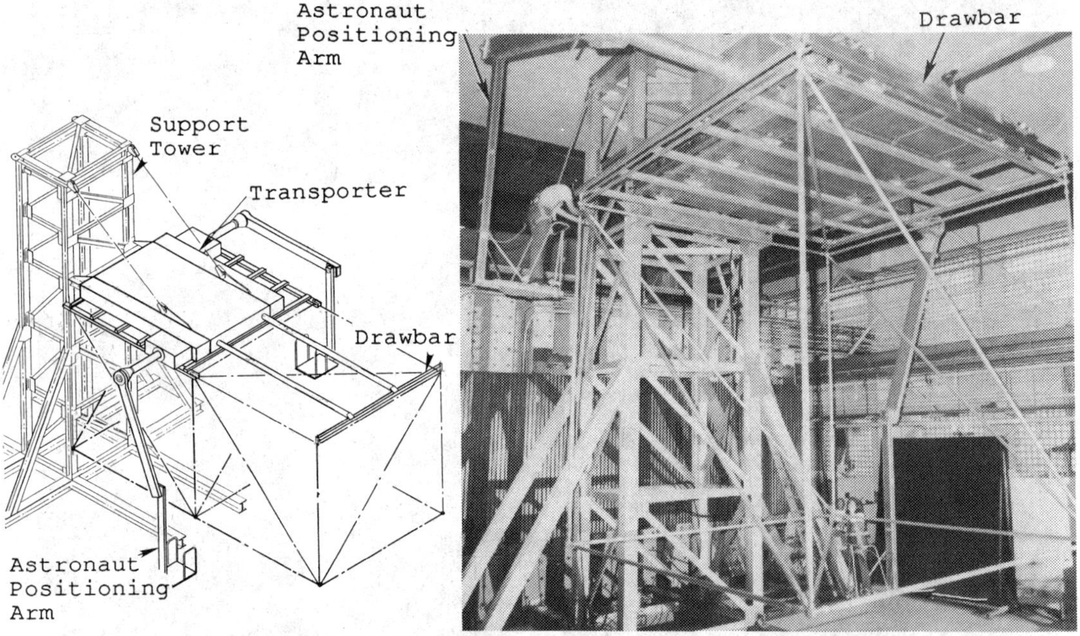

(a) Test setup (b) Hardware

Fig. 15 Transporter concept for truss assembly ground tests.

This simulates the relative motion which occurs on-orbit where the truss would be immobile and the transporter would move. Fig. 15(b) shows the transporter hardware in the laboratory. This same test setup will be used in neutral buoyancy tests scheduled for May-June 1988 in the Neutral Buoyancy Simulator (NBS) at MSFC.

Although strictly an LaRC program, assistance was requested of JSC to aid in evaluating the concept in neutral buoyancy tests performed by astronauts in the Weightless Environment Training Facility (WETF) at JSC. However, because of the limited size of the WETF, the same hardware cannot be used. Instead, a simulated transporter was designed and fabricated at LaRC specifically for the WETF. A sketch of the test setup is shown in Fig. 16(a) and a photograph of the hardware during fabrication is shown in Fig. 16(b). To simulate the relative motion that occurs between the transporter and the truss during assembly, the simulated transporter is held stationary at the bottom of the WETF, and the truss is moved as each bay is completed. The astronaut positioning

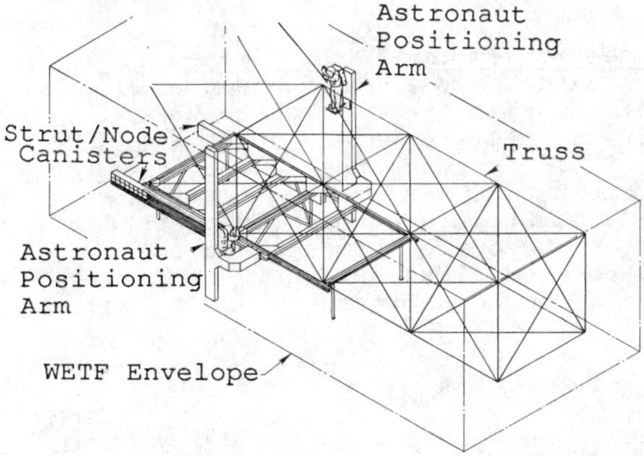

(a) Test setup

(b) Hardware fabrication

Fig. 16 Simulated transporter concept for truss assembly ground tests

devices are towers that move horizontally one bay-length, and the astronaut foot restraints move vertically on the towers one bay-length. Neutral buoyancy testing using this hardware in the WETF is scheduled to begin in December 1987.

CURRENT SHUTTLE EVA FLIGHT RULES[12,13]

To enable the extensive EVA required for Space Station assembly, Orbiter modifications as well as reevaluation and revision of flight rules that currently apply to EVA and Shuttle operations may be required. The current rules are based on Shuttle capabilities, available consumables, crew and Orbiter safety, and the particular mission requirements. Some of the current rules which directly impact EVA construction are listed below.

EVA Time

A major concern for Space Station construction is the EVA time required for assembly. The following rules will impact the allowable EVA hours:

1. For planning purposes, the maximum duration for an EVA is 6 hours.

2. The maximum duration for a Shuttle mission is 10 days.

3. For planning purposes, no scheduled EVA is permitted until flight day four. However, on a mission by mission basis, waivers can be granted to plan and conduct an EVA on flight day three.

4. A preflight planned EVA may be scheduled no later than two days before the end of the mission. This rule reserves one standard flight day for preparation for return to Earth.

5. For planning purposes, at least one flight day must separate two scheduled EVAs for a given crewman. This rule was waived in real time on STS 51-I during which LEASAT was repaired by EVA crewmen James Van Hoften and William H. Fisher. The first EVA was performed on August 31, 1985; a second EVA was performed the following day, September 1, 1985.

Crew and Orbiter Safety

<u>EVA tools will be tethered to the crewmember at all times during their use</u>. This rule prevents loss of hardware and the possibility of free-floating hardware to collide with the Orbiter on the current or subsequent orbits. This rule was waived for the ACCESS experiment on Shuttle mission STS 61-B because the ACCESS components were small and sufficient time was available for the structure to be disassembled and the Orbiter maneuvered to take it out of a collision course with lost hardware. Space Station, however, will not be maneuvered to evade lost hardware, thus improved tethering techniques that reduce astronaut hand fatigue should be developed.

Astronaut Appraisal

During the STS 61-B mission, astronauts Sherwood Spring and Jerry Ross performed EVA space construction tasks associated with the EASE and ACCESS experiments. The following comments, taken from their postflight EVA debriefing, address Space Station technology transfer.

[unrestrained assembly]

1. Ross: "I think the way to go is in foot restraints of some sort."

[tethering]

2. Ross: "We made some inputs to the safety committee very early requesting that we not tether the ACCESS components. They were originally designed to have tethers on everything, and we were very thankful that they gave us clearance not to have to tether all that...we did not tether, by the way, either the completed EASE or the ACCESS [structures during manipulation]...we didn't want to have the crewman as the bungee in between two massive objects should we have some kind of a problem with the Orbiter and we wanted to separate."

3. Ross: "If we have to tether the hardware that we're going to be assembling, number one, that ought to be eliminated, if at all possible, and, if not, then it ought to be in such a manner that it's easy--like a harpoon. Doing the wrist tether kind of bit is very fatiguing to the hands, the way you've got to squeeze both buttons and squeeze the hook to open it. That gets you real quick."

[pressure suits (EMUs)]

4. Ross: "...this is the first time we've had completely interchangeable suits ...I think the overhead, in terms of being able to resize the suits, is minimal...I think we gave the Crew Systems folks the confidence that the crew can do that kind of thing on-orbit, with a high degree of confidence it'll be done properly. And, quite frankly, I don't know why we don't make that the baseline."

[handling and material supply]

5. Spring: "I think we could have handled bigger structures. It's not inconceivable [that] we could be working 12 to 15-foot sections with about the same ease...as long as he's [the astronaut] got something he can work against, whether it's a foot restraint or a three-point hardmount that's attached to his belly...as long as he's got something he can work from, I don't think you're going to have a particular problem maneuvering pieces within reason."

6. Spring: "I think the biggest problem you're going to have when doing something like this is presenting new work material to the astronaut so you can keep him working...not the astronaut being able to manhandle and manipulate things into position."

[EVA time]

7. Spring: "I don't think we're at the limit yet [on how long an EVA crew can perform this type of work without a break], although I thought that the 6-hour EVA every other day for quite a long period of time would probably be pretty good."

8. Ross: "We went nearly 7 hours on the second EVA, and I think we both felt like we had put in a pretty decent day's work. But we could have gone longer...8 hours, I think, is not a bad time in a suit. We ended up being in the suit almost 8 hours on that second one [EVA]."

9. Ross: "I think 7 or 8 days [of EVA by an individual, every other day] would be a pleasure."

10. Spring: "I would sign up for that [one set of two guys doing three or four EVAs, every other day, on a single Shuttle flight]."

CONCLUSIONS AND RECOMMENDATIONS

Two general conclusions can be drawn from the EVA assembly tests and astronaut comments summarized herein:

1. EVA is a viable means to assemble Space Station truss structure.

2. Use of positionable foot restraints by the EVA astronauts is required for efficient EVA assembly.

To reduce risks associated with EVA and instill confidence in the astronaut corps for carrying out the assembly activities required for Space Station, the following recommendations are made:

1. A new neutral buoyancy facility, larger than the WETF or NBS, should be built. It should be large enough to accommodate mockups of interim Space Station structure and utilities being assembled on-orbit and simultaneously accommodate mockups that enable astronaut training for future flights.

2. Future Shuttle flights, where practical, should schedule EVA's. It is important for as many astronauts as possible to obtain EVA experience before Space Station construction begins.

3. The capability to accommodate two EVA crews per Shuttle flight during Station build-up should be provided.

4. An integrated system flight test that focuses on key Space Station assembly tasks should be conducted.

REFERENCES

1. Mikulas, M. M., Jr.; et al: Space Station Truss Structures and Construction Considerations. NASA TM 86338, Jan. 1985.

2. Mikulas, M. M., Jr.; et al: Deployable/Erectable Trade Study for Space Station Truss Structures. NASA TM 87573, July 1985.

3. Mikulas, M. M., Jr. and Bush, H. G.: Design, Construction, and Utilization of a Space Station Assembled from 5-Meter Erectable Struts. NASA TM 89043, Oct. 1986.

4. Stinson, R. G. and Thomas, L. C.: Space Station Truss Assembly Fixture Simulation Final Test Report. NASA JSC-22125, Mar. 1987.

5. Thomas, L. C.: Space Station Node/Strut Cluster Evaluation Final Test Report. NASA JSC-32045, Aug. 1987.

6. Heard, Walter L., Jr.; Bush, Harold G.; Wallsom, Richard E.; and Jensen, J. Kermit: A Mobile Work Station Concept for Mechanically Aided Astronaut Assembly of Large Space Trusses. NASA TP 2108, Mar. 1983.

7. Watson, Judith J.; Heard, Walter L., Jr.; and Jensen, J. Kermit: Swing-Arm Beam Erector (SABER) Concept for Single Astronaut Assembly of Space Structure. NASA TP 2379, Mar. 1985.

8. Akin, David L.: EASE, Experimental Assembly of Structures in EVA, Overview of Selected Results. NASA CP 2490, pp. 199-218, Oct. 1987.

9. Heard, Walter L. Jr.: ACCESS (Assembly Concept for Construction of Erectable Space Structure). IEEE 18th Electronics and Aerospace Systems Conference, EASCON '85, pp. 157-165.

10. Heard, Walter L. Jr.; Watson, Judith J.; Ross, Jerry L.; Spring, Sherwood C.; and Cleave, Mary L.: Results of the ACCESS Space Construction Shuttle Flight Experiment. AIAA Space Systems Technology Conference, A Collection of Technical Papers, San Diego, CA, pp. 118-125. Copyright 1986.

11. Bush, Harold G.; Mikulas, Martin M., Jr.; Wallsom, Richard E.; and Jensen, J. Kermit: Conceptual Design of a Mobile Remote Manipulator System. NASA TM-86262, July 1984.

12. STS Operational Flight Rules. JSC-12820, Sept. 1, 1987.

13. Crew Scheduling Constraints. JSC-22354, Mar. 1987.

DISCUSSION

<u>Dr. Edwin E. Speaker</u>: Let me see if our discussion panel has any specific questions for Walter or his co-authors?

<u>Question</u>: What's going to be the largest technological hurdle to overcome, the assembly of the truss itself, the deployment of the utilities, the rotary joint? What is the weak link?

<u>Walter L. Heard, Jr.</u>: Well, there are a lot of weak links there. Assembling the truss, I think if you ask Jerry and Woody, is probably no problem. That's probably the simple thing to do. The hard thing to do is put in utilities and put in the connected payloads. I showed a scenario, where you might deploy utility trays out ahead of the build, which will be a good thing to do if you can do that. With that method the trays could be longer than one bay in length, too. And it's nice to have fewer joints; the fewer joints you have, the better off you are. But you still have to connect up the fluid joints inside those trays, and they have to be leak-proof. So, how do you do that? That's a big challenge right there. There may be some other problems, too, with thermal control of the transporter, if you're talking about the transporter concept -- that's crawling out on the truss. We just have to see which way is the best way to go. We like the transporter a lot at Langley. Now, I can't say that anybody else likes it.

<u>Dr. Edwin E. Speaker</u>: Questions, anyone? Here's a question to Walter Heard.

<u>Question</u>: If the assembly tasks are really repetitive and boring, have we looked at totally automating the process? Are we looking for tasks for manned operations versus automating?

<u>Walter L. Heard, Jr.</u>: Well, I think there is a role for automation. Whether we want to wait for it and spend the money for it right now is an option. I think we have the technology, to build a Space Station, now. We can do it. EVA and as we go along, automation technology will develop. I also think EVA technology will develop. I don't think we are going to stand still there. For the dangerous tasks, we need to develop the automation technology. But if we want to build a Space Station, we need to go with what we can do now. I'm glad our early pioneers didn't wait for the chainsaw to be invented.

<u>Robert F. Thompson</u>: Gentlemen, I would just like to make a comment, that I think we ought to be a little careful about taking on a group of tests on hardware and conclude that building the real Space Station at the real point in time would be boring. I don't think it would be boring at all. I think we shouldn't jump to that conclusion based on these kinds of tests.

<u>Jerry L. Ross</u>: Bob, I think Woody and I couldn't agree more with you. We certainly weren't bored on our mission. So, I don't think boredom is going to be a problem.

<u>Answer?</u>: My own feeling, too, is you wind up using some combination of techniques, a mixture of erectables and deployables and there will be some others that are more automated and others less automated.

<u>Dr. Edwin E. Speaker</u>: Any other questions? Over here, Lyle?

Question: Do you feel like we need more flight testing before we commit to the FEL? FDL?

Walter L. Heard, Jr.: If I were an astronaut, I would want more training.

We don't have a big enough neutral buoyancy facility to do adequate training. Now, maybe we are going to get one; we need one. Yes, I think we need a flight test to address those key elements. We have to identify what the key elements are. I would like to have Jerry and Woody address this. Nobody has to tell the astronaut to go train, they do extra training because they have to perform up there. Nobody wants to fail, least of all the astronauts. They want to have the confidence that they are not going to do something that they have not done before.

Sherwood C. Spring: I think Doug mentioned the most important point on that. We really need to see what we are going to build when we name the contractor and know what the hardware looks like. Then we will see what the problem areas are going to be. Hopefully, we will try those things in the water tank, and then we would be able to get a better feel for the tasks. But I do believe there will be some type of a space flight verification before we go for the final construction.

Dr. Edwin E. Speaker: Are there any other questions?

Question: I was wondering whether Jerry and others have any sensitivity about the physical dimensions of the structure and whether that changes any of the work, for instance, if we went from 12 to 15 to even maybe 18 Ft. Strut assemblies and the total distances away from the Orbiter -- just their own projection of how it might affect some of the thinking.

Jerry L. Ross: I don't think that the physical dimensions of the hardware will be much more challenging compared with what we've already covered. The masses of the hardware that we will be dealing with on Station, to my understanding, will be significantly smaller than what we have already dealt with, since the beams that we were handling on EASE are around 64 pounds each. As far as getting farther away from the Orbiter, that only improves the view, as far as I'm concerned. There's less vision that's being blocked. One concern that Woody and I have expressed to our crew systems people at JSC is that we were putting out some fairly heavy BTU levels. On the first space walk, I was putting out an average of over 1600 BTUs an hour and Woody was up over a thousand. Yet when the sun went down, as we got above the cargo bay on the end of the manipulator arm or even sometime as we went over the sill, we could feel a very definite cooling trend in the suit. Most of the time, even with our high BTU output levels, we were operating our cooling systems in the suit at fairly low demanding levels. So, once you get above the Orbiter and you're radiating to deep space from all aspects (and hopefully the task may not be quite as physically demanding as what we demonstrate with EASE and ACCESS), then you have a very strong potential problem of not being able to maintain proper body warmth inside the suit. So that's something we are going to look at. And I know that the guys back home are looking at that right now and investigating whether or not we have to somehow modify the cooling system in the suit or provide some kind of heating capabilities as well. That is something, that very definitely will have to be looked at also for the Space Station suits of the future.

Dr. Edwin E. Speaker: Any other questions? Walter, Kitty, Jerry, and Woody, I want to thank you. You have one more in closing.

Jerry L. Ross: Ted, I'd like to make one more comment. I'm sorry I wasn't here this morning, but I understand that you got a chance to see some video of the EASE activity from our mission. I hope you saw, from Doug's presentation here today, that EASE was specifically designed for guys to be in a free-floating manner, to see what that meant in the performance of tasks at a desirable level. ACCESS was designed for both guys to be in foot restraints so that they could do the job in a proper fashion. We do not advocate the EASE method of doing work except for very specific applications of short duration, where you might want to bridge a gap between two larger structures and the only way to get out there is to shinny up the pole. We told the principal investigators from MIT repeatedly, as they were developing their hardware, that this is not the proper way to attack this task. All the management at JSC asked, why are we doing this? But I think we did get some very valuable engineering data out of it. We did demonstrate that it is capable of being done, but for very specific applications. You certainly want to have your feet in foot restraints so that you don't get the fatigue factor. When you're free-floating up there and clutching very securely the structure with one hand while you're manipulating these 64-pound beams around with the other hand, trying to align those and maintain body orientation while you're doing that, and then making the connection, it's not an easy thing. And the BTU levels that were being expended were pretty much from the elbows through the finger tips. So you can understand why they become fatigued very quickly.

Likewise, if the structures that we assemble for the Space Station happen to be composed of graphite-epoxy beams that have very fragile outer surfaces, first, we have to remember if there is a way for those guys to crawl around those beams when they are out there assembling the station or maintaining it in years to come. They are going to do it. So you have to have some integrity in that system that people going across it in a space suit will not cause deterioration. And second, we must put the guys in foot restraints; they will then be able to handle materials in a controlled fashion and they won't be beating them, one against the other, and you will have a station that will come out in good shape.

Dr. Edwin E. Speaker: Thank you very much, gentlemen. I appreciate your contribution.

Our next paper, titled, Deployable Truss Support Structures for Extremely Accurate Reflectors, is to be presented by Dr. John Hedgepeth, who is a consultant with Astro Aerospace Corporation, located in California and he has come a long way. John, the podium is yours.

AAS 87-687 to -690

SPACE STATION/LARGE STRUCTURES - PANEL DISCUSSION

Panel Moderator: Edwin E. Speaker
Panelists: S. Z. Rubenstein*[*]
Robert F. Thompson[†]
Allen J. Louviere[‡]
Charlie Blankenship[**]

Dr. Edwin E. Speaker: I'm going to ask our distinguished panel and discussion group if they want to make any general statements about the topic that we have been discussing today. Let me start with Sy Rubenstein.

S.Z. Rubenstein: We have seen in both discussions that the structural problem is an interesting problem. You have to consider it coupled overall with all the equivalent secondary systems that have to go aboard. Depending on the application, it could be just electrical feeds or electrical, thermal, and fluid systems, and I think that is the way you have to really look at it. For the Space Station application, the interesting aspect which always intrigues me is that what you are putting together is a system of partially configured systems. This brings up some interesting problems in analysis and certification and check-out between system stages and then how it operates, either in structural failure modes or system failure modes and where we start out with one system, we may wind up with 20 or 30 for analysis. That should be a consideration as you go through the design process and more important than that, is finding a way of keeping these systems separated or uncoupled, whether it is in a frequency spectrum or in an electrical sense. I think what we have seen earlier in the EVA world is not something fully deployed or fully erected but some combination of mixed capability between a partially assembled or preassembled system that will mix, most likely deployable utilities and erectable structures. I have never seen, in my experience, a system that is purely one form. So that is a set of initial comments.

* President, Space Station System Division, Rockwell International Corporation, 12214 Lakewood Blvd., Downey, California 90241.

† Vice President - General Manager, Space Station Division, McDonnell Douglas Space Systems Company, 5301 Bolsa Ave., Huntington Beach, California 92647.

‡ Senior Vice President, Engineering and Operations, Space Industries, Inc., 711 West Bay Area Blvd., Webster, Texas 77598.

** Director of Structures, NASA Langley Research Center, Langley Station, Hampton, Virginia 23665.

Dr. Edwin E. Speaker: Let me ask Bob Thompson if he has any comment.

Robert F. Thompson: Let me just make an observation. It's boot-strapping a little bit on what Sy was saying. I think, after listening to some of the discussions here, both this morning and also this afternoon, the purpose of some of these symposia is to get together as a group and discuss things and try to assemble our collective thoughts and then learn from these kinds of meetings. I think we have to be careful in taking any one of these subjects and then try to push it into an "either/or" kind of a confrontation. For example, the structure as part of the Space Station system has to really be looked at in context with the total Space Station system. You have to get the utilities distributed properly. You have to get the various tanks in the right place. You have to get your control modes and gyros in the right place. You obviously have to get your structure in the right place. And, I think, certainly from a program manager's point of view, you don't want to get boxed in to these "either/or" kinds of arguments. I think putting the structure that we have identified for Space Station in place is really not a constraining factor for Space Station. I'm convinced it can be done in a very straightforward manner with a very acceptable EVA and I think you get a lot of versatility having to develop only that one basic way of doing things. I think it's going to be very important that we retain the versatility, because the structure in Space Station is basically the land, if you will, that we are buying for this facility. We are not sure exactly what we want to build in the back or what we want to put on the front of the lot, and so forth. I think we have to leave ourselves a lot of versatility in the way we go about this structure so that we can let it grow it in the right direction, at the right time. We can make either one of these things work. There is a place somewhere in our future for pack trusses. There is a place in our future for EVA systems, certainly, even on Space Station, for things like some antenna booms or some RCS thruster pods, some kind of deployable structure that would make a lot more sense. I think we have to avoid manned versus unmanned, erectable versus deployable, and come at these things with a little bit more of a systems aspect. Retain the versatility you need to manage a program effectively; then the answers will fall out pretty easily.

Dr. Edwin E. Speaker: I would like to ask Al Louviere, if he wants to comment.

Allen J. Louviere: I am presently not associated directly with the Space Station, but I will tell you that I am still an avid fan and a loyal supporter, so you ought take these comments as those of an interested spectator. I don't spend much time thinking about Space Station anymore, but at one time it occupied a great deal of my day. In particular, this problem occupies a great deal of your day if you have to do systems engineering because it is the place where all the systems come together and intersect. It is an enormous task. We treat it rather glibly here, but it has literally hundreds of pieces in it that all have to fit together at some place. It's very sobering. You think about it for a moment and you find all the places and all the pieces that have to fit just right. I used to have a list I carried around with me of everything I would check every day, and I can still remember the list. I want to just mention a few of them: deploy versus erect, now it's going to be both of those things The other factors were two really interesting technical factors. One gentleman this morning said he worried about the thermal, so did I, and the other gentleman, a minute ago, said stiffness. As far as I was concerned, those two were the two driving factors on those two things, plus crew utilization. Our crew is a very precious commodity when you get up there, and I have a feeling that is going to enter into it. The one that used to frustrate me the most, though, was STS performance. When I saw the mission rules, they haven't changed a bit, it's going to be an interesting chore. The one that has not been discussed is the transition from the man-tended version to the manned version. We keep forgetting that we are going to this station untended for a while. So all the gentlemen who talked earlier today about robotics and automation, you ought to listen to them because you are going to have to do a lot of things remote, whether you want to or not. And the list goes on, verification was the other one. So you might say that this particular aspect of Space Station is the intersection of the engineering, the operations, and the logistics, and that's a real clash. It's a driving factor in this whole program and I think in the end NASA and the contractors will have to get greater resolve and higher commitment just to do this. This will be the largest engineering task we have taken on in this country, probably since the Brooklyn bridge or the San Francisco bridge. I want to thank you for inviting me. I was flattered when I was invited, until I realized you sat me between these two gentlemen as a referee. Thank you.

Dr. Edwin E. Speaker: Now, our last panel discusser is Charlie Blankenship, who has been very patient. Charlie, I pass you the microphone and you can make the bottom line wrap-up here.

Charlie Blankenship: Thank you, Ted. John Hedgepeth, for your information, to let you know that your perseverance over the last ten years on a large deployable reflector has not gone unnoticed, it is now officially part of the Civilian Space Technology Initiative Program that Stan Sadin told us about this morning.

We have finally changed the name to get us out of a box and it's now called the Precision Segmented Reflector and that's one of our benchmark programs now in the Civilian Space Technology Initiative. And we already have people dreaming of how we might use the Space Station once it's in orbit, one of the many uses we envision involves constructing a Precision Segmented Reflector about 30-meters in diameter off the Space Station.

I think one other way of looking at what we have seen here today, and again not at an issue of erectable versus deployable, but I think we have seen some progress, due to the perseverance, and ingenuity of very capable people in this country with some foresight over the past ten years, who have diligently conducted research and asked "How would I build a structure in space, would I construct it, would I deploy it, would I erect it?" And they set about in a laboratory and also in neutral buoyancy tanks to learn how to do that, and answer a lot of questions, such that today we have the technology to look at both of those options and decide which makes the best sense. I think that is what we have seen today. Thank you.

Dr. Edwin E. Speaker: Thank you, Charlie, and I would like to just summarize with a comment that I have seen two very interesting papers this afternoon: One was by Walter Heard which recapped, with the assistance of the two mission specialists, who actually participated in the validation of assembly procedures for large structures, and an extremely fascinating way and that, of course, was combined with the films we saw this morning, which I thought were very informative. Then the creative and imaginative work that John Hedgepeth has done to develop new designs for compact packaging and then very stiff and precise structures once they are opened up. I think it has been very, very informative.

I want to thank you, John, and I would like to thank our discussion group. Are there any questions, for either, John Hedgepeth or for our panel of experts from the audience? Very well, and I will declare this session closed. Thank you all very much for your attendance and support.

CONCLUSION: WHERE FROM HERE?

CLOSING PANEL

Chairman:	David C. Honhart Captain, U.S. Navy Office of the Chief of Naval Research
Host Chairman:	Donald K. Slayton President Space Services Inc. of America

The following paper numbers were not assigned:

AAS 87-696 to -699

AAS 87-691 to -695

SPACE - A NEW COMMUNITY OF OPPORTUNITY - CLOSING RECAP: WHERE FROM HERE? -

David C. Honhart*, Donald K. Slayton†, H. Norman Abramson‡, Gentry Lee** and Willis Hawkins††

Donald K. (Deke) Slayton: Ladies and gentlemen, the purpose of this session is to close and recap, and our mission is to skim the cream off all of the previous sessions that have been run. We are going to try to do that, not quite as structured as that, but I think the speakers will be addressing and summarizing most of the major issues that have been brought up. If we have a little time at the conclusion we will do a clean-up if we discover we have not really addressed issues adequately. Our first speaker this afternoon is your president, I know you've seen him on a few other occasions here, so he probably doesn't need much introduction from me, but Capt. Honhart is the director for Space Oceanography, the Naval Research Laboratory Office at the Chief of Naval Research and is president of the AAS and he would like to make a few introductory comments.

David C. Honhart: Thanks, Deke. It seems to me that there is a tremendous amount of energy being spent in this country on space. A lot of engineering effort, a lot of scientific effort, a lot of money being spent, and it's my thought, based on my experience in Washington, and based on my experience of sitting in symposium workshops like this and listening, that perhaps we have got to step back and think about doing business in a little different manner. So I have written a short piece here that I want to read to you for the sake of time, because we want to hear from our distinguished panelists here this afternoon.

---

* Captain Honhart is President of the American Astronautical Society. He is affiliated with the Office of the Chief of Naval Research, 800 N. Quincy Street, Code 13D, Arlington, Virginia 22217.

† "Deke" Slayton was selected as one of the original Mercury Astronauts and served as flight crew operations director 1963-1972. He flew on the Apollo-Soyuz Test Project (ASTP), in July 1975. Currently he is President, Space Services Inc. of America, 7015 Gulf Freeway, Suite 140, Houston, Texas 77087.

‡ Dr. Abramson is Executive Vice President, Southwest Research Institute, 6220 Culebra Road, San Antonio, Texas 78284.

** Chief Engineer of the Galileo Project, Jet Propulsion Laboratory, 4800 Oak Grove Drive, Pasadena, California 91109.

†† Senior Advisor, Lockheed Corporation, 4500 Park Granada Blvd., Calabasas, California 91399.

My treatise starts as follows: For three days we have listened to a cavalcade of distinguished speakers, engineers, scientists, and managers. We have talked about new reference systems, the benefits of manned versus unmanned programs, trips to Mars, bulky, costly, outdated guidance systems, international cooperation, the massive bureaucratic acquisition process, dreams of outer galactic exploration, the quest for better understanding for life sciences, the need for a national space plan and other topics.

Regarding the national space plan, there is no doubt in my mind that the need exists. It is not unlike another project I'm involved in for the Navy. Today, there does not exist an operational computer model for forecasting oceanographic conditions on a global basis. Within the Navy we have decided that the problem is so complex, the funding resource is so scarce, and the scientific experts so few, that we are spearheading a national effort, involving the brightest minds from the Navy Research Centers, the college and university systems, and private industry to develop a master research and development plan which is coordinated, prioritized, and parceled out to eliminate the elements of destructive competition, the duplication of effort and the wasting of limited scientific and engineering talent and funding.

In a similar vein, a national space program must be developed, which involves the cognizance of all appropriate federal agencies and departments, defense, commerce, interior, and NASA, and the cooperation of the National Academy of Science and Engineering and the National Science Foundation. The plan must be focused on clear operational objectives, but allowing also for the healthy advance of science and exploration. It has got to have both operational and science exploration objectives. The plan must have near-term and long-term objectives to ensure continuity with the constant change of elected officials, political appointees, career civil servants and military officers. Above all else, the plan must be prioritized to ensure that our most important space programs, both civil and military, stay on track and are protected from the whimsical slash of a bureaucratic pen. It is my opinion, this represents or presents an opportunity to work hand-in-glove with the Government, to develop the best plan, which takes advantage of the right technology, not necessarily the most sophisticated technology. We have a propensity in this country to strive for complex, sophisticated solutions when, perhaps, less complex systems would get the job done. One of the admirals of the Soviet Navy, maintains, "Better is the enemy of good enough, at a price we can afford". I think there is a lesson there for all of us.

On another subject, I am not convinced that competition always delivers the best product, or that it delivers the most cost-effective product. In fact, it is my personal viewpoint, based on almost seven years of working in Washington, that literally hundreds of millions of dollars and years of engineering talent are wasted as a result of the often cumbersome, competitive, procurement process. I am reminded of a comment of a former office-mate of mine, Dr. Richard James, who is an oceanographer of some national renown. One day Dr. James walked into our office, slammed the door, and shouted, "We have three feet of paperwork, and after four years we still don't have a system." Perhaps, rather than wasting the time and expense of required full-scale, open competition for every procurement, we should use the time to work with industry and our legislative bodies to develop a national space plan, including common goals and a method to equitably assign the elements of the plan to all of industry to execute as a team. There are those who say that we can't trust industry. Whatever happened to the Yankee spirit of trust. Are we not all Americans pulling together for the common good of our country, those within the Government and those within industry?

In these past few days comments were made about the national press. My personal thought is that we made a mistake following the *Challenger* accident. After it was determined that the booster joint was the cause of the accident, we should have taken the position that the problem was not an engineering problem, but rather an operations problem. And rather than looking for an engineering fix, we should have looked for an operations fix. That is, we will not operate the Shuttle Launch System below a certain temperature and surely not, with icicles hanging from the gantry, nor will we launch ELV's into thunder and lightning storms. I think it is important for the American people to realize that space travel is a dangerous business, more so than flying on an airplane or driving in a car.

Other accidents will occur, and all of the fancy formal reviews, and all of the finger pointing, and all of the twenty-some approval signatures, will not alter the statistical fact of accidents to come. There is risk involved. When accidents happen, we must pick up the ball and start running again. In the aftermath of *Challenger*, I hold the national press and the Government at fault for being too preoccupied with the pointing of fingers, and not occupied enough with educating the American population with the necessity of picking up the ball and running.

As a closing thought, I also think it is necessary for our professional astronautical societies, including the AAS, to accept some responsibilities for not doing a more effective job in promoting the necessity of a national space plan and through its membership, working with the governmental leaders to ensure a healthy United States space program, which would restore the program to the world prominence it has known before. To this degree, it is my intention to take the results of this national meeting and ensure that national leaders within the Administration and the Congress are cognizant of the collected opinion of this esteemed national society.

Donald K. Slayton: Thank you, David. There's a lot of food for thought in that message. The next speaker is Gentry Lee. Gentry is from JPL. He's the chief engineer on the Galileo Program. He was a member of the small committee that prepared the, so-called, Ride Report. Another point that I didn't know about him, he's also a novelist. He has done some novels in collaboration with Arthur Clarke. Mr. Lee.

Gentry Lee: Thank you, Deke. The title of this panel is, "Where from Here?" and I will, first of all, tell you that I do not know much about the military space program, so I'm going to address my comments to the civilian space program. I began my work as an engineer, a navigation manager and engineer, and there are two problems in navigation: One is, "Where are you trying to go", and the other one is, "Where are you?" I think that we should start by recognizing where we are and trying to define some sort of a metric that allows us to assess where we are from a reasonable point of view. Now, there are two different kinds of metrics that you might use. One metric you might use is where you compare where you are to where you wanted to be, which is a very normal thing you do when you're doing navigation. I happened, before I came here, to pick up a detailed description of the five-year plan for NASA in 1983. You may or may not know that, as of today, we are supposed to be flying 20 to 25 space shuttles a year. Galileo has passed an asteroid and is on its way to Jupiter. The space telescope has been launched and is probing ten times deeper than any spacecraft has ever probed into the heavens. So by one metric we can use to determine where we are, we're not very far by our own admission.

Now another metric you can use just to figure out where you are is to look at what your competitors or cooperators are doing and where they are. Let's try to do that. Let's ask ourselves, what are three measurable quantities by which we can compare ourselves, for example, to the Soviets. (1) Access to space, how are we doing in comparison with them? We are, for all practical purposes, dead in the water while they have recently upgraded their best launch vehicle. They now have a new one, it's roughly the equivalent of the Saturn. They have more access to space than anybody else has ever had, and they show no signs of slowing down. So by any measure, they are way ahead of us there. (2) Another one is the length of time spent in space by human beings. I don't need to tell you that we're rapidly approaching the day when 90% of all the time spent in space is by someone who says thank you in Russian. We did not think it was going to be that way ten years ago, but that is the way it is now. (3) How about exploration and science programs? Once we said we were the best. We were way ahead of everybody else. I even had a colleague who stood up in a meeting like this five years ago and said, no one will ever catch us. I'll remind you that John of Portugal, after he took over for Henry the Navigator, also said the Portuguese will be the first in exploration forever; they were so far ahead. It looks now as if we have a chance to become the Portugal of the 21st century. For as I sit here, the United States has no national space plan despite the fact that many have been developed. The Russians have announced that they are going to Mars in 1988. They are going to blow some surface off Phobos and also land there. They are going back again in 1992 to put a rover on the surface and may bring a sample back in 1996 or 1998, depending on how many other people they can get to cooperate with them and how long it takes to do it.

I was asked recently, what is your plan? And I said, my personal plan is embodied in the work that I did with Dr. Sally Ride last year and the Ride Report. But what the United States plan is no one knows. And we stand here today after having the Paine Commission Report, the Ride Report, the Report of the National Advisory Council and many other reports on what we ought to be doing. And nothing has happened. Why has nothing happened? Let's call spades, spades, and nonleaders, nonleaders. The civilian space program has been in a period of benign neglect for the last decade. The current president has no idea where space is, much less a policy. Now, we were told, those of us who were blindly spending our time working on the Ride Report, that it had to get into the White House by April. April came, May came, June came, and July came, and to this date, and I've checked very carefully, I even asked Aaron just a minute ago, I do not think anyone has even been able to get past the screener who decides what goes on the calendar for Ronald Reagan.

This is not leadership. What are examples of leadership? People are asking me this all the time. I'll give you just a couple of examples. In the second century, A.D., there was a man named Hadrian, who took over the Roman Empire at it's fullest, and was told the provinces were ungovernable. If you don't take some steps, the Roman Empire is going to collapse. He said, I can't understand what their problems are. They are too far away. They said, you have got to solve that problem. In the next decade he learned 20 languages, went and spent three months at a time in each of the provinces and promulgated laws that lasted for five years; that's leadership.

A millennium later, in the 12th century, Henry II, Henry Plantagenet ascended to the throne of England. Since he came from France, a lot of people in England wondered what was going to happen when this man from Anjou became king. He said, they are not going to follow me, are they? Some people said, "You had better get in touch with your subjects right away." Do you know what he did, other than learn English very quickly? He got himself a troop of people and he went from city to city and castle to castle, day after day, until he got to know everybody.

Leadership is someone in the White House saying, this is important. It doesn't matter what the rest of us do, it is somebody at the top who has to say what we are to do. Now, let's just for a moment talk about the two elements of policy that need to be decided upon: (1) Where we are going? and (2) What kind of international work are we going to do in the future? Those are two nice meaty issues. On the international scene, the Soviets have made it very clear, they intend to be preeminent in all phases of space exploration. They intend to lead, both in terms of what they do and in terms of the international cooperation that they espouse. The 92 to 94 mission, whenever it is they launch it, is about to become a 25 to 30 nation venture, in which people are going to be supplying instruments for them.

The United States currently has no international space policy and needs to address that issue as well. On this issue of leadership, I want to pass on to you a story that was told to me and I will, first of all, say this is a second-hand story. Therefore, you must weigh it, but it's nevertheless a very powerful story. I was told by a friend of mine, that in a conversation with Academician Sagdeev who basically runs the major part of the Russian space program, one night, Academician Sagdeev told about how he briefed Gorbachev on the Halley Armada, the five spacecraft that went by Halley's Comet, two Russian, two Japanese, and one European. They were supposed to be in Gorbachev's office for thirty minutes, but the chairman kept asking questions, and they ended up being there for three hours. And when he was done, Gorbachev said, this is sensational, we should brief the Politburo. Sagdeev said, I'll prepare it whenever you want me to. No, Gorbachev said, I'm going to do it myself. Now, can you imagine anybody in the White House briefing the cabinet on anything that's been talked about for the past three days? It is grist for thought, but what should we do?

I'm going to say that those of us in this room have been too timid. I understand that all of you work for industry and in a certain sense for NASA, and have to make sure you don't alienate the people in power, because they're the ones who give you the contracts. I'm sorry, though, a large part of a very small pie, is not as significant as a small part of a big pie. And there are more compelling issues. The fate of the nation, in terms of its 21st century high technology is at stake here. And I'll call upon this group, just like I've called upon the AIAA, to take off your timidity and to have the temerity to take a stand, to say to the nation, we are floundering. We are falling apart in our civilian space program. Here is a plank, a policy, a set of issues that should be addressed, and should be addressed now. It is fine to get together and talk about what the major issues are in the future. But if we don't solve the nonleadership issue, we're all smoking opium. Thank you very much.

Donald K. Slayton: I didn't see anybody dosing off during that speech. Our next speaker is Norman Abramson. He is executive vice president for Southwest Research Institute. His background is fluid structure interaction problems. He's a member of the Air Force Scientific Advisory Board. Norman.

Dr. H. Norman Abramson: If I'm going to get any kind of attention at all after Gentry's talk, I think I had better get up here. When he's sitting down and I'm standing up, I have the advantage of height. I can't be quite as global as Gentry has been, nor can I be as dynamic. But I do have two issues that I thought that I would bring before you, which I think are, each in itself, quite significant.

In both, Panel 1 and Panel 8, questions arose regarding basic research and its role in the space program, whether or not basic research can indeed, survive in the arena of large science. As a matter of fact, let me just expand on the theme.

Let me talk about the question of whether there's room for small science in the national scheme of things, much less the space program. First of all, Gentry referred to the Ride Report. The Ride Report names four major goals. These goals are a manned mission to Mars, an unmanned program of solar exploration, a manned lunar base, and space-base study of Earth. All of those are significant goals, which we should pursue, but they're big goals. The super-conducting, super-collider is equal to anyone of those four goals. And you could go on and name countless other large national programs. Each of these large national programs involves what we now call big science. And there are people who are concerned that big science leads to the obscuration of small science. Each of these major programs, in its own right, will do good science. Each of these major programs will lead to advances in technology that will, in itself, be important and will help other related programs. But each of these major programs also is characterized by stretch-outs, by greatly increased cost, by the emergence of political issues and conflicting objectives by principal agencies like NASA and DOD.

Yet, there's still a role for small science. A role which we must be vigorous in pursuing to ensure that it does not ever become eliminated. In space, for example, we still have a very important small science program involving suborbital sounding rockets of the atmosphere and the ionosphere, a very important scientific program that must not be allowed to die but which forever faces extinction in the face of rising budgets for big science programs. The Polar Arches Program, for example, is a good example of what can be done even with relatively low budgets and short lead times if there are people who are dedicated to their survival.

The nation to some extent has embarked on a dangerous course, if we ignore small science. Basic research in the United States is funded, essentially, first of all by the National Science Foundation, secondly by the basic agencies of the Department of Defense, thirdly and fourthly, by the Department of Energy and the Department of Transportation. All agencies are now embarked head-long in the development of a new concept, the concept of center-oriented research. The National Science Foundation is setting the tone with its program of engineering research centers, and now forthcoming, science and technology research centers. Each of the other agencies that I mentioned is anxiously competing to play a second role to the National Science Foundation and seeing how many such centers it can set up, all of this to the detriment of individual investigator initiated research and development work.

Let me remind you that in the good old days not so many years ago, when the leading basic research agencies, besides the National Science Foundation, were the Office of Naval Research, the Air Force Office of Scientific Research and the Army Research Office, that even those highly competitive programs accepted 30% to 50% of the proposals that were submitted. Today, we have competition for engineering research centers and now forthcoming, science and technology centers and so on. Each of those centers is characterized by an institutional investment, sometimes of many hundreds of thousands of dollars, admittedly, not very important in comparison with a major aerospace firm proposal for a major new system development, but still important when you think in terms of the man hours, the labor, the level of the labor that's involved in the preparation of such proposals. And what's the acceptance right now on proposals submitted for engineering research centers? Under five percent. I ask you if this is a wise use of some of the greatest resource that we have available in the country, namely, the experience, the capabilities of our most senior and distinguished scientists? We must at all cost protect the role of basic research and its funding in all of these major programs. You and the aerospace industry, those of you who are involved in large missions, large programs, must never forget that the ultimate success of your programs will still depend on the science and the technologies that come from the basic research program of the country which must not be allowed to falter.

Now if I can take my remaining few minutes to embark on a totally different subject, which has also emerged during the course of the discussions here in the last three days. And this relates to the issue of large launch vehicles. As one who used to be involved a number of years ago with large launch systems, I'm appalled at the problem that we face today. When SDI became an issue of some importance and when it became likely that there could be some space-based SDI hardware and that the Air Force would look for an opportunity to put that hardware into space, both NASA and the Air Force realized that there might be an opportunity for a new launch vehicle. And both started scrambling very, very hard to come up with the right concepts and sell their ideas to Congress and to the Executive Branch. It happened very quickly, just the end of last year, that it became apparent there might be such an opportunity for a new vehicle design. The Air Force has gone ahead with its plans for a version of a heavy lift vehicle, which they are now calling the Advanced Launch System.

Now I'm going to quote from *Science* magazine on a recent analysis of what has happened since both NASA and the Air Force announced their plans for moving ahead with new heavy lift vehicle concepts. Congress voted on the 1987 Supplemental Appropriation Bill on the 1st of July, While prohibiting early deployment of any SDI hardware and it reduced the Administrations request for funds, but at the same time the Congress empowered the Air Force to go on with the Advanced Launch System Program and provided 75 million dollars in R&D money, specifying that 38 million of it should be spent in NASA propulsion facilities. The Air Force was allowed to use 12 million dollars from an earlier budget to start their work on the Advanced Launch System. Simultaneously NASA was allowed to spend 5 million dollars of its own funds on cargo vehicle studies known as the Shuttle-C Program. NASA and the Air Force were ordered to find a way to cooperate on the Advanced Launch System and to report back to Congress before spending the 75 million dollars. Also the Air Force has begun making commitments to spend the 75 million, it has not agreed on an R&D plan with NASA and no report has been filed. On the 10th of July, just four short months ago, the Air Force awarded contracts worth 5 million dollars a piece to seven big aerospace firms, each of which must come back in a year with a design for a system that could fly at the rate of about 30 times a year and have a large cargo capacity and reduce operational cost in 1998 by a factor of ten. And the Air Force also asked bidders to prepare an Interim Advanced Launch System that would be not quite as efficient but would be able to lift huge payloads, presumably for deployment of SDI by 1994. One month later, NASA invited bidders to prepare designs for the large Shuttle-C cargo vehicle that would carry the same size payload and be ready for the first flight, at roughly the same time as the Interim ALS or perhaps a year earlier. Secretary of the Air Force Aldridge, asked NASA to submit its plans for the joint R&D Program and NASA declined.

I'm not going to say which is right, and which is wrong, but I think that it's almost criminal to think that here we have the two major agencies of the Government concerned with launch vehicles who are really not speaking to each other and not working cooperatively. And the cost is the dollars that you and I provide, the dollars that could be used for other significant programs and which were being used to compete with each other, rather than for the national good. I can hardly think of a better issue for this society to address than that of demanding that the Government come to grips with its requirement to establish a suitable policy and force NASA, the Air Force, to develop the heavy launch capability that the country needs to compete with the Soviets, and to reemerge as the world power in space. Thank you.

Donald K. Slayton: Thank you, Norm. Our clean-up speaker is Willis Hawkins. Willis is a senior advisor for Lockheed Corporation, an association of about 50 years. He's a retired officer and director of Lockheed, ex-Assistant Secretary of the Army for Research and Development, chairman of NASA's Aerospace Safety Advisory panel for many years. Most of you in the audience have had some personal experiences with Willis, and will look forward to hearing from him.

Willis Hawkins: If this audience likes a challenge, I suggest you consider the task of clean-up speaker for a panel like you have just heard. We had an enthusiastic explorer in space who knew what he was talking about, an enthusiastic researcher who knew what he was talking about. I'm just a dirty-fingered engineer who has been trying to figure what I should say that would be equally important. I did take a bunch of notes during these meetings and a lot of the issues that you folks addressed are the issues that we must address in the United States, and the sooner the better.

I have elected to concentrate on one single issue. But it is an issue that would be applicable to almost every other issue we have been talking about, and that's the business of getting low-cost access to space. If we don't get low-cost, we are not going to get access at all, unless we do it on a panic basis, one of these years, after the Russians have done something even more spectacular than they are doing every day. When we take a look at what is going on in the booster business, we see that the needs are being added up, demanding something that is really a new booster. If you take a look at the SDI build-up that is proposed in some of the plans, the build-up necessary to put the Space Station in orbit and add that to the routine loads we are planning, we discover a total of very large loads to orbit. Designing for very large payloads produces low-cost in dollars per pound, but pretty high-cost per launch, in fact, magnificent cost per launch. I'd like to suggest that there are a lot of sustaining requirements we're going to face, and are facing today, where researchers have loads that they'd like to launch, and they hate to have to wait for the other 200,000 pounds to come along. So we must also take a look at how we approach low-cost for modest payloads. I see no national effort toward both the high and low end of our payload requirement and I see no real cooperation between DOD and NASA in taking a look at what they ought to do with a total spectrum.

I believe that this is a national issue in its own right, and I second what was suggested by Norm Abramson. There is a way of fulfilling this need that would save us a lot of dollars. We seem to be caught in the habit of how we did the boosters in the old days. I'd like to see somebody put together a booster proposal that consists of a cluster of identical units. We do that almost every day, when we put the circle of solids around the bottom of one of our old boosters. But those aren't identical with the core, and I think you can make the core identical. With this concept one can assemble just enough of the elements to meet the payload that you want, and you don't have to have everything different every time you launch. I have seen a couple of glimmers of this that are running around in proposal form, but nobody seems to be officially interested, and I think the nation ought to be.

There is a parallel problem also lurking about. Many believe, including me, that commercial boosters, might be a step function in low-cost. If you have looked at the recovery on the Shuttle system, and what they are doing to be ready for the next launch, we're building a system of paperwork that we are going to have one devil of a time tearing back down again to save money. We have to convert to something like commercial practice and the only way you can really do that is to let the commercial operators have the responsibility.

There is still another problem, even if you wanted to go "commercial", and that's how to mesh with the Military/NASA Launch Facility Complex, how to share the cost of those complexes, how to share the cost of the tracking and launching. I would like to suggest that maybe commercial boosters would begin to appear if payload loss insurance were available at a reasonable cost, if the commercial operator had control of his own schedule and his own launch procedures and if the public, including the international public, liability burdens were reduced or eliminated. This can only take place when some firm licensing rules are established, and certification standards of safety are defined. I think if we solved these problems, commercial boosters would appear.

As I thought about this, just for fun I created a kind of imaginary heaven to see where it might lead. The first thing that happens in this heaven of mine is that the United States buys a Pacific island, and makes launch areas available for lease. There are no neighboring lands, so omni-directional launch is possible, and there is land to build launch pads. Second, the United States organizes an insurance program for payloads, and launch-base damage, and liability for third parties. I think that perhaps the United States wouldn't really have to put up the money; it could provide a loan guarantee for commercial insurance people to tackle it and see what reasonable rates might be offered. Thirdly, I think the United States would have to determine fee schedules for tracking and communication services if they weren't otherwise available, but make them public so everybody knows how much they would cost. And finally the U.S. affirms a policy, that they will, indeed, hire commercial launch services for scientific public service, and even Space Station supply missions. Then, I think, the U.S. must also affirm a willingness to be bought out when experience and market rates, or market futures, stabilize. The United States might even make available, for lease and purchase, older track and communication facilities when the state-of-the-art or some national need, like a military requirement, demands new Government facilities.

A plan like this, if somebody put his mind to it, would get the commercial adrenalin really flowing, and we would see what commerce could do, their way.

Before I stop, I would like to discuss the creation of national policy, just the mechanism of it. When the Government tries to create a policy, it has a lot of trouble. There are many, many, too many, parties, all with their turf jealousies, that somehow, have to be brought to vote the same way. We know how difficult that is. If it were floated on the public, it would be subject to multiple critics, most of whom have a bias. As a matter of fact, we have been sitting here for three or four days, criticizing quite a few things that somebody, I think, floated as a potential policy. I would like to suggest that maybe we ought to turn this process around. I think, perhaps, you as a society, could get together with other non-Government entities and societies and see if you could put together an activity which actually created rational goals and policy. I know that efforts have been made like this with big names. They have done good jobs, but they didn't address the steps needed to reach their lofty goals. Maybe, if we had a bigger, more transition oriented, population working on this, we might be able to get more attention. Let the Government be the critic, and keep forcing, the Government through the leaders of this activity -- those that really contributed to it -- nagging them, if you will, to respond. If we finally got their attention, I'm quite sure that what was left over may be a sound foundation for the future. Maybe it's worth a try, and you in this association can figure out whether you want to be the lead agent. Thank you.

Donald K. Slayton: Thank you, Willis. Some people in this audience who are familiar with our Conestoga family of vehicles would think I wrote that speech for you, because you defined it perfectly. We also subscribe completely to the parameters you talked about that are relative to dealing with the Government and how to do things. I think we have picked up almost all of the major issues that were brought to our attention from the panels in one form or another.

DISCUSSION

Donald K. Slayton: I've got one package of issues here, questions that relate pretty well to one subject and let me just touch on that briefly, because I don't think it was addressed. That's the risk assessment, and primarily related to the launch systems and it's a series of questions about liquids versus solids, what's an acceptable reliability number, and how you improve on it, those kinds of things. My opinion is that I don't care what anybody inside this room or outside of it does, you are never going to get a 100% reliable transportation system. I would like to remind you that back in the days when we flew the Mercury on Atlas the reliability number was 80%, and that was good enough, because that was the best we knew how to do. You wouldn't accept that today. Today you'd like to see a Shuttle that is about 99.99% reliable and if it's 99%, that's probably pretty good. The bottom line is its going to be up in there somewhere. Whatever it is, it's acceptable. There will be other failures, there's no question about it. The key point is that you do exactly what we've done with the past failures; you go back and analyze what happened, fix it and get on with the program. You don't sit around and contemplate your navel and decide whether you ought to quit or not, that's the bottom line. I think for the unmanned expendable vehicles probably a number more like 95% would normally be considered acceptable, somewhere between that and 97%. That's my take on that series of questions.

We could take time to pick up a couple of questions off the floor here, if anybody has any questions. This question is to Gentry Lee.

Question: What do you think would be the most effective way of getting Congress and the White House leadership to pay attention, short of writing letters to elected officials? Who wants to be the Portugal of the 21st century with our greatest launch vehicle rusting away in Texas, Alabama, and Florida Sun?

Gentry Lee: Well, first of all, in the best of all possible worlds, to quote * Dr. Pangloss, what would happen if the leaders of the companies that you all work for, plus the leaders of the various centers, would do something emphatic to make it clear to the White House that benign neglect is not a satisfactory way to run a civilian space program? Now, I recognize that this is counter to the nature of many of the people in the industrial establishment to do such things. But that would certainly make people stand up and take notice. I don't know. I do know that in January of 1989, when the Russians are hovering over Phobos, blowing stuff off the surface, and collecting it, that there will be a rush-around like there was during Sputnik, as if no one had told them that this was all going to happen. I have to say, that it was in-

* In Voltaire's Candide.

credible to me, that we spent all the time to study and report. The Ride group was commissioned to produce the Ride Report, and all of us worked under the idea that somebody was just pantingly waiting for the output. As you can see in the six months since we finished the job, I don't believe it has even been read. There was even a suggestion at one stage that we not publish it. So I think what it is going to take is a combined community action, that societies and companies say to the White House, "You may have problems somewhere else, but if you don't address this one, you're going to go down in history as the guy who allowed the world's first space-faring nation to fall into a position of nonparity with the other great space powers of the world." He might pay attention to that.

Question: What is status of the current ELV efforts, in relation to Government interfaces and where do we have remaining problems?

Donald K. Slayton: We've been working this area for quite a number of years, and we've got a fair amount of experience. Considering where we started from, we think we're in great shape. When we started, we had to deal with 18 different Government agencies, all independently. Today, there is an ELV bill that's been in place now since 1984 that gives us a single-stop shopping point, if you will, within the Department of Transportation. We think that bill, as it is written, is very good.

There is one remaining problem that I see from a small vehicle point of view. Maybe the big guys have some different problems than we have -- but we are operating our own system from Government ranges and we don't have some of the same problems they do. The biggest problem, I think, is the insurance issue, and that is being worked on. We have requirements to provide liability insurance, not only for our problems, but for the Government's problems when we're on the Government range. We are liable for anybody that is doing anything relative to our system. As an example, on Wallops Island, Virginia, where we have an agreement to launch if a Government driver in a Government vehicle runs off the bridge while delivering our mail we are liable. We think that is not fair and nobody else does either. That's something that's probably going to take legislation to fix. Part of that is putting a cap on the total insurance, which we are required to carry. That's one of the things that has many people uptight because there's basically, unlimited liability at the present time. We can only insure for up to a half a billion, and the rest becomes uninsurable. Those are the big issues, but we don't believe any of that is insurmountable and I don't think the big guys do either. We will figure out a way to make it work.

Gentry Lee: If I might jump back, Deke, the person that wrote this question had a very good suggestion, which I meant to mention, that the candidates who are running for president next year should be required at some stage to state what their space policy is. I would be interested if some of you had heard something that I haven't, but I haven't heard a single candidate from either party make a statement about space yet.

I'll tell you a funny story. We might like to end up on a piece of humor. There was a time when I did consider running for office and I discovered two things. First of all, that there was no way I was going to be able to get enough money together to run. And then my wife pointed out that there was at least one significant reason why I shouldn't run. And I said, "What is that, darling?" She said, "The way you talk, you'll get shot."

<u>**Donald K. Slayton**</u>: Your president, Capt. Honhart, would like to have an opportunity to make a couple of closing comments, and I would just like to thank you for the opportunity to run this session today. I think it has been very stimulating and a lot of good ideas have come out of it. I hope somebody can make something happen. I think we're really not dealing with a lack of a plans and that most of us would agree the problem is lack of implementation. Thank you.

<u>**David C. Honhart**</u>: Well, let me make one comment, to put the thought back in your mind, before we give some thanks and close the meeting and then get on to other things. There is a tremendous amount of money being wasted in this country on this space program, on this whole competitive basis. If I said hundreds of millions of dollars, it's probably billions of dollars, and if you think for a minute, instead of putting the onus on the Government, that you are a cartel in yourselves, that industry is a cartel, and all that money that we spend in competition, including the second, third, fourth, and fifth company that lose the competition, and if you had taken that money and that engineering effort, we could put a lot more things into space because the money's there to do it. So it's food for thought, but I suggest to you that we've got to come up with a different way of doing business.

Anyway, at this time the meeting here in Houston is over, but before we go I want to point out to you, once again, that the whole success of the American Astronautical Society has to do with the companies that are its corporate members and the individuals who are individual members. It's really their effort through which we have meetings like this, our annual meeting once a year, our military space symposium which occurred this year in June, and our Goddard Symposium which occurs in March. I want to acknowledge especially Aaron Cohen from NASA, Hank Bowes from Lockheed, and Bill Straight, I'd really like a loud hand of applause for them and for their effort. And on the part of the national organization, I would like to say, thank you very much.

APPENDIX

PUBLICATIONS OF THE AMERICAN ASTRONAUTICAL SOCIETY

Following are the principal publications of the American Astronautical Society:

JOURNAL OF THE ASTRONAUTICAL SCIENCES (1954-)

Published quarterly and distributed by AAS Business Office, 6212-B Old Keene Mill Court, Springfield, VA 22152. Back issues available from Univelt, Inc., P.O. Box 28130, San Diego, CA 92128.

SPACE TIMES (1986-)

Published bi-monthly and distributed by AAS Business Office, 6212-B Old Keene Mill Court, Springfield, VA 22152., Virginia 22152

AAS NEWSLETTER (1962-1985)

Incorporated in *Space Times*. Back issues available from AAS Business Office, 6212-B Old Keene Mill Court, Springfield, VA 22152.

ASTRONAUTICAL SCIENCES REVIEW (1959-1962)

Incorporated in *Space Times*. Back issues still available from Univelt, Inc., P.O. Box 28130, San Diego, CA 92128.

ADVANCES IN THE ASTRONAUTICAL SCIENCES (1957-)

Proceedings of major AAS technical meetings. Published and distributed for the American Astronautical Society by Univelt, Inc., P.O. Box 28130, San Diego, CA 92128.

SCIENCE AND TECHNOLOGY SERIES (1964-)

Supplement to *Advances in the Astronautical Sciences*. Proceedings and monographs, most of them based on AAS technical meetings. Published and distributed for the American Astronautical Society by Univelt, Inc., P.O. Box 28130, San Diego, CA 92128

AAS HISTORY SERIES (1977-)

Supplement to *Advances in the Astronautical Sciences*. Selected works in the field of aerospace history under the editorship of R. Cargill Hall. Published and distributed for the American Astronautical Society by Univelt, Inc., P.O. Box 28130, San Diego, CA 92128.

AAS MICROFICHE SERIES (1968-)

Supplement to *Advances in the Astronautical Sciences*. Consists principally of technical papers not included in the hard-copy volume. Published and distributed for the American Astronautical Society by Univelt, Inc., P.O. Box 28130, San Diego, CA 92128.

Subscriptions to the *Journal* and the *Space Times* should be ordered from the AAS Business Office. Back issues of the *Journal* and all books and microfiche should be ordered from Univelt, Inc.

ADVANCES IN THE ASTRONAUTICAL SCIENCES SERIES (1957-)

ISSN 0065-3438, LIBRARY OF CONGRESS CARD NO. 57-43769

Proceedings of Major AAS Technical Meetings

Vol. 1 Third Annual AAS Meeting, Dec. 6-7, 1956, New York, NY, 1957, 184p., ed. Norman V. Petersen, Microfiche only, $20 *(ISBN 0-87703-002-2)*

Vol. 2 Fourth Annual AAS Meeting, Jan. 29-31, 1958, New York, NY, 1958, 440p., eds. Norman V. Petersen, Horace Jacobs, Microfiche only, $20 *(ISBN 0-87703-003-0)*

Vol. 3 First Western National AAS Meeting, Aug. 18-19, 1958 530p., eds. Norman V. Petersen, Horace Jacobs, Microfiche only, $20 *(ISBN 0-87703-004-9)*

Vol. 4 Fifth Annual AAS Meeting, Dec. 27-31, 1958, Washington, D.C., 1959, 462p., ed. Horace Jacobs, Microfiche only, $20 *(ISBN 0-87703-005-7)*

Vol. 5 Second Western National AAS Meeting, Aug. 4-5, 1959, Los Angeles, CA, 1960, 364p., ed. Horace Jacobs, Microfiche only, $20 *(ISBN 0-87703-006-3)*

Vol. 6 Sixth Annual AAS Meeting, Jan. 18-21, 1960, New York, NY, 1961, 968p., eds. Horace Jacobs and Eric Burgess, Hard Cover $45 *(ISBN 0-87703-007-3)*

Vol. 7 Third Western National AAS Meeting, Aug. 4-5, 1960, Seattle, WA, 1961, 464p., eds. Horace Jacobs and Eric Burgess, Microfiche only, $20 *(ISBN 0-87703-008-1)*

Vol. 8 Seventh Annual AAS Meeting, Jan. 16-18, 1961, Dallas, TX, 1963, 602p., ed. Horace Jacobs, Microfiche only, $20 *(ISBN 0-87703-009-X)*

Vol. 9 Fourth Western Regional AAS Meeting, Aug. 1-3, 1961, San Francisco, CA, 1963, 910p., ed. Eric Burgess, Hard Cover $45 *(ISBN 0-87703-010-3)*

Vol. 10 Manned Lunar Flight (AAS/AAAS Symposium) Dec. 19, 1961, Denver, CO, 1963, 310p., eds. George W. Morgenthaler and Horace Jacobs, Hard Cover $35 *(ISBN 0-87703-011-1)*

Vol. 11 Eighth Annual AAS Meeting, Jan. 16-18, 1962, Washington, D.C., 1963, 808p., ed. Horace Jacobs, Hard Cover $45 *(ISBN 0-87703-012-X)*

Vol. 12 Scientific Satellites - Mission and Design (AAS/AAAS Symposium), Dec. 27, 1962, Philadelphia, PA, 1963, 262p., ed. Irving E. Jeter, Hard Cover $25 *(ISBN 0-87703-013-8)*

Vol. 13 Interplantetary Missions, 9th Annual AAS Meeting, Jan. 15-17, 1963, Los Angeles, CA, 1963, 690p., ed. Eric Burgess, Hard Cover $45 *(ISBN 0-87703-014-6)*

Vol. 14 Second AAS Symposium on Physical and Biological Phenomena under Zero G Conditions, Jan. 18, 1963, Los Angeles, CA, 1963, 382p., eds. Elliot T. Benedikt and Robert W. Halliburton, Hard Cover $30 *(ISBN 0-87703-015-4)*

Vol. 15 Exploration of Mars Symposium, Jun. 6-7, 1963, Denver, CO, 1963, 634p., ed. George W. Morgenthaler, Hard Cover $45 *(ISBN 0-87703-016-2)*

Vol. 16 Space Rendezvous, Rescue, and Recovery Symposium, Sept. 10-12, 1963, Edwards, CA 1963, 1408p., ed. Norman V. Petersen, Hard Cover, **Part 1**, 1028p., $45 *(ISBN 0-87703-017-0)*; **Part 2**, 380p., $30 *(ISBN 0-87703-018-9)*

Vol. 17 Bioastronautics - Fundamental and Practical Problems (AAS/AAAS Symposium), Dec. 30, 1963, Cleveland, OH, 1964, 128p., ed. William C. Kaufman, Microfiche only, $10 *(ISBN 0-87703-019-7)*

Vol. 18 Lunar Flight Programs, 10th Annual AAS Meeting, May 4-7, 1964, New York, NY, 1964, 630p., ed. Ross Fleisig, Hard Cover $45 *(ISBN 0-87703-020-0)*

Vol. 19 Unmanned Exploration of the Solar System Symposium, Feb. 8-10, 1965, Denver, CO, 1965, 1000p., eds. George W. Morgenthaler, Robert G. Morra, Hard Cover $45 *(ISBN 0-87703-021-9)*

Vol. 20 Post Apollo Exploration, 11th Annual AAS Meeting, May 3-6, 1965, Chicago, IL, 1966, 1220p., ed. Francis Narin, Microfiche only, **Part I**, 572p., $30 *(ISBN 0-87703-022-7)*; **Part 2**, 648p., $35 *(ISBN 0-87703-023-5)*

Vol. 21 Practical Space Applications Symposium, Feb. 21-23, 1966, San Diego, CA, 1967, 508p., ed. Lawrence L. Kavanau, Hard Cover $40 *(ISBN 0-87703-024-3)*

Vol. 22 The Search for Extraterrestrial Life, 12th Annual AAS Meeting, May 23-25, 1966, Anaheim, CA, 1967, 388p., ed. James S. Hanrahan, Microfiche only $30 *(ISBN 0-87703-025-1)*; Microfiche Suppl. (Vol. 1 AAS Microfiche Series) $12 *(ISBN 0-87703-132-0)*

Vol. 23 Commercial Utilization of Space, 13th Annual AAS Meeting, May 1-3, 1967, Dallas, TX, 1968, 512p., eds. J. Ray Gilmer, Alfred M. Mayo, Ross C. Peavey, Hard Cover *(ISBN 0-87703-026-X)*; plus Microfiche Suppl. (Vol. 3 AAS Microfiche Series) $60 *(ISBN 0-87703-216-5)*

Vol. 24 Exploitation of Space for Experimental Research, 14th Annual AAS Meeting, May 13-15, 1968, Dedham, MA, 1968, 363p., ed. Harry Zuckerberg, Hard Cover $30 *(ISBN 0-87703-027-8)*

Vol. 25 Advanced Space Experiments, Sept. 16-18, 1968, Ann Arbor, MI, 1969, 530p., eds. O. Lyle Tiffany and Eugene M. Zaitzeff, Hard Cover $40 *(ISBN 0-87703-028-6)*

Vol. 26 Planning Challenges of the 70's in Space, 15th Annual AAS Meeting, Jun. 17-20, 1969, Denver, CO, 1970, 470p., eds. George W. Morgenthaler and Robert G. Morra, Hard Cover $35 *(ISBN 0-87703-053-7)*; Microfiche Suppl. (Vol. 14 AAS Microfiche Series) $20 *(ISBN 0-87703-130-4)*

Vol. 27/28 Space Stations (**v27**) and Space Shuttles and Interplanetary Missions (**v28**), 16th Annual AAS Meeting, Jun. 8-10, 1970, Anaheim, CA, 1970, **Vol. 27**, eds. Lewis Larmore and Robert L. Gervais, 6060p., Hard Cover $45 *(ISBN 0-87703-054-5)*; **Vol. 28**, eds. Lewis Larmore and Robert L. Gervais, 488p., Hard Cover $35 *(ISBN 0-87703-055-3)*

Vol. 29 The Outer Solar System, 17th Annual AAS Meeting, Jun. 28-30, 1971, Seattle, WA, 1971, 1358p., $85; ed. Juris Vagners, Hard Cover, **Part 1**, 618p., $40 *(ISBN 0-87703-059-6)*; **Part 2**, 740p., $45 *(ISBN 0-87703-060-X)*

Vol. 30 International Congress of Space Benefits, 19th Annual AAS Meeting, Jun. 19-21, 1973, Dallas, TX, 1974, 528p., ed. Francis S. Johnson, Hard Cover $40 *(ISBN 0-87703-065-0)*

Vol. 31 The Skylab Results, 20th Annual AAS Meeting, Aug. 20-22, 1974, Los Angeles, CA, 1975, 1174p., eds. William C. Schneider and Thomas E. Hanes, Microfiche only (ISBN 0-87703-072-3); Plus Microfiche Suppl. (Vol. 22 AAS Microfiche Series) $60 *(ISBN 0-87703-043-6)*

Vol. 32 Space Shuttle Missions of the 80's, 21st Annual AAS Meeting, Aug. 26-28, 1975, Denver, CO, 1977, 1364p., eds. William J. Bursnall, George W. Morgenthaler, Gerald E. Simonson, Hard Cover, **Part 1**, 598p., $40 *(ISBN 0-87703-078-2)*; Hard Cover, **Part 2**, 766p., $55 *(ISBN 0-87703-087-1)*; Microfiche Suppl. (Vol. 25 AAS Microfiche Series $65 *(ISBN 0-87703-133-9)*

Vol. 33 AAS/AIAA Astrodynamics Conference, July 28-30, 1975, Nassau, Bahamas, 1976, 390p., eds. William F. Powers, Herbert E. Rauch, Byron D. Tapley, Carmelo E. Velez, Hard Cover $35 *(ISBN 0-87703-079-0)*; Microfiche Suppl. (Vol. 26 AAS Microfiche Series) $40 *(ISBN 0-87703-142-8)*

Vol. 34 Apollo Soyuz Mission Report, 1977, 336p., ed. Chester M. Lee, Hard Cover $35 *(ISBN 0-87703-089-8)*

Vol. 35 The Bicentennial Space Symposium - New Themes for Space: Mankind's Future Needs and Aspirations, 22nd AAS Meeting, Oct. 6-8, 1976, Washington, D.C., 1977, 242p., ed. William C. Schneider, Hard Cover $25 *(ISBN 0-87703-090-1)*

Vol. 36 The Industrialization of Space, 23rd Annual AAS Meeting, Oct. 18-20, 1977, San Francisco, CA, 1978, 1160p., eds. Richard A. Van Patten, Paul Siegler, Edward V.B. Stearns, Hard Cover, **Part 1**, 610p., $55 *(ISBN 0-87703-094-4)*; Hard Cover, **Part 2**, 550p., $45 *(ISBN 0-87703-095-2)*; Microfiche Suppl. (Vol. 28 AAS Microfiche Series) $15 *(ISBN 0-87703-121-5)*

Vol. 37 Space Shuttle and Spacelab Utilization, What are the Near-Term and Long-Term Benefits for Mankind?, 16th Goddard Memorial Symposium, 24th Annual AAS Meeting, March 8-10, 1978, Washington, D.C., 1978, 865p., eds. George W. Morgenthaler and Manfred Hollstein, Hard Cover, **Part 1**, 400p., $40 *(ISBN 0-87703-096-0)*; Hard Cover, **Part 2**, 465p., $45 *(ISBN 0-87703-097-9)*

Vol. 38 The Future U.S. Space Program, 25th Anniversary Conference, Oct. 20 - Nov. 2, 1978, Houston, TX, 1979, 880p., eds. Richard S. Johnston, Albert Naumann, Jr., Clay W. G. Fulcher, Hard Cover, **Part 1**, 444p., $45 *(ISBN 0-87703-098-7)*; Hard Cover, **Part 2**, 436p., $40 *(ISBN 0-87703-099-5)*; Microfiche Suppl. (Vol. 30 AAS Microfiche Series) $15 *(ISBN 0-87703-129-0)*

Vol. 39 Guidance and Control 1979, Feb. 24-28, 1979, Keystone, CO, 1979, 492p., ed. Robert D. Culp, Hard Cover $45 *(ISBN 0-87703-100-2)*; Microfiche Suppl. (Vol. 31 AAS Microfiche Series) $10 *(ISBN 0-87703-128-2)*

Vol. 40 AAS/AIAA Astrodynamics Conference, Jun. 25-27, 1979, Provincetown, MA, 1980, 996p., eds. Paul A. Penzo, Bernard Kaufman, Louis Friedman, Richard Battin, Hard Cover, **Part 1**, 494p., $45 *(ISBN 0-87703-107-X)*; Soft Cover $35 *(ISBN 0-87703-108-8)*; Hard Cover, **Part 2**, 502p., $45 *(ISBN 0-87703-109-6)*; Soft Cover $35 *(ISBN 0-87703-110-X)*; Microfiche Suppl. (Vol. 32 AAS Microfiche Series) $20 *(ISBN 0-87703-139-8)*

Vol. 41 Space Shuttle: Dawn of an Era, 26th Annual AAS Meeting, Oct. 29-Nov. 1, 1979, Los Angeles, CA, 1980, 980p., eds. William F. Rector, III and Paul A. Penzo, Hard Cover, **Part 1**, 452p., $45 *(ISBN 0-87703-111-8)*; Soft Cover $35 *(ISBN 0-87703-112-6)*; Hard Cover, **Part 2**, 528p., $55 *(ISBN 0-87703-113-4)*; Soft Cover $40 *(ISBN 0-87703-114-2)*; Microfiche Suppl. (Vol. 33 AAS Microfiche Series) $10 *(ISBN 0-87703-136-3)*

Vol. 42 Guidance and Control 1980, Feb. 17-21, 1980, Keystone, CO, 1980, 738p., ed. Louis A. Morine, Hard Cover $60 *(ISBN 0-87703-137-1)*; Soft Cover $45 *(ISBN 0-87703-138-X)*

Vol. 43 Shuttle/Spacelab - The New Transportation System and its Utilization, (3rd DGLR/AAS Symposium), Apr. 28-30, 1980, Hannover, Germany, 1981, 342p., eds. Dietrich E. Koelle and George V. Butler, Hard Cover $45 *(ISBN 0-87703-144-4)*; Soft Cover $35 *(ISBN 0-87703-146-0)*

Vol. 44 Space--Enhancing Technological Leadership, 27th Annual AAS Meeting, Oct. 20-23, 1980, Boston, MA, 1981, 580p., ed. Lawrence P. Greene, Hard Cover $65 *(ISBN 0-87703-147-9)*; Soft Cover $50 *(ISBN 0-87703-148-7)*; Microfiche Suppl. (Vol. 35 AAS Microfiche Series) $10 *(ISBN 0-87703-164-9)*

Vol. 45 Guidance and Control 1981, Jan. 31- Feb. 4, 1981, Keystone, CO, 1981, 506p., ed. Edward J. Bauman, Hard Cover $60 *(ISBN 0-87703-150-9)*; Soft Cover $50 *(ISBN 0-87703-151-7)*; Microfiche Suppl. (Vol. 36 AAS Microfiche Series) $15 *(ISBN 0-87703-156-8)*

Vol. 46 AAS/AIAA Astrodynamics Conference, Aug. 3-5, 1981, North Lake Tahoe, NV, 1982, 1124p., eds. Alan L. Friedlander, Paul J. Cefola, Bernard Kaufman, Walt Williamson, G.T. Tseng, Hard Cover, **Part 1**, 552p., $55 *(ISBN 0-87703-159-2)*; Soft Cover $45 *(ISBN 0-87703-160-6)*; Hard Cover, **Part 2**, 572p., $55 *(ISBN 0-87703-161-4)*; Soft Cover $45 *(ISBN 0-87703-162-2)*; Microfiche Suppl. (Vol. 37 AAS Microfiche Series) $40 *(ISBN 0-87703-163-0)*

Vol. 47 Leadership in Space - For Benefits on Earth, 28th Annual AAS Meeting, Oct. 26-29, 1981, San Diego, CA, 1982, 310p., ed. William F. Rector, III, Hard Cover $45 *(ISBN 0-87703-168-1)*; Soft Cover $35 *(ISBN 0-87703-169-X)*

Vol. 48 Guidance And Control 1982, Jan. 30 - Feb. 3, 1982, Keystone, CO, 1982, 558p., eds. Robert D. Culp, Edward J. Bauman, W. E. Dorroh, Jr., Hard Cover $65 *(ISBN 0-87703-170-3)*; Soft Cover $50 *(ISBN 0-87703-171-1)*; Microfiche Suppl. (Vol. 38 AAS Michrofiche Series) $10 *(ISBN 0-87703-180-0)*

Vol. 49 Spacelab, Space Platforms, and the Future, Fourth AAS/DGLR Symposium and 20th Goddard Memorial Symposium, Mar. 17-19, 1982, Greenbelt, MD, 1982, 502p., eds. Peter M. Bainum, Dietrich E. Koelle, Hard Cover $55 *(ISBN 0-87703-174-6)*; Soft Cover $45 *(ISBN 0-87703-175-4)*; Microfiche Suppl. (Vol. 42 AAS Microfiche Series) $15 *(ISBN 0-87703-181-9)*

Vol. 50 Proceedings on an International Symposium on Engineering Sciences and Mechanics, Dec. 29-31, Tainan, Taiwan, 1983, **two parts**, 1570p., eds. Han-Min Hsia, Richard W. Longman, You-Li Chou, Hard Cover $120 *(ISBN 0-87703-176-2)*; Microfiche Suppl. (Vol. 43 AAS Microfiche Series) $10 *(ISBN 0-87703-215-7)*

Vol. 51 Guidance and Control 1983, Feb. 5-9, 1983, Keystone, CO, 1983, 494p., eds. Edward J. Bauman, Zubin W. Emsley, Hard Cover $60 *(ISBN 0-87703-182-7)*; Soft Cover $50 *(ISBN 0-87703-183-5)*; Microfiche Suppl. (Vol. 44 AAS Microfiche Series) $10 *(ISBN 0-87703-214-9)*

Vol. 52 Developing the Space Frontier, 29th Annual AAS Meeting, Oct. 25-27, 1982, Houston, TX, 1983, 436p., eds. Albert Naumann, Grover Alexander, Hard Cover $55 *(ISBN 0-87703-189-4)*

Vol. 53 Space Manufacturing 1983, May 9-12, 1983, Princeton, NJ, 1983, 496p., eds. James D. Burke, April S. Whitt, Hard Cover $60 *(ISBN 0-87703-188-6)*; Soft Cover $50 *(0-87703-189-4)*

Vol. 54 AAS/AIAA Astrodynamics Conference, Aug. 22-25, 1983, Lake Placid, NY, 1984, **two parts**, 1370p., eds. G.T. Tseng, Paul J. Cefola, Peter M. Bainum, David A. Levinson, Hard Cover $120 *(ISBN 0-87703-190-8)*; Soft Cover $90 *(ISBN 0-87703-191-6)*; Microfiche Suppl. (Vol. 45 AAS Microfiche Series) $40 *(ISBN 0-87703-192-4)*

Vol. 55 Guidance and Control 1984, Feb. 4-8, 1984, Keystone, CO, 1984, 500p., eds. Robert D. Culp, Parker S. Stafford, Hard Cover $60 *(ISBN 0-87703-199-1)*; Soft Cover $50 *(ISBN 0-87703-200-9)*; Microfiche Suppl. (Vol. 48 AAS Microfiche Series $15 *(ISBN 0-87703-201-7)*

Vol. 56 From Spacelab to Space Station, Fifth DGLR/AAS Symposium, Oct. 3-5, 1984, Hamburg, Germany, 1985, 270p., eds. H. Stoewer, Peter M. Bainum, Microfiche Only $30 *(ISBN 0-87703-209-2)*

Vol. 57 Guidance and Control 1985, Feb. 2-6, 1985, Keystone, CO, 1985, 618p., eds. Robert D. Culp, Edward J. Bauman, Charles A. Cullian, Hard Cover $65 *(ISBN 0-87703-211-4)*; Soft Cover $50 *(ISBN 0-87703-212-2)*; Microfiche Suppl. (Vol. 50 AAS Microfiche Series) $15 *(ISBN 0-87703-213-0)*

Vol. 58 AAS/AIAA Astrodynamics Conference, Aug. 12-15, 1985, Vail, CO, 1986, **two parts**, 1556p., eds. Bernard Kaufman, Joseph J.F. Liu, Robert A. Calico, Felix R. Hoots, Hard Cover $140 *(ISBN 0-87703-245-9)*; Soft Cover $110 *(ISBN 0-87703-246-7)*; Microfiche Suppl. (Vol. 51 AAS Microfiche Series); $60 *(ISBN 0-87703-247-5)*

Vol. 59 Space Station Beyond IOC, 32nd Annual AAS Meeting, Nov 6-7, 1985, Los Angeles, CA, 1986, 188p., ed. M. Jack Friedenthal, Hard Cover $40 *(ISBN 0-87703-252-1)*; Soft Cover $30 *(ISBN 0-87703-253-X)*

Vol. 60 Space Exploitation and Utilization, First AAS/JRS Symposium, Dec. 15-19, 1985, Honolulu, HI, 1986, 740p., eds. Gayle L. May, Peter M. Bainum, Kenji Ikeda, Tamiya Nomura, Tatsuo Yamanaka, Ryojiro Akiba, Hard Cover $70 *(ISBN 0-87703-254-8)*; Soft Cover $55 *(ISBN 0-87703-255-6)*; Microfiche Suppl. (Vol. 52 AAS Microfiche Series) $10 *(ISBN 0-87703-256-4)*

Vol. 61 Guidance and Control 1986, Feb. 1-5, 1986, Keystone, CO, 1986, 460p., eds. Robert D. Culp, John C. Durrett, Hard Cover $60 *(ISBN 0-87703-257-2)*; Soft Cover $50 *(ISBN 0-87703-258-0)*; Microfiche Suppl. (Vol. 53 AAS Microfiche Series) $10 *(ISBN 0-87703-259-9)*

Vol. 62 Tethers in Space, Proceedings of First International Conference on Tethers in Space (NASA & PSN Sponsors; AIAA, AAS, & AIDAA Co-Sponsors), Sept. 17-19, 1986, Arlington, VA, 1987, 784p., eds. Peter M. Bainum, Ivan Bekey, Luciano Guerriero, Paul A. Penzo, Hard Cover $80 *(ISBN 0-87703-264-5)*; Soft Cover $70 *(ISBN 0-87703-265-3)*

Vol. 63 Guidance and Control 1987, Jan. 31 - Feb. 4, 1987, Keystone, CO, 1987, 638p., eds. Robert D. Culp, Terry J. Kelly, Hard Cover $75 *(ISBN 0-87703-268-8)*; Soft Cover $60 *(ISBN 0-87703-269-6)*

Vol. 64 Aerospace Century XXI, 33rd AAS Annual Meeting, Oct. 26-29, 1986, Boulder, CO, 1987, **all three parts**, Hard Cover $225 *(ISBN 0-87703-276-9)*; Soft Cover $180 *(ISBN 0-87703-277-7)*; **Part I**, Space Missions and Policy, 686p., eds. George W. Morgenthaler, Gayle L. May, Hard Cover $75 *(ISBN 0-87703-279-3)*; Soft Cover $60 *(ISBN 0-87703-282-3)*; **Part II**, Space Flight Technologies, 608p., eds. George W. Morgenthaler, W. Kent Tobiska, Hard Cover $75 *(ISBN 0-87703-280-7)*; Soft Cover $60 *(ISBN 0-87703-283-1)*; **Part III**, Space Sciences, Applications, and Commercial Developments, 724p., eds. George W. Morgenthaler, Jean N. Koster, Hard Cover $75 *(ISBN 0-87703-281-5)*; Soft Cover $60 *(ISBN 0-87703-284-X)*; Microfiche Suppl. (Vol. 54 AAS Microfiche Series) $25 *(ISBN 0-87703-278-5)*

Vol. 65 AAS/AIAA Astrodynamics Conference, Aug. 10-13, 1987, Kalispell, MT, 1988, **two parts**, 1774p., eds. John K. Soldner, Arun K. Misra, Robert E. Lindberg, Walton Williamson, Hard Cover $180 *(ISBN 0-87703-285-8)*; Soft Cover $150 *(ISBN 0-87703-286-6)*; Microfiche Suppl. (Vol. 55 AAS Microfiche Series); $70 *(ISBN 0-87703-287-4)*

Vol. 66 Guidance and Control 1988, Jan. 30 - Feb. 3, 1988, Keystone, CO, 1988, 576p., eds. Robert D. Culp, Paul L. Shattuck, Hard Cover $75 *(ISBN 0-87703-288-2)*; Soft Cover $60 *(ISBN 0-87703-289-0)*; Microfiche Suppl. (Vol. 56 AAS Microfiche Series) $10 *(ISBN 0-87703-290-4)*

Vol. 67 Space - A New Community of Opportunity, 34th AAS Annual Meeting, Nov. 3-5, 1987, Houston, TX, 1989, approx. 472p., ed. William G. Straight, Henry N. Bowes, Hard Cover $70 *(ISBN 0-87703-297-1)*; Soft Cover $55 *(ISBN 0-87703-298-X)*

Order from Univelt, Inc., P.O. Box 28130, San Diego, California 92128

INDEX

NUMERICAL INDEX

VOLUME 67 — **ADVANCES IN THE ASTRONAUTICAL SCIENCES**, SPACE -- A NEW COMMUNITY OF OPPORTUNITY

(34th AAS Annual Conference, November 3-5, 1987, Houston, Texas)

AAS 87-601 to -602 Space: A New Community of Opportunity -- Welcoming Remarks, H. N. Bowes, A. Cohen, K. J. Whitmire, G. Griffin

AAS 87-603 to -604 Space: A New Community of Opportunity -- Opening Remarks, D. C. Honhart, A. Cohen

AAS 87-605 Space: A New Community of Opportunity -- Keynote Address, W. Hawkins

AAS 87-606 Governing Policies and Issues: Panel Discussion, A. Cohen, W. Hawkins, C. Stadd, J. K. Manber, J. Allen, R. H. Truly, R. C. Duncan

AAS 87-607 Not Assigned

AAS 87-608 Governing Policies and Issues -- Introduction, W. G. Straight

AAS 87-609 Governing Policies and Issues: Department of Transportation Perspective, C. Stadd

AAS 87-610 Governing Policies and Issues: An Office of Commercial Space Department of Commerce Perspective, J. K. Manber

AAS 87-611 Governing Policies and Issues: Space Industries Perspective, J. Allen

AAS 87-612 Governing Policies and Issues: U.S. Navy/NASA Perspective, R. H. Truly

AAS 87-613 Governing Policies and Issues: Department of Defense Perspective, R. C. Duncan

AAS 87-614 Space Utilization/Application -- Introduction, E. Young, D. S. Lilly

AAS 87-615 Space Utilization/Application -- Panel Discussion, E. Young, R. E. Berry, L. J. Evans, Jr., R. Filep, H. Hudson

AAS 87-616 Return to Eminence: Issues for U.S. Communications Research, Development, and Applications, R. Filep

AAS 87-617 Transfer of Satellite Applications and Technology: The Need for a U.S. Initiative, H. E. Hudson

AAS 87-618 International Space Applications: Where Competition and Cooperation Co-Exist, R. E. Berry

AAS 87-619 Industrial Applications of the Micro-Gravity Environment, L. J. Evans, Jr.

AAS 87-620	Astrodynamics and Planetary Missions -- Introduction, J. J. F. Liu
AAS 87-621	Expandable Launch Vehicle and Upper Stage Modernization -- Introduction, R. L. Gates
AAS 87-622	Not Assigned
AAS 87-623	Astrodynamic Standards, V. Szebehely, J. J. F. Liu
AAS 87-624	Planetary Missions, W. I. McLaughlin
AAS 87-625	Avionics for Today's Expendable Launch Vehicles, M. J. Davis, D. A. Willis
AAS 87-626	Avionics for Future Expendable Launch Vehicles, J. R. Allder
AAS 87-627	Expandable Launch Vehicle and Upper Stage Modernization -- Dual Versus Three-String Trade-Offs, J. C. Karas
AAS 87-628	Not Available
AAS 87-629	Expandable Launch Vehicle and Upper Stage Modernization -- Better Utilization of Cross Programs, L. Thayne
AAS 87-630	Guidance and Control Requirements for Future Launch Vehicles, J. W. Hardtla
AAS 87-631	Rocket Propulsion -- Introduction
AAS 87-632	Rocket Propulsion -- Panel Discussion, M. T. Constantine, W. M. Piland, J. R.Thompson, Jr., J. R. Wormington
AAS 87-633	Rocket Propulsion -- The National Aerospace Plane, W. M. Piland
AAS 87-634	Rocket Propulsion -- The Shuttle-C, J. R. Thompson, Jr.
AAS 87-635	Not Available
AAS 87-636	Not Available
AAS 87-637	Astronomy, Astrophysics & Solar System Exploration -- Introduction, G. Newton, W. Phinney
AAS 87-638	Astronomy, Astrophysics & Solar System Exploration -- Panel Discussion, G. Newton, J. Aaron, D. Blanchard
AAS 87-639	Planetary Exploration in the 1990's, D. Blanchard
AAS 87-640	Future Exploration Activities -- Astronomy, Astrophysics & Solar System Exploration, J. Aaron
AAS 87-641	Great Observatories in Space, G. Newton
AAS 87-642	Life Sciences -- Introduction, A. E. Nicogossian
AAS 87-643	Not Assigned
AAS 87-644	Physiological Effects of Space Flight, C. L. Huntoon
AAS 87-645	Medical Care Delivery in Space, D. F. Stewart

AAS 87-646 Personality and Organizational Influences on Aerospace Human Performance, R. L. Helmreich

AAS 87-647 Bio-Regenerative Life Support, R. D. MacElroy, T. Wydeven, Jr.

AAS 87-648 to -649 Tracking and Data Systems -- Introduction, P. Hopkins, C. T. Force

AAS 87-650 Tracking Requirements to Support Space -- Very Long Baseline Interferometry, J. F. Jordan

AAS 87-651 Electrophoresis Experiment Command and Data Handling, D. W. Richman

AAS 87-652 Advanced Technology for Space Communications, Tracking, and Robotic Sensors, K. Krishen

AAS 87-653 Space Station Communications, A. W. Weinrich

AAS 87-654 Tracking and Data Systems Aspects of Telerobotics, A. G. Stephenson

AAS 87-655 Honors and Awards Banquet -- Master of Ceremonies Presentation, A. Cohen

AAS 87-656 Honors and Awards Banquet -- Keynote Address, T. O. Paine

AAS 87-657 Honors and Awards Presentations, D. C. Honhart

AAS 87-658 Structures and Composite Materials -- Introduction, A. Amos, D. C. Wade, H. N. Bowes

AAS 87-659 Not Assigned

AAS 87-660 Not Available

AAS 87-661 Structural Design and Validation Challenges of Large Space Systems, K. Soosaar

AAS 87-662 Large Scale Computational Issues in the Dynamics of Large Space Structures, E. J. Haug

AAS 87-663 Ground and Flight Experimentation in Future Large Space Systems Development, W. P. Witt, III

AAS 87-664 to -665 Automation and Robotics -- Introduction, J. D. Erickson, R. A. Cliff

AAS 87-666 Automation and Robotics for the Space Station -- An ATAC Perspective, R. R. Nunamaker

AAS 87-667 Space Station Initial Operational Concept (IOC) Operations and Safety View -- Automation and Robotics for Space Station, W. V. Bates, Jr.

AAS 87-668 The Flight Telerobotic Servicer Project, H. McCain

AAS 87-669 Research and Development View -- Automation and Robotics for Space Station, J. Yang

AAS 87-670 Systems Integration View -- Automation and Robotics for Space Station, A. Nease

AAS 87-671	Space Station Growth: University Views -- Automation and Robotics for Space Station, R. A. Volz
AAS 87-672	NASA View -- Automation and Robotics for Space Station, D. L. Pivirotto
AAS 87-673	Automation and Robotics -- Discussion, R. A. Cliff, R. Nunamaker, W. V. Bates, Jr., H. McCain, J. Yang, A. Nease, R. A. Volz, D. L. Pivirotto, S. R. Sadin,T. Kanade, S. Weiss
AAS 87-674	The Promise of Automation and Robotics for the Moon, Mars and Beyond, R. A. Cliff
AAS 87-675	OAST Research Program -- Automation and Robotics for the Moon, Mars and Beyond, S. R. Sadin
AAS 87-676	University Research -- Automation and Robotics for the Moon, Mars and Beyond, T. Kanade
AAS 87-677	Industry Research -- Automation and Robotics for the Moon, Mars and Beyond, S. Weiss
AAS 87-678 to -679	Space Station/Large Structures -- Introduction, E. E. Speaker M. Mickulas
AAS 87-680	Not Assigned
AAS 87-681 to -684	EVA Space Construction: Experience and Fundamental Issues, W. L. Heard, Jr., J. J. Watson, J. L. Ross, S. C. Spring, K. A. Havens
AAS 87-685	Not Available
AAS 87-686	Not Available
AAS 87-687 to -690	Space Station/Large Structures -- Panel Discussion, E. E. Speaker, S. Z. Rubenstein, R. F. Thompson, A. J. Louviere, C. Blankenship
AAS 87-691 to -695	Space -- A New Community of Opportunity -- Closing Recap: Where From Here?, D. C. Honhart, D. K. Slayton, H. N. Abramson, G. Lee, W. Hawkins
AAS 87-696 to -699	Not Assigned

AUTHOR INDEX*

* For each author the paper number is given. The page numbers refer to Volume 67, Advances in the Astronautical Sciences.